Delmar's Standard Guide to Transformers

DELMAR'S STANDARD GUIDE TO TRANSFORMERS

Stephen L. Herman

Donald E. Singleton

Australia • Brazil • Japan • Korea • Mexico • Singapore • Spain • United Kingdom • United States

Delmar's Standard Guide to Transformers
Stephen L. Herman, Donald E. Singleton

Publisher: Rober D. Lynch

Acquisitions Editor: Mark Huth

Developmental Editor: Jeanne Mesick

Project Editor: Thomas Smith

Project Editor: Karen Smith

Art and Design Coordinator: Cheri Plasse

Cover Designer: Betty Kodela

Cover Photo Courtesy: Magne Tek

Library of Congress Control Number: 96-16475

ISBN-13: 978-0-8273-7209-2

ISBN-10: 0-8273-7209-4

Delmar
Executive Woods
5 Maxwell Drive
Clifton Park, NY 12065
USA

Cengage Learning is a leading provider of customized learning solutions with office locations around the globe, including Singapore, the United Kingdom, Australia, Mexico, Brazil, and Japan. Locate your local office at:
international.cengage.com/region

Cengage Learning products are represented in Canada by
Nelson Education, Ltd.

For your lifelong learning solutions, visit **delmar.cengage.com**

Visit our corporate website at **www.cengage.com**

Printed in the United States of America
6 7 8 9 10 20 19 18 17 16

Contents

Preface

Delmar's Standard Guide to Transformers is a practical approach to understanding transformers. This text provides the student with a unique blend of the theoretical and the practical. It is assumed that the student has a knowledge of basic alternating current and concepts of inductive and capacitive loads, as well as phase angle and power factor.

This text begins with a study of magnetism and the principles of magnetic induction. The basic types of single-phase transformers; isolation, auto, and current, are covered. Calculations of voltage current and turns ratio are illustrated for each transformer type.

A general discussion of three phase power is followed by coverage of three phase transformer connections. Calculations are shown for each type of three phase transformer connection plus calculations for single phase loads connected to three phase systems. Sections on transformer installation, maintenance, and cooling are also included. Transformer installation is presented in accord with National Electrical Code® requirements.

Delmar's Standard Guide to Transformers includes a set of laboratory experiments using standard control transformers and incandescent lamps. These experiments present basic transformer characteristics, polarity, single phase and three phase connections and calculations from a practical hands-on perspective.

Acknowledgements

The author and *Delmar Publishers* gratefully acknowledge the comments and suggestions from the reviewers of this book. Their time and effort is appreciated.

Thomas Pickren, Albany Technical Institute, Albany, GA 31708

R. A. Smith, Red Rocks Community College, Evergreen, CO 80439

Richard E. Hoover, Owens Community College, Millbury, OH 43447

1

Magnetism

Objectives

After studying this unit, you should be able to

- Discuss the properties of permanent magnets
- Discuss the difference between the axis poles of the earth and the magnetic poles of the earth
- Discuss the operation of electromagnets
- Determine the polarity of an electromagnet when the direction of the current is known
- Discuss the different systems used to measure magnetism
- Define terms used to describe magnetism and magnetic quantities

Magnetism is one of the most important phenomena in the study of electricity. It is the force used to produce most of the electrical power in the world. The force of magnetism has been known for over 2000 years. It was first discovered by the Greeks when they noticed that a certain type of stone was attracted to iron. This stone was first found in Magnesia in Asia Minor and was named magnetite. In the Dark Ages, the strange powers of the magnet were believed to be caused by evil spirits or the devil.

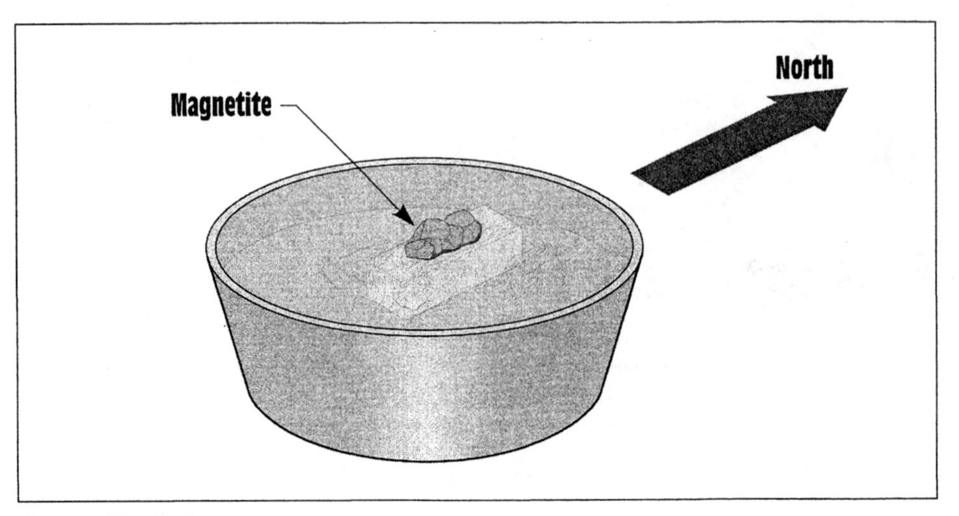

Figure 1-1 The first compass.

1-1 The Earth Is a Magnet

The first compass was invented when it was noticed that a piece of magnetite, a type of stone that is attracted to iron, placed on a piece of wood floating in water always aligned itself north and south *(Figure 1-1)*. Because they are always able to align themselves north and south, natural magnets became known as "leading stones" or **lodestones**. The reason that the lodestone aligned itself north and south is because the earth itself contains magnetic poles. *Figure 1-2* illustrates the position of the true North and South poles, or the axis, of the earth and the position of the magnetic poles. Notice that *magnetic* north is not located at the true North Pole of the earth. This is the reason that navigators must distinguish between true north and magnetic north. The angular difference between the two is known as the angle of declination. Although the illustration shows the magnetic lines of force to be only on each side of the earth, the lines actually surround the entire earth like a magnetic shell.

Also notice that the magnetic north pole is located near the southern polar axis and the magnetic south pole is located near the northern polar axis. The reason that the *geographic* poles (axis) are called north and south is because the north pole of a compass needle points in the direction of the north geographic pole. Since unlike magnetic poles attract, the north magnetic pole of the compass needle is attracted to the south magnetic pole of the earth.

1-2 Permanent Magnets

Permanent magnets are magnets that do not require any power or force

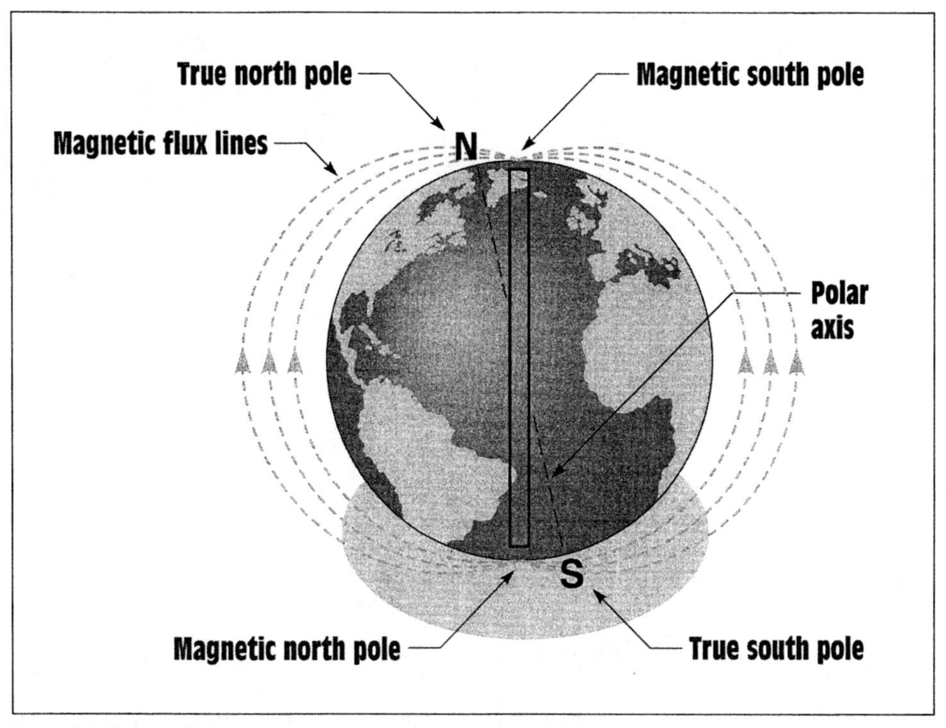

Figure 1-2 The earth is a magnet.

to maintain their field. They are an excellent example of one of the basic laws of magnetism, which states that **Energy is required to create a magnetic field, but no energy is required to maintain a magnetic field.** Man-made permanent magnets are much stronger and can retain their magnetism longer than natural magnets.

1-3 The Electron Theory of Magnetism

There are actually only three substances that form natural magnets: iron, nickel, and cobalt. Why these materials form magnets has been the subject of complex scientific investigations, resulting in an explanation of magnetism based on **electron spin patterns**. It is believed that electrons spin on their axis as they orbit around the nucleus of the atom. This spinning motion causes each electron to become a tiny permanent magnet. Although all electrons spin, they do not all spin in the same direction. In most atoms, electrons that spin in opposite directions tend to form pairs *(Figure 1-3)*. Since the electron pairs spin in opposite directions, their magnetic effects cancel each other out as far as having any effect on distant objects. In a similar manner two horseshoe magnets connected together would be strongly

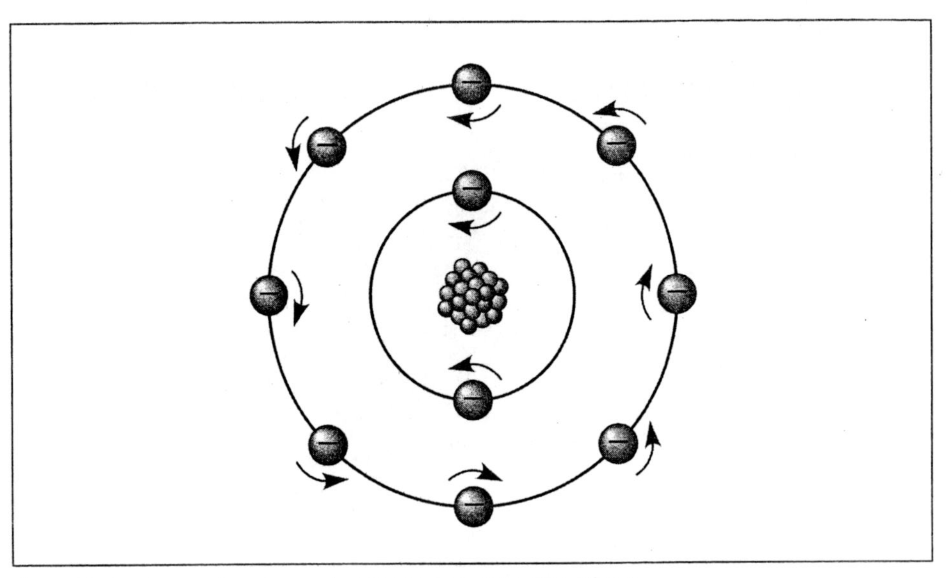

Figure 1-3 Electron pairs generally spin in opposite directions.

attracted to each other, but would have little effect on surrounding objects *(Figure 1-4)*.

An atom of iron contains 26 electrons. Of these 26, 22 are paired and spin in opposite directions, canceling each other's magnetic effect. In the next to the outermost shell, however, four electrons are not paired and spin in the same direction. These four electrons account for the magnetic properties of iron. At a temperature of 1420°F, or 771.1°C, the electron spin patterns rearrange themselves and iron loses its magnetic properties.

When the atoms of most materials combine to form molecules, they arrange themselves in a manner that produces a total of 8 valence electrons. The electrons form a spin pattern that cancels the magnetic field of the mate-

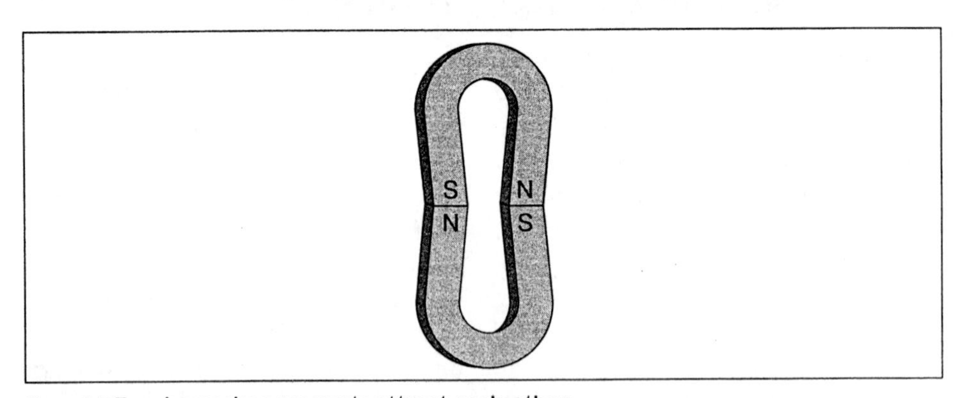

Figure 1-4 Two horseshoe magnets attract each other.

rial. When the atoms of iron, nickel, and cobalt combine, however, the magnetic field is not canceled. Their electrons combine so that they share valence electrons in such a way that their spin patterns are in the same direction, causing their magnetic fields to add instead of cancel. The additive effect forms regions in the molecular structure of the metal called **magnetic domains** or **magnetic molecules**. These magnetic domains act like small permanent magnets.

A piece of nonmagnetized metal has its molecules in a state of disarray as shown in *Figure 1-5*. When the metal is magnetized, its molecules align themselves in an orderly pattern as shown in *Figure 1-6*. In theory, each molecule of a magnetic material is itself a small magnet. If a permanent magnet were cut into pieces, each piece would be a separate magnet *(Figure 1-7)*.

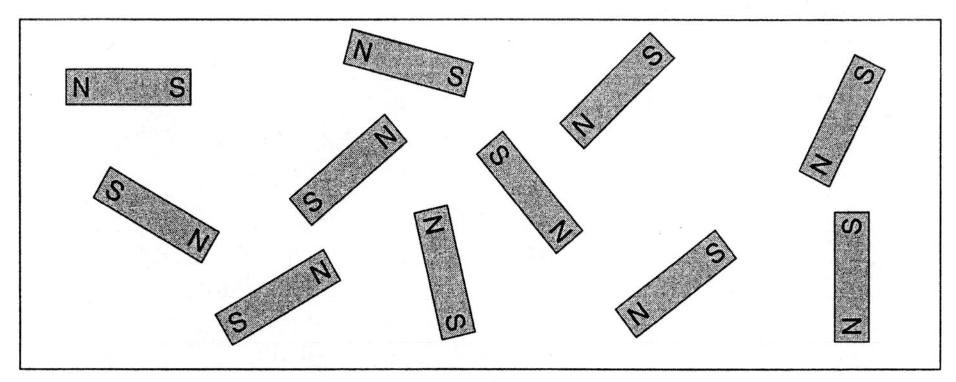

Figure 1-5 The molecules are disarrayed in a piece of nonmagnetized metal.

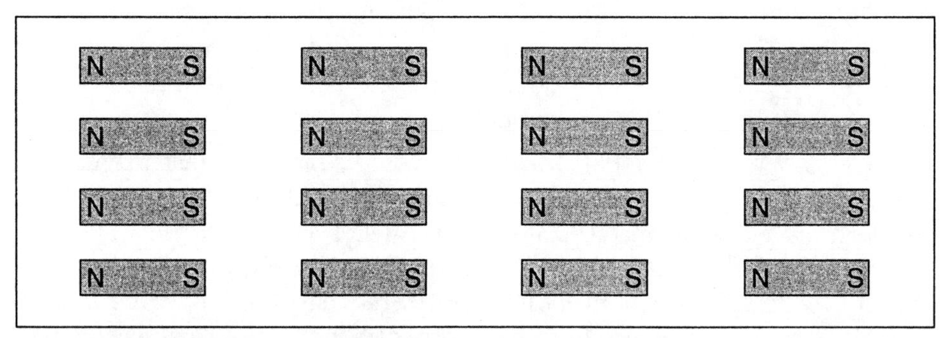

Figure 1-6 The molecules are aligned in an orderly fashion in a piece of magnetized metal.

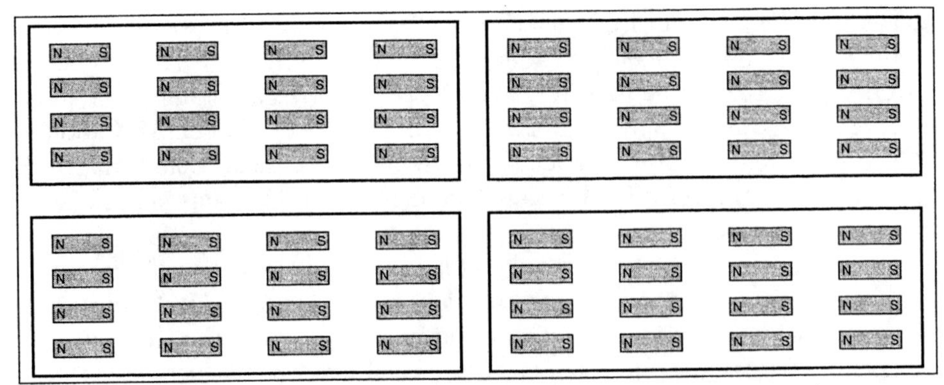

Figure 1-7 When a magnet is cut apart, each piece becomes a separate magnet.

1-4 Magnetic Materials

Magnetic materials can be divided into three basic classifications. These are:

Ferromagnetic materials are metals that are easily magnetized. Examples of these materials are iron, nickel, cobalt, and manganese.

Paramagnetic materials are metals that can be magnetized, but not as easily as ferromagnetic materials. Some examples of paramagnetic materials are platinum, titanium, and chromium.

Diamagnetic materials are either metal or nonmetal materials that cannot be magnetized. The magnetic lines of force tend to go around them instead of through them. Some examples of these materials are copper, brass, and antimony.

Some of the best materials for the production of permanent magnets are alloys. One of the best permanent magnet materials is Alnico 5, which is made from a combination of aluminum, nickel, cobalt, copper, and iron. Another type of permanent magnet material is made from a combination of barium ferrite and strontium ferrite. Ferrites can have an advantage in some situations because they are insulators and not conductors. They have a resistance of approximately 1,000,000 Ω per centimeter. These two materials can be powdered. The powder is heated to the melting point and then rolled and heat-treated. This treatment changes the grain structure and magnetic properties of the material. The new type of material has a property more like stone than metal and is known as a ceramic magnet. Ceramic magnets can be powdered and mixed with rubber, plastic, or liquids. Ceramic magnetic materials mixed with liquids can be used to make magnetic ink, which is used on checks. Another frequently used magnetic material is iron oxide, which is used to make magnetic recording tape and computer diskettes.

1-5 Magnetic Lines of Force

Magnetic lines of force are called **flux**. The symbol used to represent flux is the Greek letter phi (Φ). Flux lines can be seen by placing a piece of cardboard on a magnet and sprinkling iron filings on the cardboard. The filings will align themselves in a pattern similar to the one shown in *Figure 1-8*. The pattern produced by the iron filings forms a two-dimensional figure, but the flux lines actually surround the entire magnet *(Figure 1-9)*. Magnetic **lines of flux** repel each other and never cross. Although magnetic lines of flux do not flow, it is assumed they are in a direction north to south.

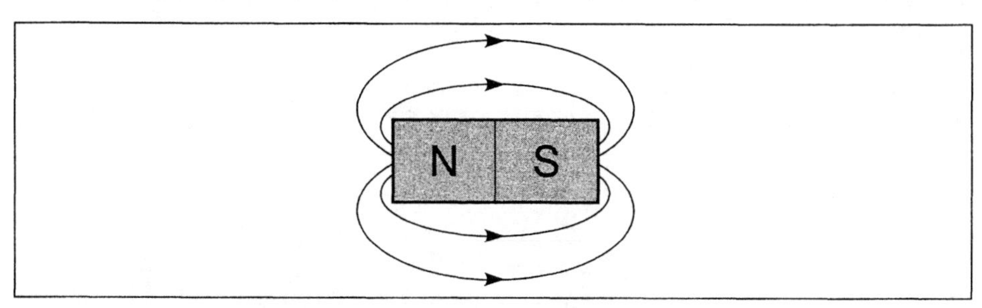

Figure 1-8 Magnetic lines of force are called flux lines.

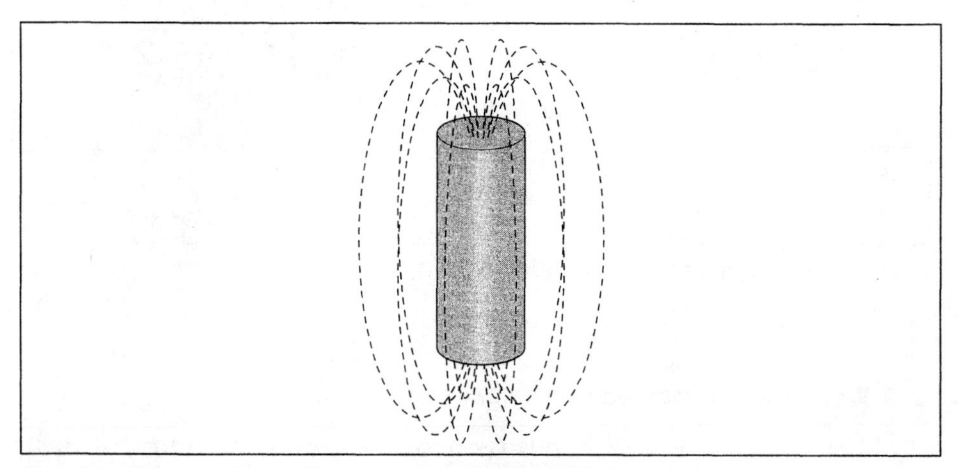

Figure 1-9 Magnetic lines of force surrounding the entire magnet.

A basic law of magnetism states that **unlike poles attract and like poles repel**. *Figure 1-10* illustrates what happens when a piece of cardboard is placed over two magnets with their north and south poles facing each other and iron filings are sprinkled on the cardboard. The filings form a pattern showing that the magnetic lines of flux are attracted to each other. *Figure 1-11* illustrates the pattern formed by the iron filings when the cardboard is placed over two magnets with like poles facing each other. The filings show that the magnetic lines of flux repel each other.

If the opposite poles of two magnets are brought close to each other, they will be attracted to each other as shown in *Figure 1-12*. If like poles of the two magnets are brought together, they will repel each other.

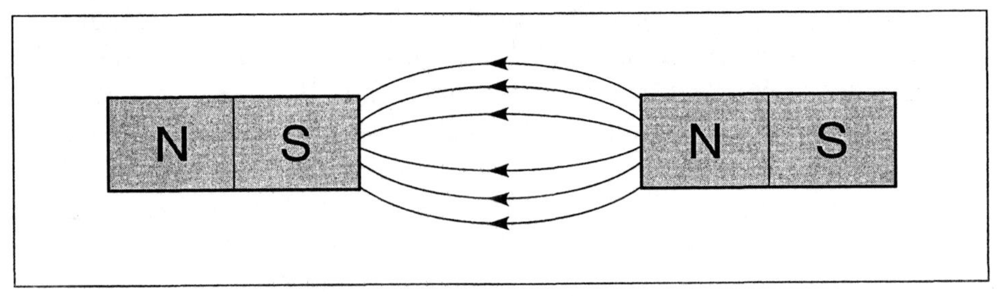

Figure 1-10 Opposite magnetic poles attract each other.

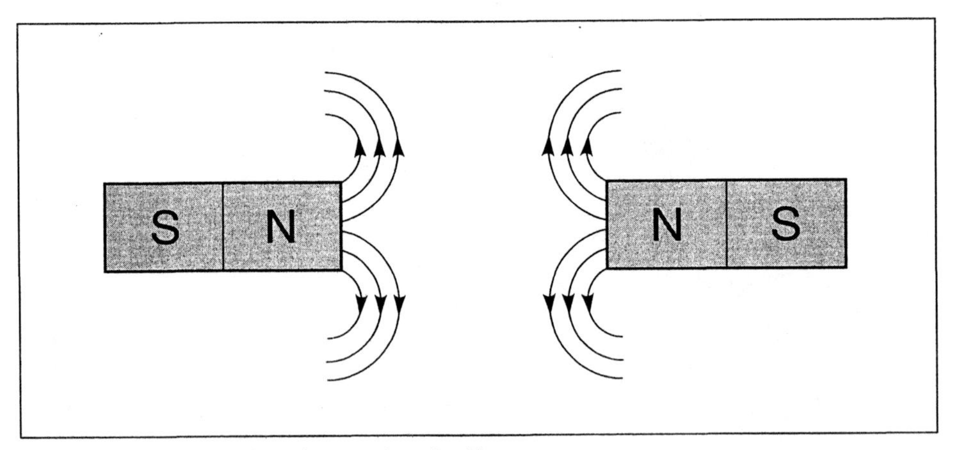

Figure 1-11 Like magnetic poles repel each other.

1-6 Electromagnetics

A basic law of physics states that **whenever an electric current flows through a conductor, a magnetic field is formed around the conduc-**

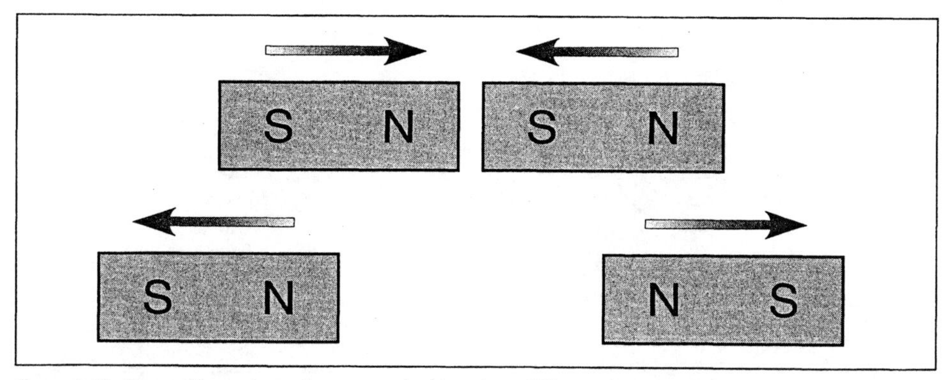

Figure 1-12 Opposite poles of a magnet attract and like poles repel.

tor. **Electromagnets** depend on electric current flow to produce a magnetic field. They are generally designed to produce a magnetic field only as long as the current is flowing; they do not retain their magnetism when current flow stops. Electromagnets operate on the principle that current flowing through a conductor produces a magnetic field around the conductor *(Figure 1-13)*. If the conductor is wound into a coil as shown in *Figure 1-14*, the magnetic lines of flux add to produce a stronger magnetic field. A coil with 10 turns of wire will produce a magnetic field that is 10 times as strong as the magnetic field around a single conductor.

Another factor that affects the strength of an electromagnetic field is the amount of current flowing through the wire. An increase in current flow will cause an increase in magnetic field strength. The two factors that determine the number of flux lines produced by an electromagnet are the number of turns of wire and the amount of current flow through the wire. The strength of an electromagnet is proportional to its **ampere-turns**. Ampere-turns are determined by multiplying the number of turns of wire by the current flow.

Core Material

Coils can be wound around any type of material to form an electromagnet. The base material is called the core material. When a coil is wound around a nonmagnetic material such as wood or plastic, it is known as an *air-core* magnet. When a coil is wound around a magnetic material such as iron or soft steel, it is known as an *iron-core* magnet. The addition of magnetic material to the center of the coil can greatly increase the strength of the magnet. If the core material causes the magnetic field to become 50 times stronger, the core material has a permeability of 50 *(Figure 1-15)*. **Permeability** is a measure of a material's willingness to become magnetized. The number of flux lines produced is proportional to the

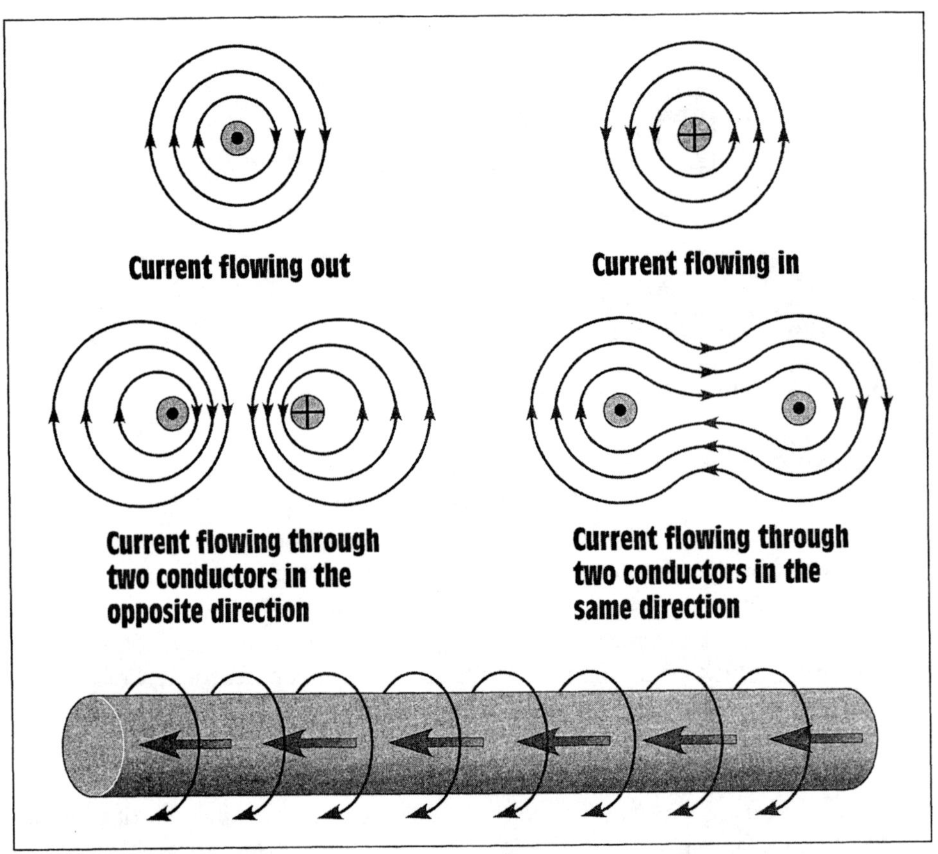

Figure 1-13 Current flowing through a conductor produces a magnetic field around the conductor.

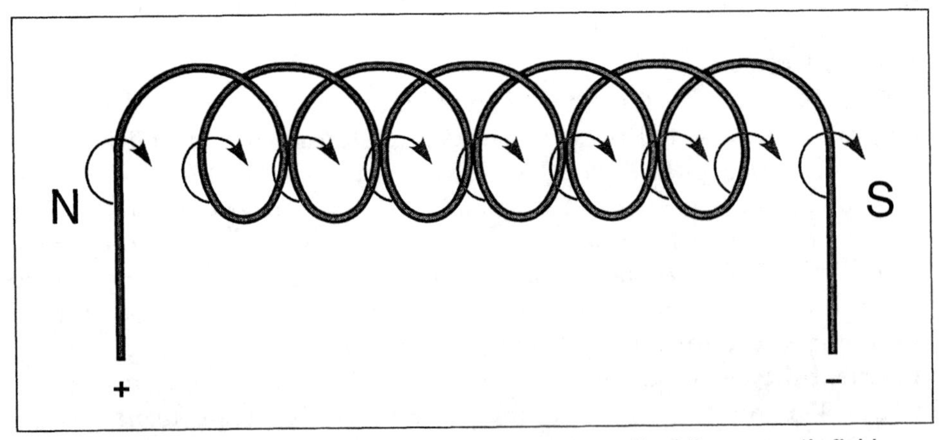

Figure 1-14 Winding the wire into a coil increases the strength of the magnetic field.

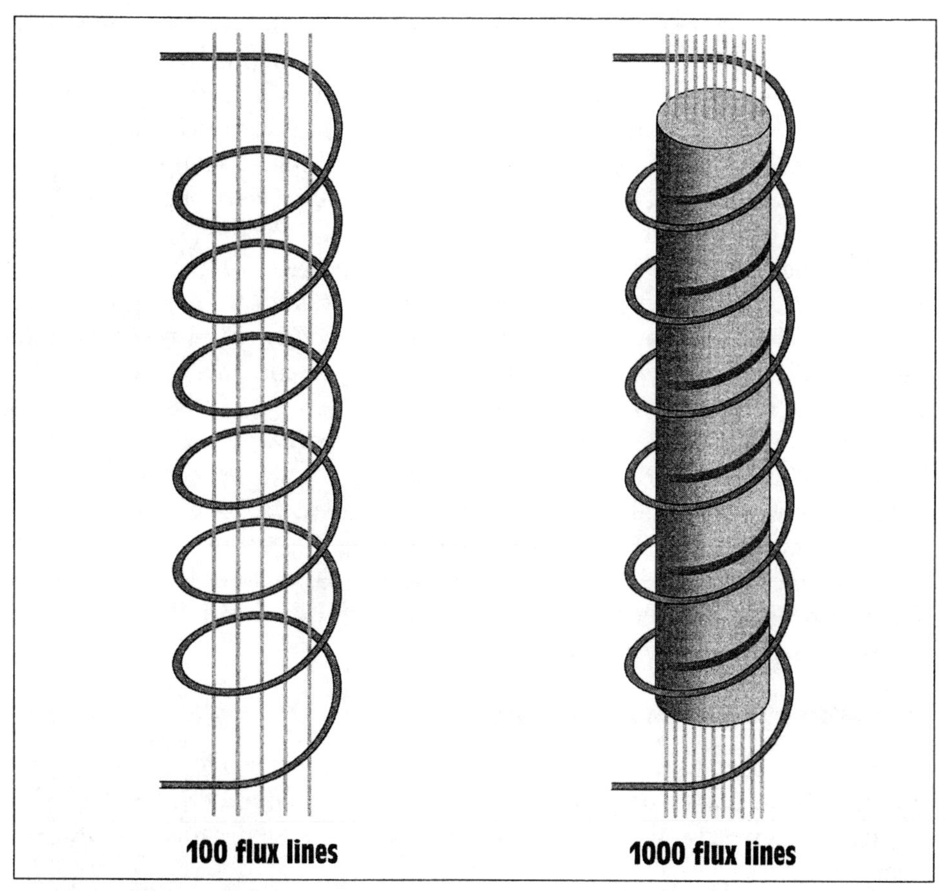

Figure 1-15 An iron core increases the number of flux line per-square -nch.

ampere-turns. The magnetic core material provides an easy path for the flow of magnetic lines in much the same way a conductor provides an easy path for the flow of electrons. This increased permeability permits the flux lines to be concentrated in a smaller area, which increases the number of flux lines per-square-inch or per-square-centimeter. In a similar manner, a person using a garden hose with an adjustable nozzle attached can adjust the nozzle to spray the water in a fine mist that covers a large area or in a concentrated stream that covers a small area.

Another common magnetic measurement is reluctance. **Reluctance** is resistance to magnetism. A material such as soft iron or steel has a high permeability and low reluctance because it is easily magnetized. A material such as copper has a low permeability and high reluctance.

If the current flow in an electromagnet is continually increased, the magnet will eventually reach a point where its strength will increase only slightly with an increase in current. When this condition occurs, the magnetic material is at a point of saturation. **Saturation** occurs when all the molecules of the magnetic material are lined up. Saturation is similar to pouring 5 gallons of water into a 5-gallon bucket. Once the bucket is full, it simply cannot hold any more water. If it became necessary to construct a stronger magnet, a larger piece of core material would be required.

When the current flow through the coil of a magnet is stopped, there may be some magnetism left in the core material. The amount of magnetism left in a material after the magnetizing force has stopped is called **residual magnetism**. If the residual magnetism of a piece of core material is hard to remove, the material has a high coercive force. *Coercive force* is a measure of a material's ability to retain magnetism. A high coercive force is desirable in materials that are intended to be used as permanent magnets. A low coercive force is generally desirable for materials intended to be used as electromagnets. Coercive force is measured by determining the amount of current flow through the coil in the direction opposite to that required to remove the residual magnetism. Another term that is used to describe a material's ability to retain magnetism is *retentivity*.

1-7 Magnetic Measurement

The terms used to measure the strength of a magnetic field are determined by the system that is being used. There are three different systems used to measure magnetism: the English system, the CGS system, and the MKS system.

The English System

In the English system of measure, magnetic strength is measured in a term called flux density. Flux density is measured in lines per-square-inch. The Greek letter phi (Φ) is used to measure flux. The letter B is used to represent flux density. The formula shown below is used to determine flux density.

$$\text{B (flux density)} = \frac{\Phi \text{ (flux lines)}}{\text{A (area)}}$$

In the English system, the term used to describe the total force producing a magnetic field, or flux, is **magnetomotive force (mmf)**. Magnetomotive force can be computed using the formula:

$$\text{mmf} = \Phi \times \text{rel (reluctance)}$$

The formula shown below can be used to determine the strength of the magnet.

$$\text{Pull (in pounds)} = \frac{B \times A}{72{,}000{,}000}$$

where B = flux density in lines per-square-inch
A = area of the magnet

The CGS System

In the CGS (centimeter-gram-second) system of measurement, one magnetic line of force is known as a maxwell. A gauss represents a magnetic force of one maxwell per-square-centimeter. In the English system, magnetomotive force is measured in ampere-turns. In the CGS system, gilberts are used to represent the same measurement. Since the main difference between these two systems of measurement is that one uses English units of measure and the other uses metric units of measure, a conversion factor can be used to help convert one set of units to the other.

1 gilbert = 1.256 ampere-turns

The MKS System

The MKS (meter-kilogram-second) system uses metric units of measure also. In this system, the main unit of magnetic measurement is the dyne. The dyne is a very weak amount of force. One dyne is equal to 1/27,800 of an ounce, or it requires 27,800 dynes to equal a force of one ounce. In the MKS system, a standard called the unit magnetic pole is used. In *Figure 1-16*, two magnets are separated by a distance of 1 cm. These magnets repel each other with a force of 1 dyne. When two magnets separated by a distance of 1 cm exert a force on each other of 1 dyne, they are considered to be a unit

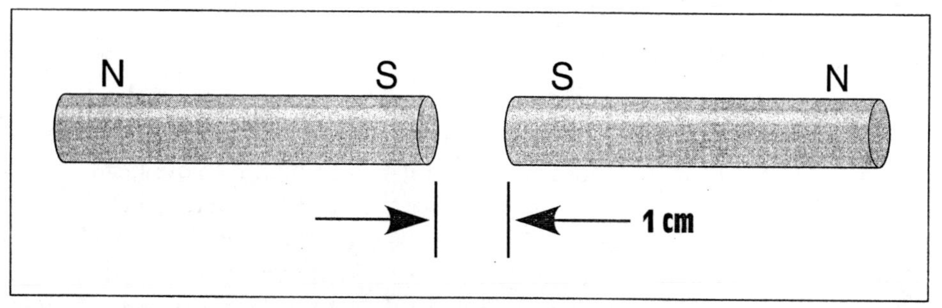

Figure 1-16 A unit magnetic pole produces a force of 1 dyne.

magnetic pole. Magnetic force can then be determined using the formula

$$\text{Force (in dynes)} = \frac{M_1 \times M_2}{D}$$

where M_1 = strength of first magnet, in unit magnetic poles
M_2 = strength of second magnet, in unit magnetic poles
D = distance between the poles, in centimeters

1-8 Magnetic Polarity

The polarity of an electromagnet can be determined using the **left-hand rule**. When the fingers of the left hand are placed around the windings in the direction of electron current flow, the thumb will point to the north magnetic pole *(Figure 1-17)*. If the direction of current flow is reversed, the polarity of the magnetic field will reverse also.

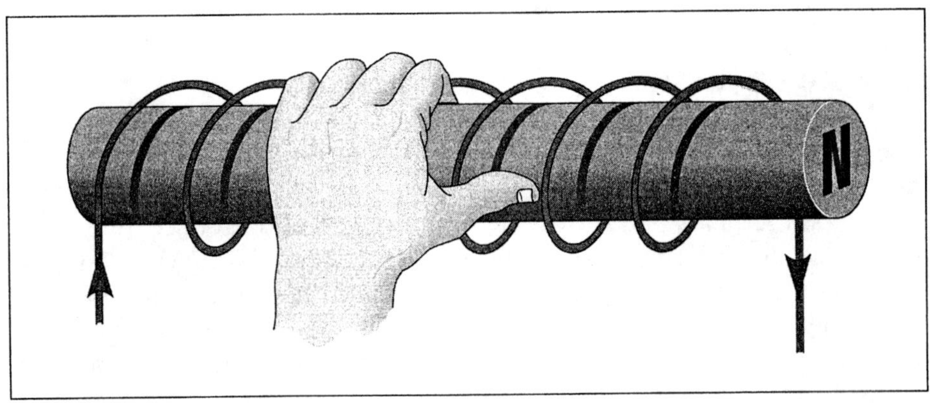

Figure 1-17 The left-hand rule can be used to determine the polarity of an electromagnet.

1-9 Demagnetizing

When an object is to be **demagnetized**, its molecules must be disarranged as they are in a nonmagnetized material. This can be done by placing the object in the field of a strong electromagnet connected to an alternating current (AC) line. Since the magnet is connected to AC current, the polarity of the magnetic field reverses each time the current changes direction. The molecules of the object to be demagnetized are, therefore, aligned first in one direction and then in the other. If the object is pulled

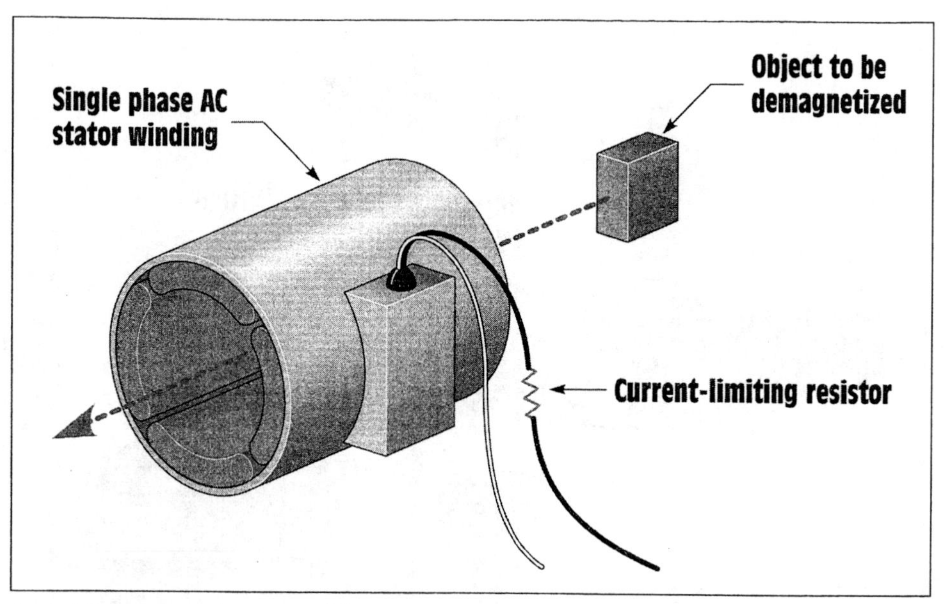

Figure 1-18 Demagnetizing an object.

away from the AC magnetic field, the effect of the field becomes weaker as the object is moved farther away *(Figure 1-18)*. The weakening of the magnetic field causes the molecules of the object to be left in a state of disarray. The ease or difficulty with which an object can be demagnetized depends on the strength of the AC magnetic field and the coercive force of the object.

An object can be demagnetized in two other ways *(Figure 1-19)*. If a magnetized object is struck, the vibration will often cause the molecules to rearrange themselves in a disordered fashion. It may be necessary to strike the object several times. Heating also will demagnetize an object. When the temperature becomes high enough, the molecules will rearrange themselves in a disordered fashion.

1-10 Magnetic Devices

A list of devices that operate on magnetism would be very long indeed. Some of the more common devices are electromagnets, measuring instruments, inductors, transformers, and motors.

The Speaker

The speaker is a common device that operates on the principle of magnetism *(Figure 1-20)*. The speaker produces sound by moving a cone; the movement causes a displacement of air. The tone is determined by how fast

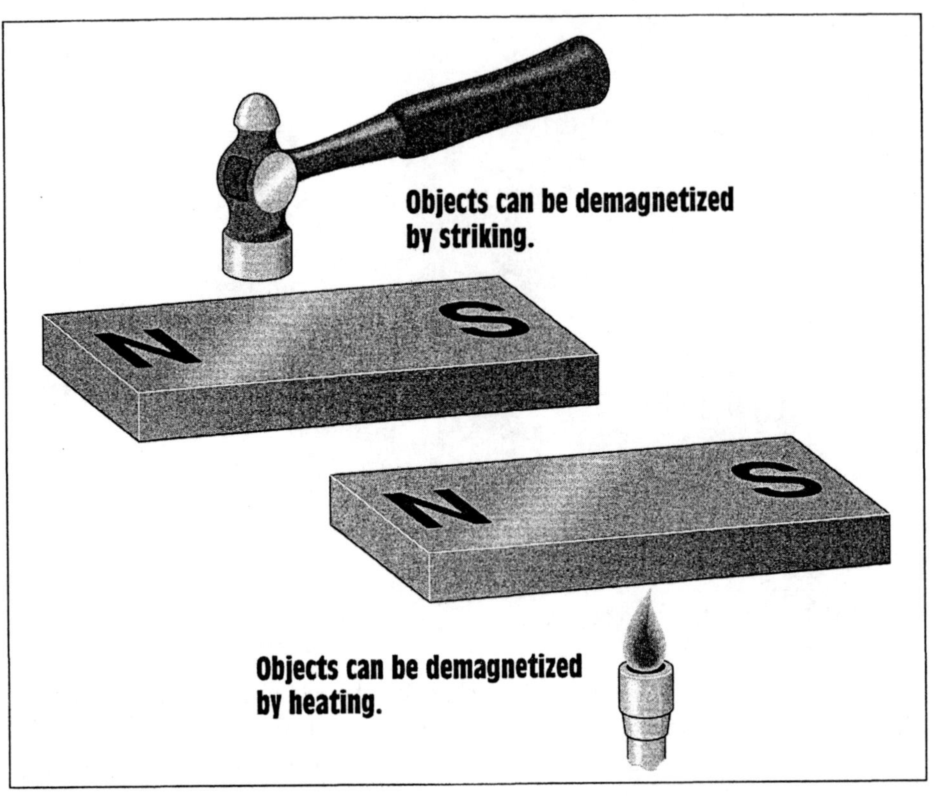

Figure 1-19 Other methods for demagnetizing objects.

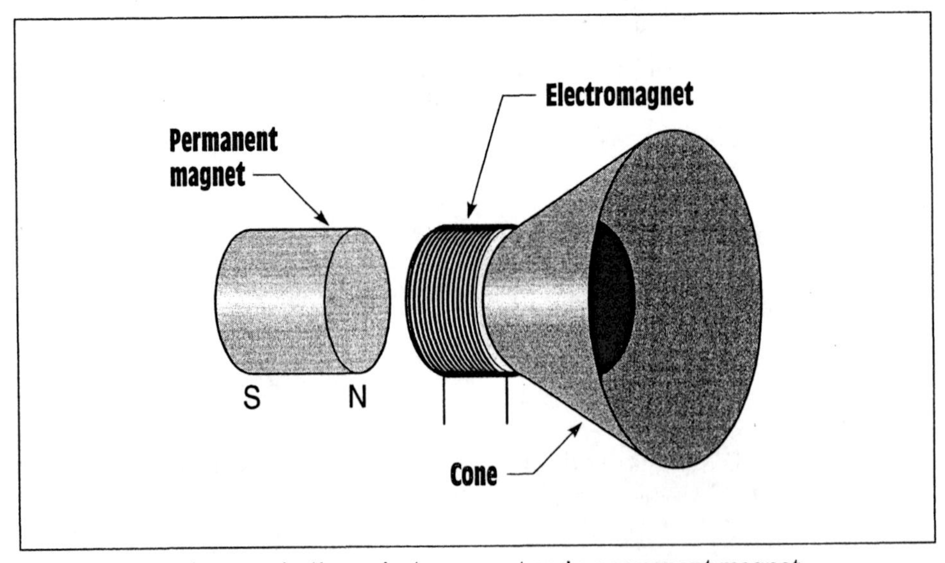

Figure 1-20 A speaker uses both an electromagnet and a permanent magnet.

the cone vibrates. Low or bass sounds are produced by vibrations in the range of 20 cycles-per-second. High sounds are produced when the speaker vibrates in the range of 20,000 cycles-per-second.

The speaker uses two separate magnets. One is a permanent magnet, and the other is an electromagnet. The permanent magnet is held stationary, and the electromagnet is attached to the speaker cone. When current flows through the coil of the electromagnet, a magnetic field is produced. The polarity of the field is determined by the direction of current flow. When the electromagnet has a north polarity, it is repelled away from the permanent magnet, causing the speaker cone to move outward and displace air. When the current flow reverses through the coil, the electromagnet has a south polarity and is attracted to the permanent magnet. The speaker cone then moves inward and again displaces air. The number of times per second that the current through the coil reverses determines the tone of the speaker.

Summary

1. Early natural magnets were known as lodestones.
2. The earth has a north and a south magnetic pole.
3. The magnetic poles of the earth and the axis poles are not the same.
4. Like poles of a magnet repel each other, and unlike poles attract each other.
5. Some materials have the ability to become better magnets than others.
6. Three basic types of magnetic material are:
   A. ferromagnetic
   B. paramagnetic
   C. diamagnetic
7. When current flows through a wire, a magnetic field is created around the wire.
8. The direction of current flow through the wire determines the polarity of the magnetic field.
9. The strength of an electromagnet is determined by the ampere-turns.
10. The type of core material used in an electromagnet can increase its strength.

11. Three different systems are used to measure magnetic values:
    A. The English system
    B. The CGS system
    C. The MKS system
12. An object can be demagnetized by placing it in an AC magnetic field and pulling it away, by striking, and by heating.

Review Questions

1. Is the north magnetic pole of the earth a north polarity or a south polarity?
2. What were early natural magnets known as?
3. The south pole of one magnet is brought close to the south pole of another magnet. Will the magnets repel or attract each other?
4. How can the polarity of an electromagnet be determined if the direction of current flow is known?
5. Define the following terms.
   Flux density
   Permeability
   Reluctance
   Saturation
   Coercive force
   Residual magnetism
6. A force of 1 ounce is equal to how many dynes?

2

Magnetic Induction

Objectives

After studying this unit, you should be able to

- Discuss magnetic induction
- List factors that determine the amount and polarity of an induced voltage
- Discuss Lenz's law
- Discuss an exponential curve.
- List devices used to help prevent inductive voltage spikes

Magnetic induction is one of the most important concepts in the electrical field. It is the basic operating principle underlying alternators, transformers, and most alternating-current motors. It is imperative that anyone desiring to work in the electrical field have an understanding of the principles involved.

2-1 Magnetic Induction

In Unit 1, it was stated that one of the basic laws of electricity is that whenever current flows through a conductor, a magnetic field is created around the conductor *(Figure 2-1)*. The direction of the current flow determines the polarity of the magnetic field, and the amount of current determines the strength of the magnetic field.

That basic law in reverse is the principle of **magnetic induction**, which states that **whenever a conductor cuts through magnetic lines of flux,**

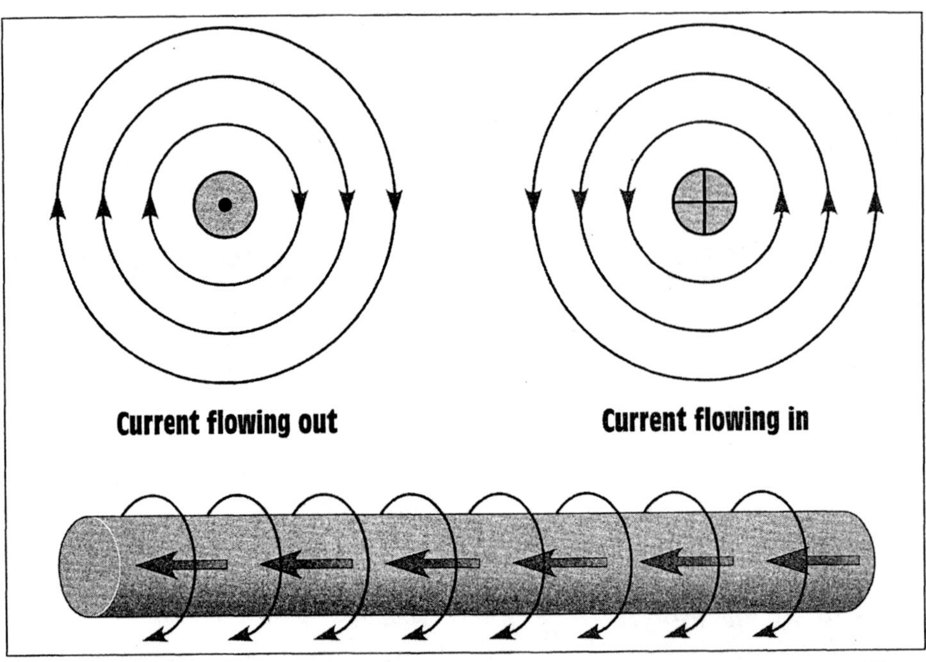

Figure 2-1 Current flowing through a conductor produces a magnetic field around the conductor.

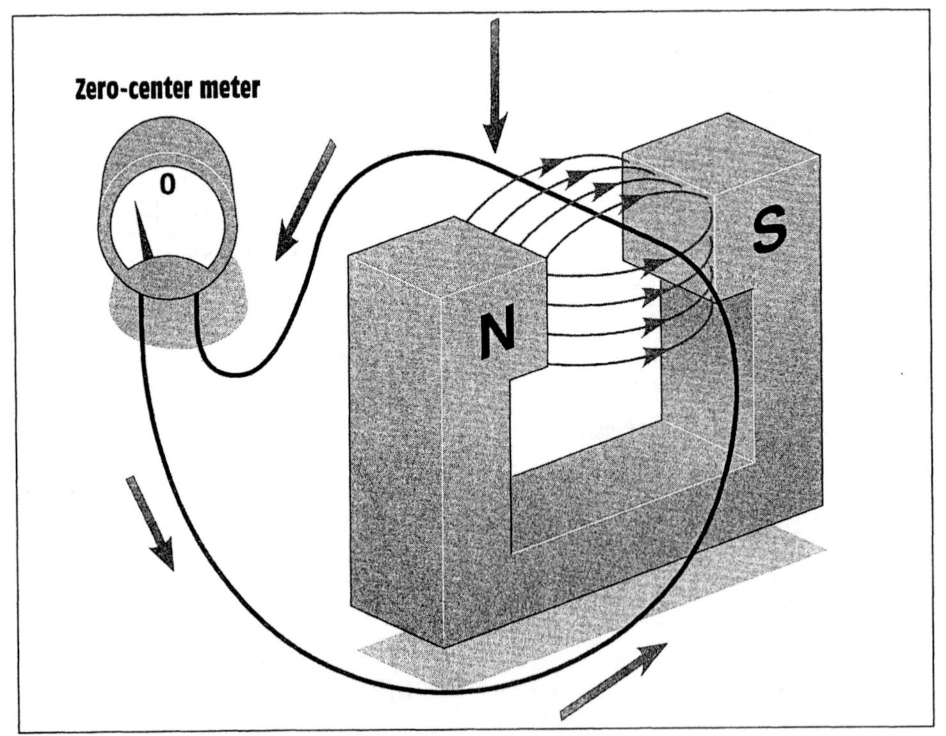

Figure 2-2 A voltage is induced when a conductor cuts magnetic lines of flux.

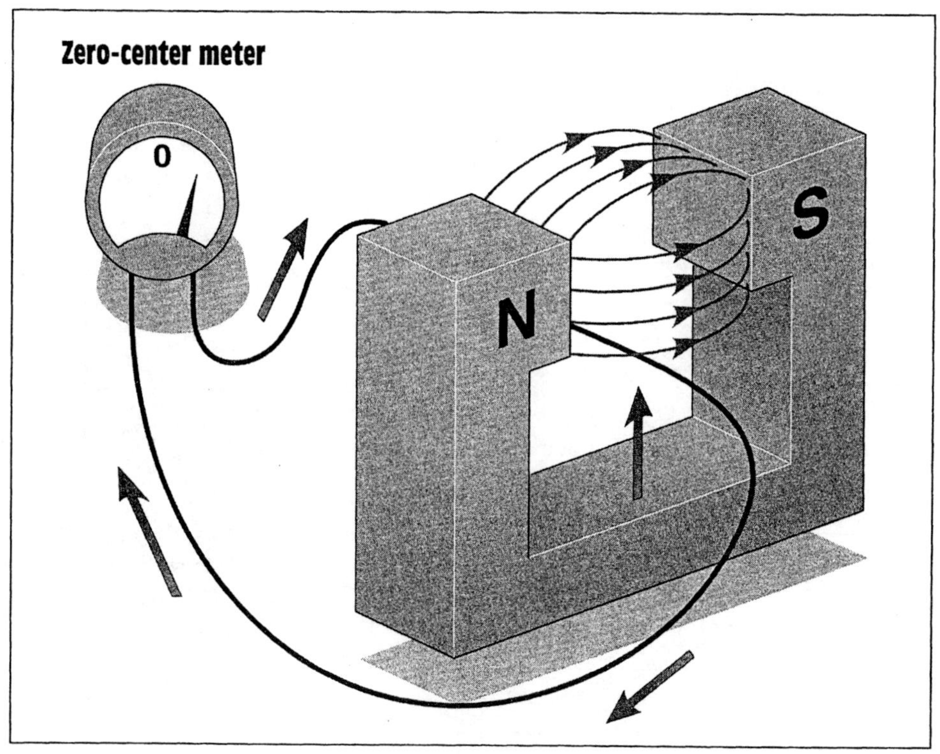

Figure 2-3 Reversing the direction of movement reverses the polarity of the voltage.

a voltage is induced into the conductor. The conductor in *Figure 2-2* is connected to a zero-center microammeter, creating a complete circuit. When the conductor is moved downward through the magnetic lines of flux, the induced voltage will cause electrons to flow in the direction indicated by the arrows. This flow of electrons causes the pointer of the meter to be deflected from the center-zero position.

If the conductor is moved upward, the polarity of induced voltage will be reversed and the current will flow in the opposite direction *(Figure 2-3)*. The pointer will be deflected in the opposite direction.

The polarity of the induced voltage can also be changed by reversing the polarity of the magnetic field *(Figure 2-4)*. In this example, the conductor is again moved downward through the lines of flux, but the polarity of the magnetic field has been reversed. Therefore the polarity of the induced voltage will be the opposite of that in *Figure 2-2,* and the pointer of the meter will be deflected in the opposite direction. It can be concluded that **the polarity of the induced voltage is determined by the polarity of the magnetic field in relation to the direction of movement**.

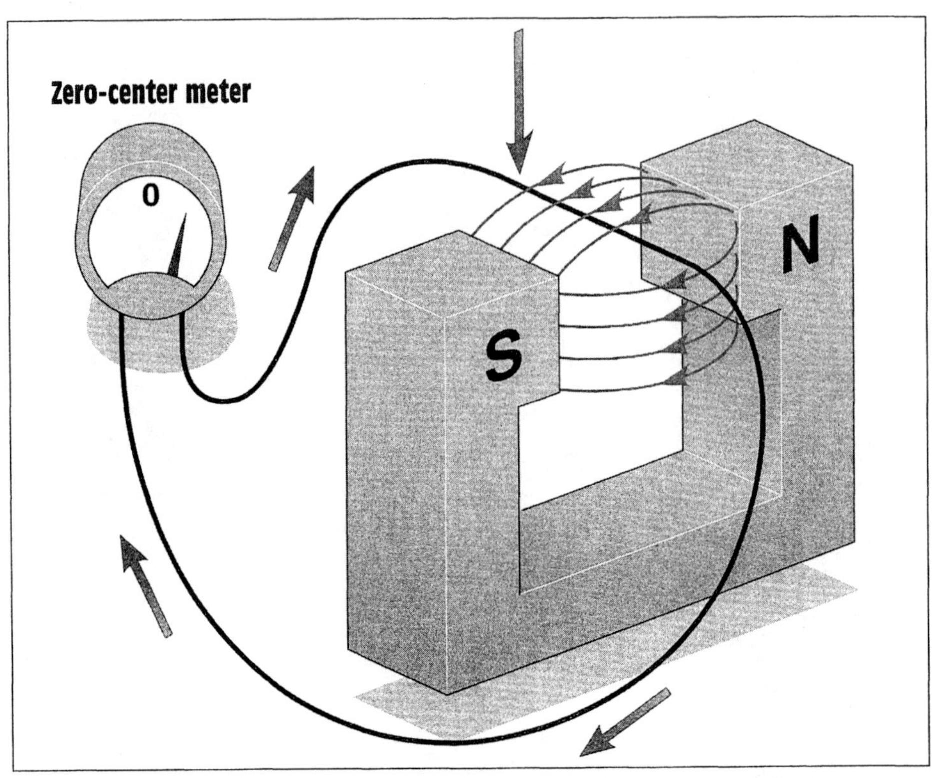

Figure 2-4 Reversing the polarity of the magnetic field reverses the polarity of the voltage.

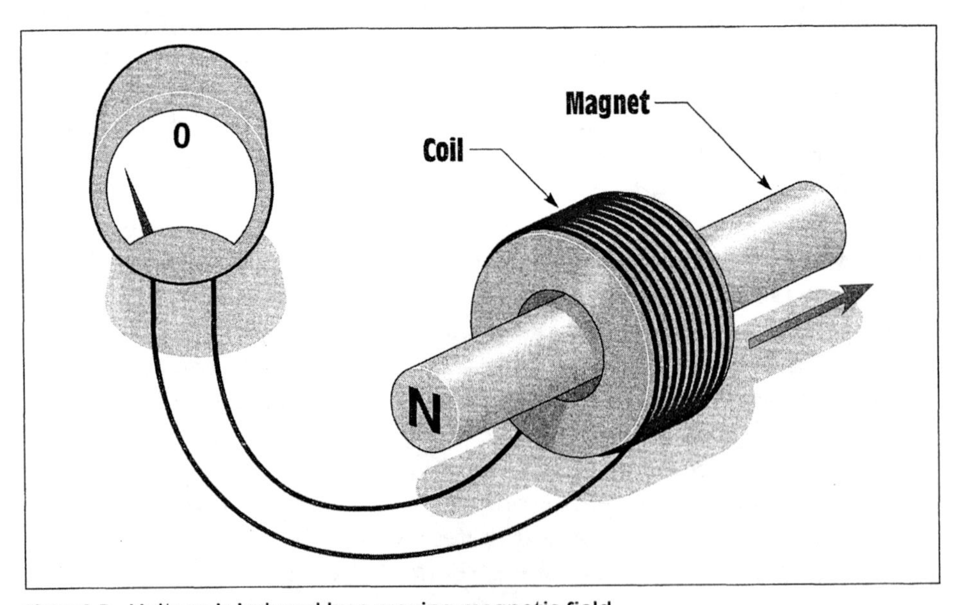

Figure 2-5 Voltage is induced by a moving magnetic field.

2-2 Moving Magnetic Fields

The important factors concerning magnetic induction are a conductor, a magnetic field, and movement. In practice, it is often desirable to move the magnet instead of the conductor. Most alternating current generators or alternators operate on this principle. In *Figure 2-5,* a coil of wire is held stationary while a magnet is moved through the coil. As the magnet is moved, the lines of flux cut through the windings of the coil and induce a voltage into them.

2-3 Determining the Amount of Induced Voltage

Three factors determine the amount of voltage that will be induced in a conductor:

1. **the number of turns of wire,**
2. **the strength of the magnetic field** (flux density), and
3. **the speed of the cutting action.**

In order to induce 1 V in a conductor, the conductor must cut 100,000,000 lines of magnetic flux in 1 s. In magnetic measurement, 100,000,000 lines of flux are equal to one **weber (Wb)**. Therefore, if a conductor cuts magnetic lines of flux at a rate of 1 Wb/s, a voltage of 1 V will be induced. A simple one-loop generator is shown in *Figure 2-6*. The loop is attached to a rod that

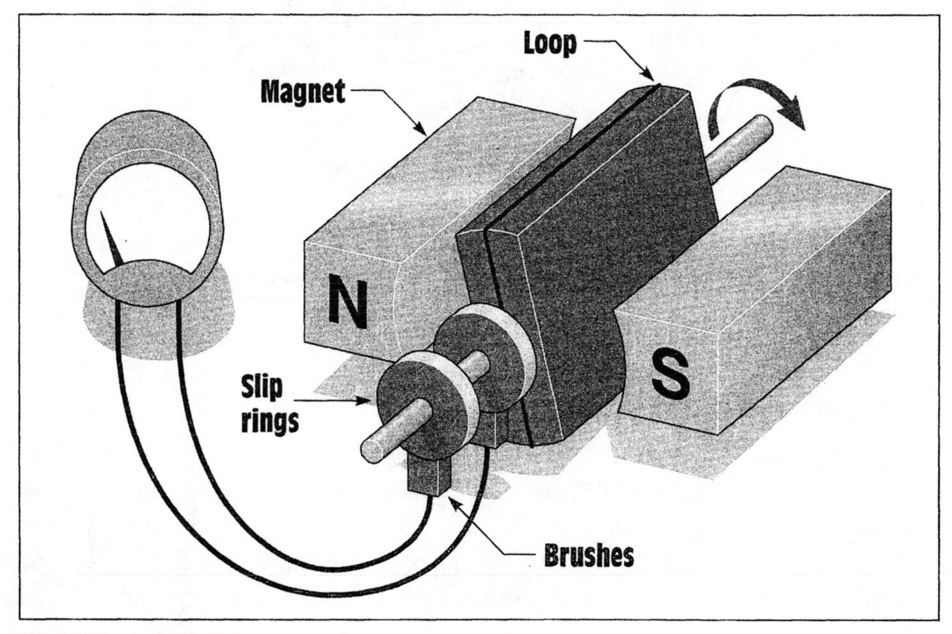

Figure 2-6 A single-loop generator

is free to rotate. This assembly is suspended between the poles of two stationary magnets. If the loop is turned, the conductor cuts through magnetic lines of flux and a voltage is induced into the conductor.

If the speed of rotation is increased, the conductor cuts more lines of flux per second, and the amount of induced voltage increases. If the speed of rotation remains constant and the strength of the magnetic field is increased, there will be more lines of flux per-square-inch. When there are more lines of flux, the number of lines cut per second increases and the induced voltage increases. If more turns of wire are added to the loop *(Figure 2-7),* more flux lines are cut per second and the amount of induced voltage increases again. Adding more turns has the effect of connecting single conductors in series, and the amount of induced voltage in each conductor adds.

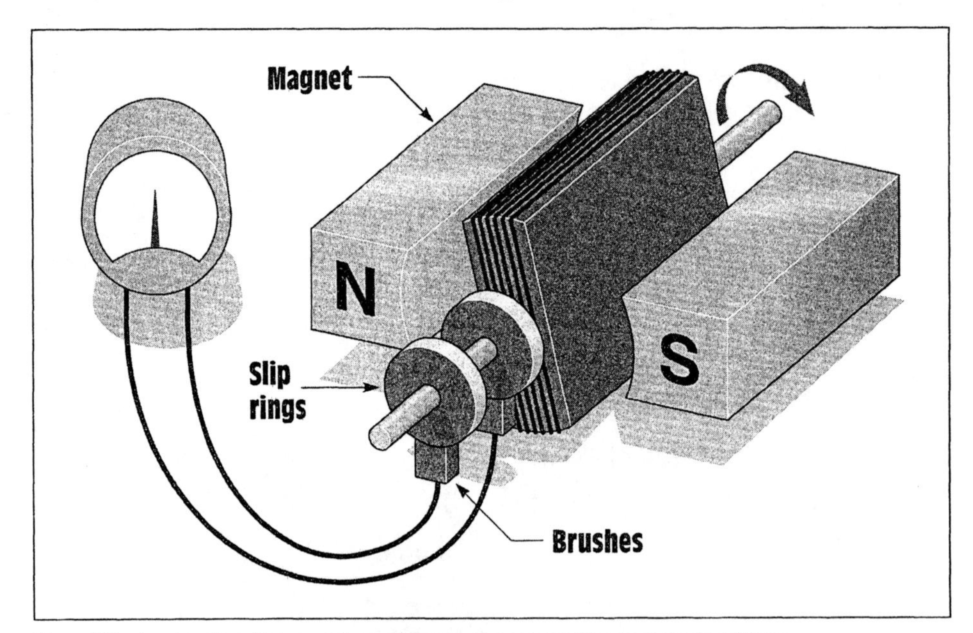

Figure 2-7 Increasing the number of turns increases the induced voltage.

2-4 Lenz's Law

When a voltage is induced in a coil and there is a complete circuit, current will flow through the coil *(Figure 2-8).* When current flows through the coil, a magnetic field is created around the coil. This magnetic field develops a polarity opposite that of the moving magnet. The magnetic field developed by the induced current acts to attract the moving magnet and pull it back inside the coil.

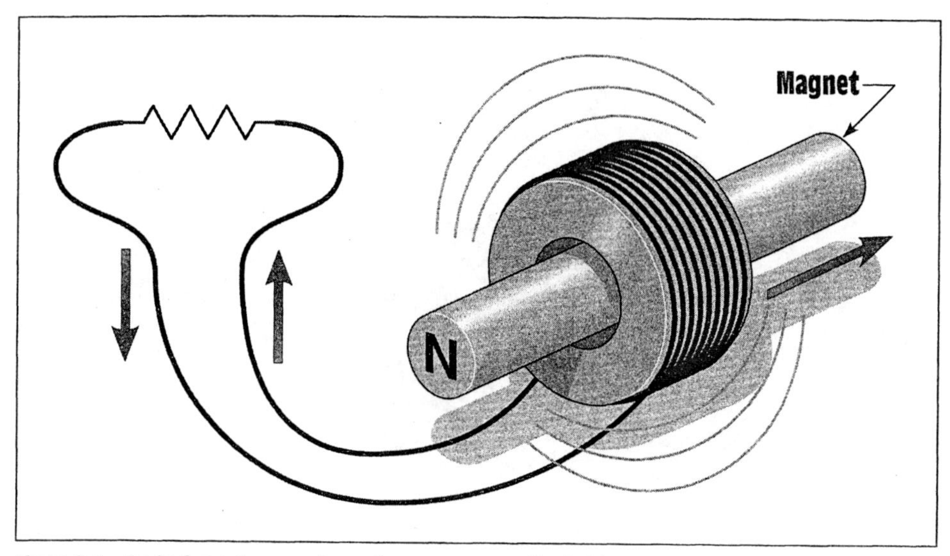

Figure 2-8 An induced current produces a magnetic field around the coil.

If the direction of motion is reversed, the polarity of the induced current is reversed, and the magnetic field created by the induced current again opposes the motion of the magnet. This principle was first noticed by Heinrich Lenz many years ago and is summarized in **Lenz's law,** which states that **an induced voltage or current opposes the motion that causes it**. From this basic principle, other laws concerning inductors have been developed. One is that **inductors always oppose a change of current**. The coil in *Figure 2-9,* for example, has no induced voltage and

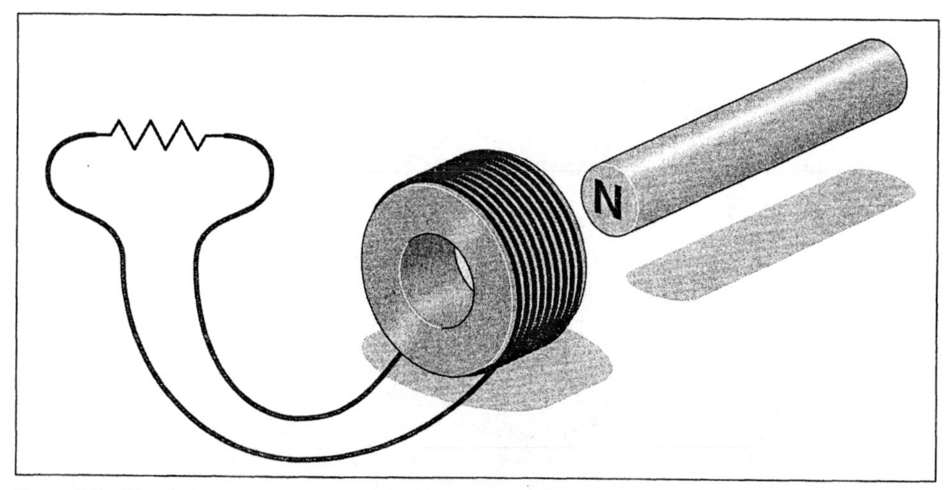

Figure 2-9 There is no current flow through the coils.

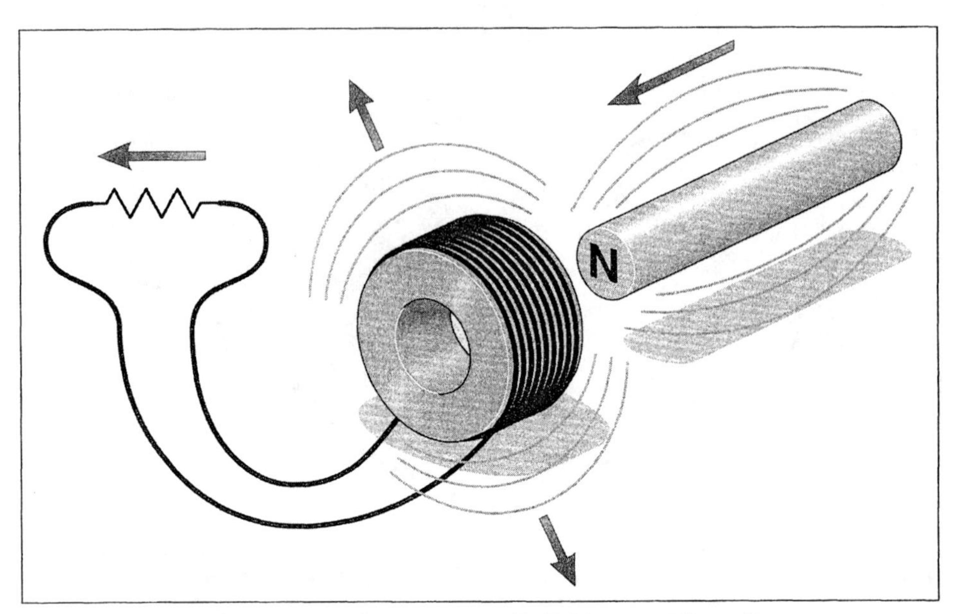

Figure 2-10 Induced current produces a magnetic field around the coil.

therefore no induced current. If the magnet is moved toward the coil, however, magnetic lines of flux will begin to cut the conductors of the coil, and a current will be induced in the coil. The induced current causes magnetic lines of flux to expand outward around the coil *(Figure 2-10)*. As this expanding magnetic field cuts through the conductors of the coil, a voltage is induced in the coil. The polarity of the voltage is such that it opposes the induced current caused by the moving magnet.

If the magnet is moved away, the magnetic field around the coil will collapse and induce a voltage in the coil *(Figure 2-11)*. Since the direction of movement of the collapsing field has been reversed, the induced voltage will be opposite in polarity, forcing the current to flow in the same direction.

2-5 Rise Time of Current in an Inductor

When a resistive load is suddenly connected to a source of direct current *(Figure 2-12)*, the current will instantly rise to its maximum value. The resistor shown in *Figure 2-12* has a value of 10 Ω and is connected to a 20-V source. When the switch is closed the current will instantly rise to a value of 2 A (20 V/10 Ω = 2 A).

If the resistor is replaced with an inductor that has a wire resistance of 10 Ω and the switch is closed, the current cannot instantly rise to its maximum value of 2 A *(Figure 2-13)*. As current begins to flow through an inductor, the expanding magnetic field cuts through the conductors, inducing a volt-

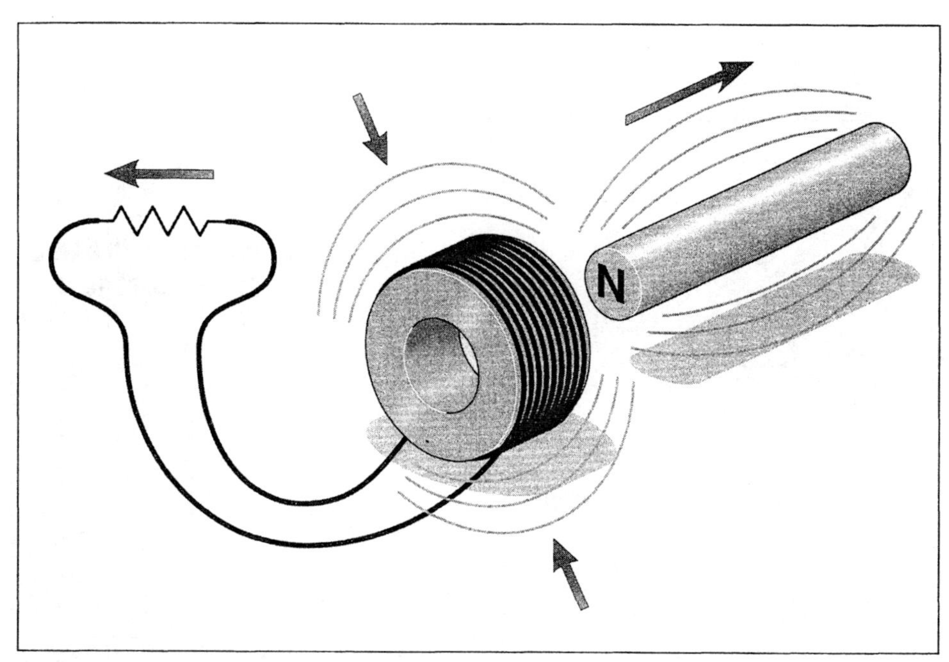

Figure 2-11 The induced voltage forces current to flow in the same direction.

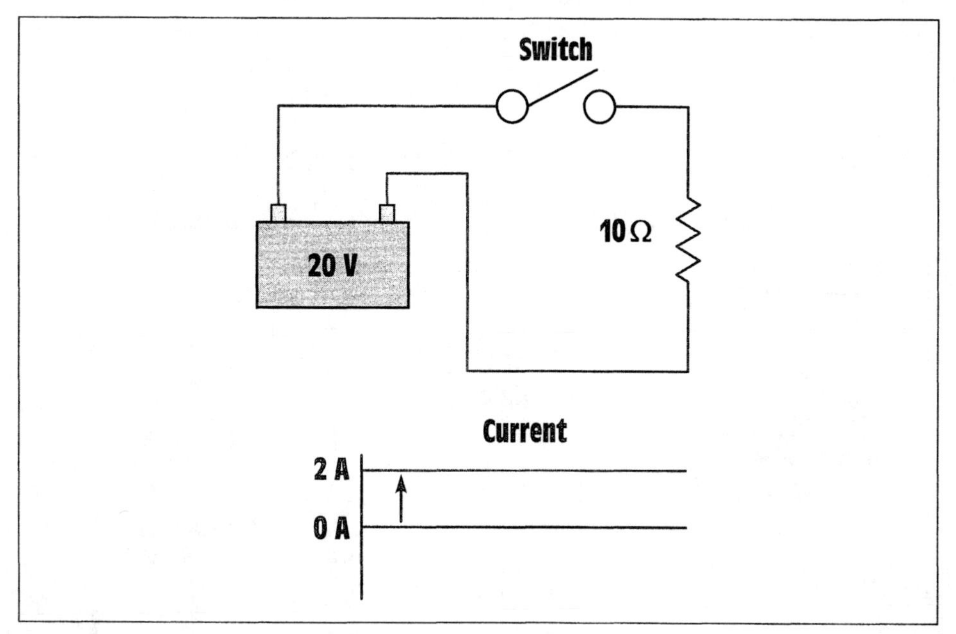

Figure 2-12 The current rises instantly in a resistive circuit.

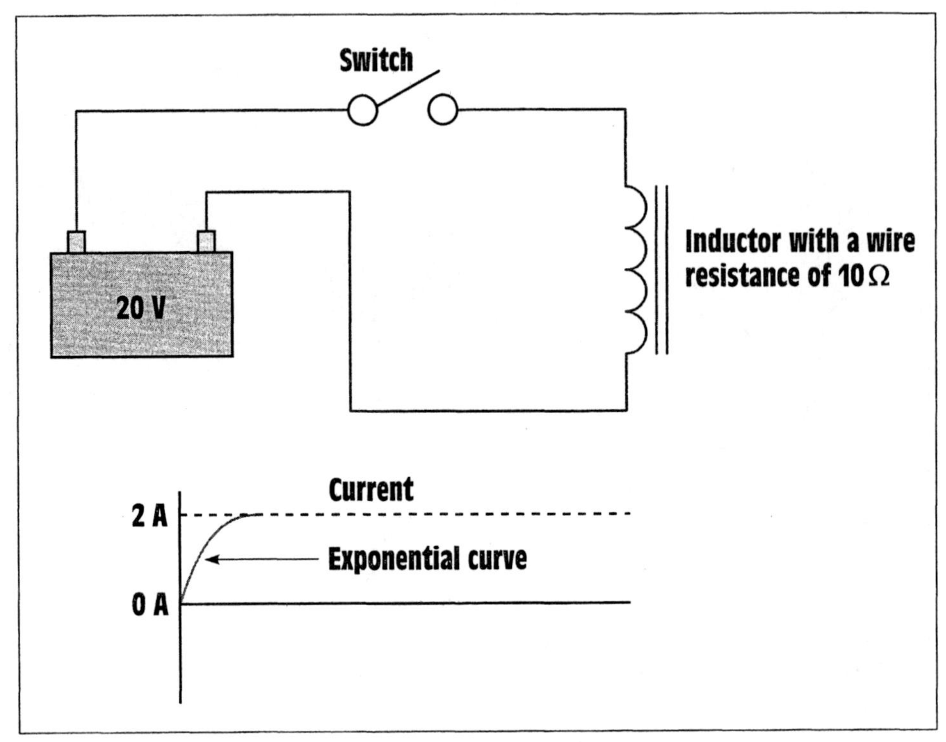

Figure 2-13 Current rises through an indicator at an exponential rate.

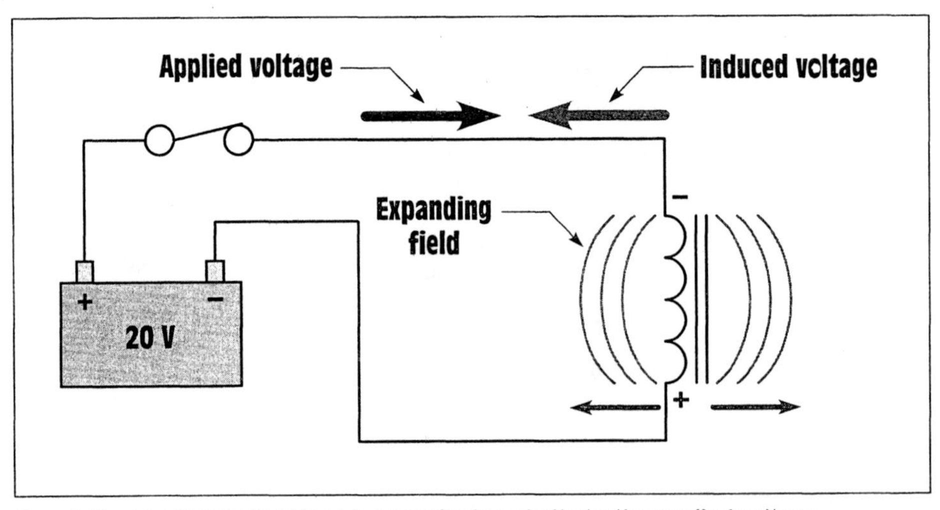

Figure 2-14 The induced voltage is opposite in polarity to the applied voltage.

age into them. In accord with Lenz's law, the induced voltage is opposite in polarity to the applied voltage. The induced voltage, therefore, acts like a resistance to hinder the flow of current through the inductor *(Figure 2-14)*.

The induced voltage is proportional to the rate of change of current (speed of the cutting action). When the switch is first closed, current flow through the coil tries to rise instantly. This extremely fast rate of current change induces maximum voltage in the coil. As the current flow approaches its maximum Ohm's law value, in this example 2 A, the rate of change becomes less and the amount of induced voltage decreases.

2-6 The Exponential Curve

The **exponential curve** describes a rate of certain occurrences. The curve is divided into five time constants. Each time constant is equal to 63.2% of some value. An exponential curve is shown in *Figure 2-15*. In this example, current must rise from zero to a value of 1.5 A at an exponential rate. In this example, 100 ms are required for the current to rise to its full value. Since the current requires a total of 100 ms to rise to its full value,

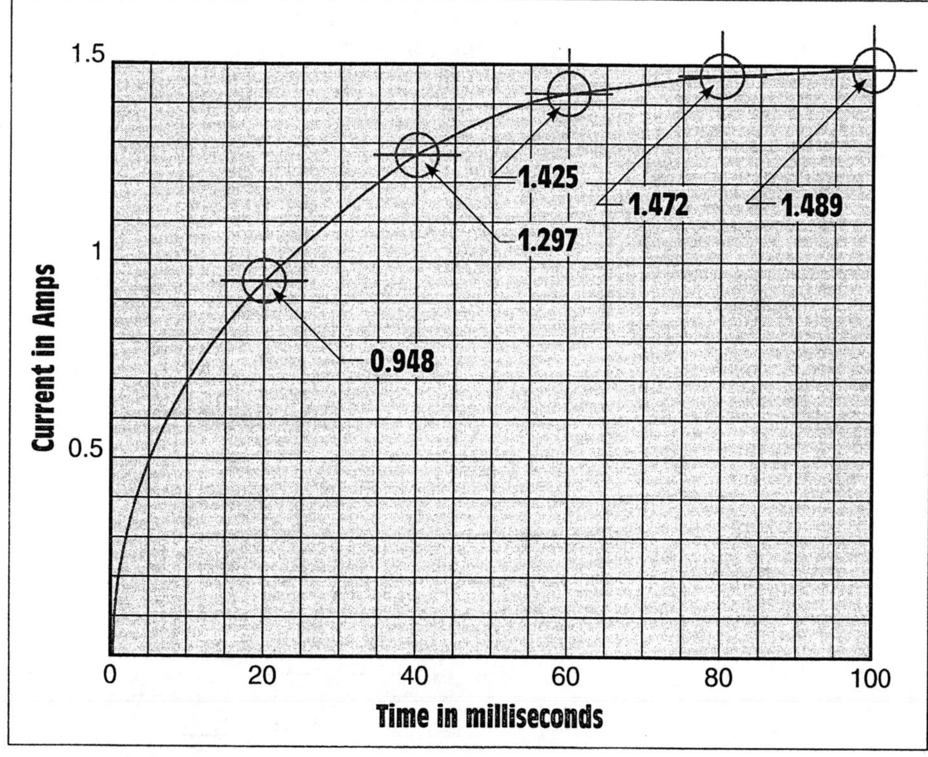

Figure 2-15 Exponential curve.

each time constant is 20 ms (100 ms/5 time constants = 20 ms per time constant). During the first time constant, the current will rise from 0 to 63.2% of its total value, or 0.948 A (1.5 x 0.632 = 0.948). During the second time constant the current will rise to a value of 1.297 A, and during the third time constant the current will reach a total value of 1.425 A.

Because the current increases at a rate of 63.2% during each time constant, it is theoretically impossible to reach the total value of 1.5 A. After five time constants, however, the current has reached approximately 99.3% of the maximum value and for all practical purposes is considered to be complete.

The exponential curve can often be found in nature. If clothes are hung on a line to dry, they will dry at an exponential rate. Another example of the exponential curve can be seen in *Figure 2-16.* In this example, a bucket has been filled to a certain mark with water. A hole has been cut at the bottom of the bucket and a stopper placed in the hole. When the stopper is removed from the bucket, water will flow out at an exponential rate. Assume, for example, it takes 5 min for the water to flow out of the bucket. Exponential curves are always divided into five time constants so in this case each time constant has a value of 1 min. In *Figure 2-17,* if the stopper is removed and water is permitted to drain from the bucket for a period of 1 min before the stopper is replaced, during that first time constant 63.2% of the water in the bucket will drain out. If the stopper is again removed for a period of 1 min, 63.2% of the water remaining in the bucket will drain out. Each time the stopper is removed for a period of one time constant, the bucket will lose 63.2% of its remaining water.

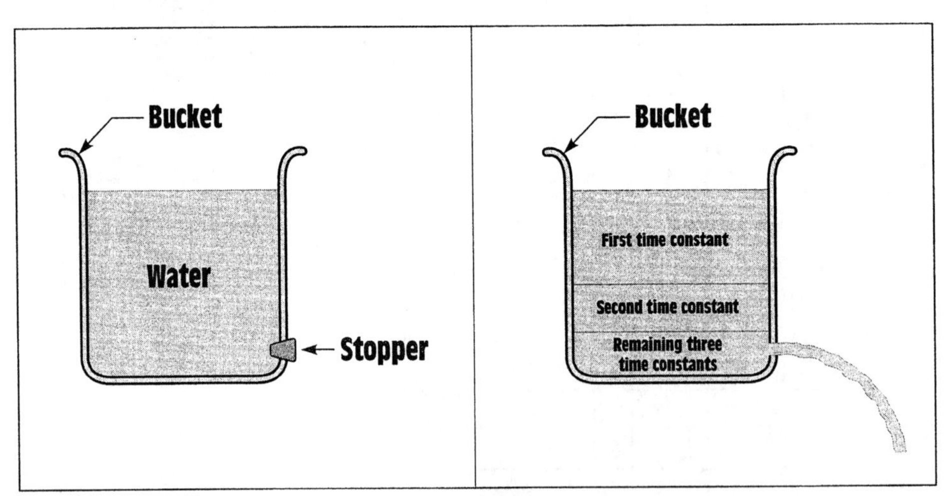

Figure 2-16 Exponential curves can be found in nature.

Figure 2-17 Water flows from a bucket at an exponential rate.

2-7 Inductance

Inductance is measured in units called the **henry (H)** and is represented by the letter *L*. **A coil has an inductance of one henry when a current change of one ampere per second results in an induced voltage of one volt.**

The amount of inductance a coil will have is determined by its physical properties and construction. A coil wound on a nonmagnetic core material such as wood or plastic is referred to as an *air-core* inductor. If the coil is wound on a core made of magnetic material such as silicon steel or soft iron it is referred to as an *iron-core* inductor. Iron-core inductors produce more inductance with fewer turns than air-core inductors because of the good magnetic path provided by the core material. Iron-core inductors cannot be used for high-frequency applications, however, because of **eddy current** loss and **hysteresis loss** in the core material.

Another factor that determines inductance is how far the windings are separated from each other. If the turns of wire are far apart they will have less inductance than turns wound closer together *(Figure 2-18)*.

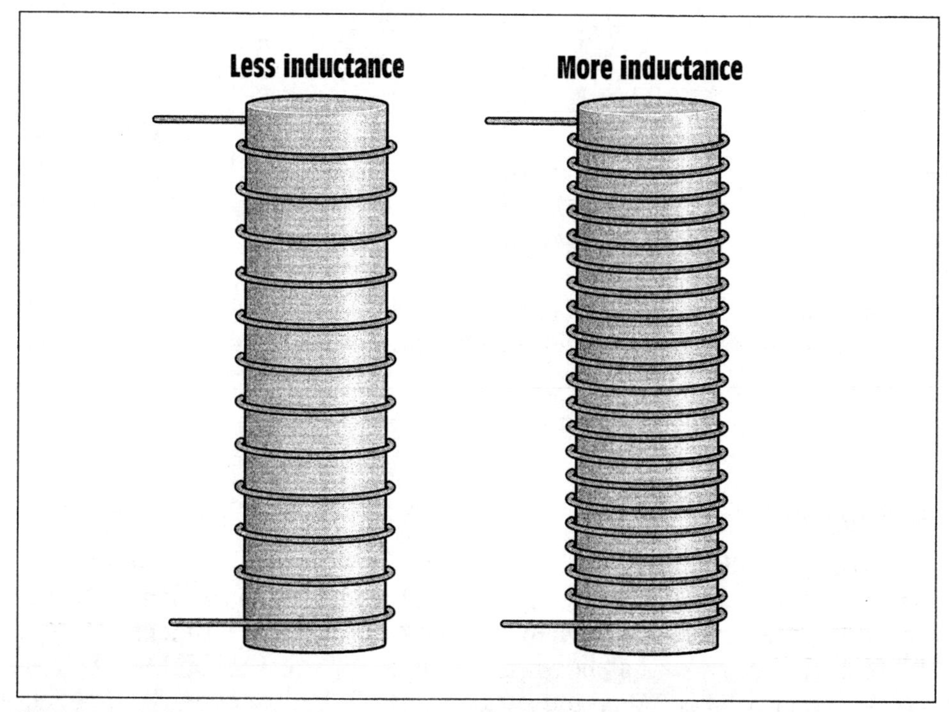

Figure 2-18 Inductance is determined by the physical construction of the coil.

2-8 R-L Time Constants

The time necessary for current in an inductor to reach its full Ohm's law value, called the **R-L time constant**, can be computed using the formula

$$T = \frac{L}{R}$$

where

T = time in seconds

L = inductance in henrys

R = resistance in ohms

This formula computes the time of one time constant.

Example 1

A coil has an inductance of 1.5 H and a wire resistance of 6 Ω. If the coil is connected to a battery of 3 V, how long will it take the current to reach its full Ohm's law value of 0.5 A (3 V/6 Ω = 0.5 A)?

Solution

To find the time of one time constant, use the formula

$$T = \frac{L}{R}$$

$$T = \frac{1.5}{6}$$

$$T = 0.25 \text{ s}$$

The time for one time constant is 0.25 s. Since five time constants are required for the current to reach its full value of 0.5 A, 0.25 s will be multiplied by 5.

$$0.25 \times 5 = 1.25 \text{ s}$$

2-9 Induced Voltage Spikes

A **voltage spike** occurs when the current flow through an inductor stops, and the current decreases at an exponential rate also *(Figure 2-19)*. As long as a complete circuit exists when the power is interrupted, there is little or no problem. In the circuit shown in *Figure 2-20,* a resistor and inductor are connected in parallel. When the switch is closed, the battery will supply cur-

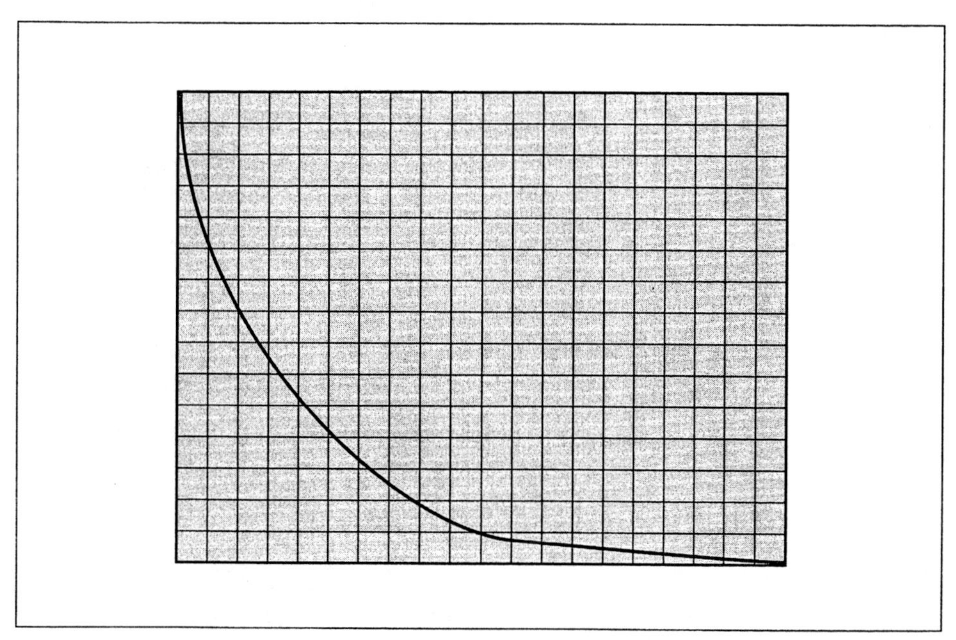

Figure 2-19 Current flow through an inductor deceases at an exponential rate.

rent to both. When the switch is opened, the magnetic field surrounding the inductor will collapse and induce a voltage into the inductor. The induced voltage will attempt to keep current flowing in the same direction. Recall that inductors oppose a change of current. The amount of current flow and the time necessary for the flow to stop will be determined by the resistor and the properties of the inductor. The amount of voltage produced by the collapsing magnetic field is determined by the maximum current in the circuit

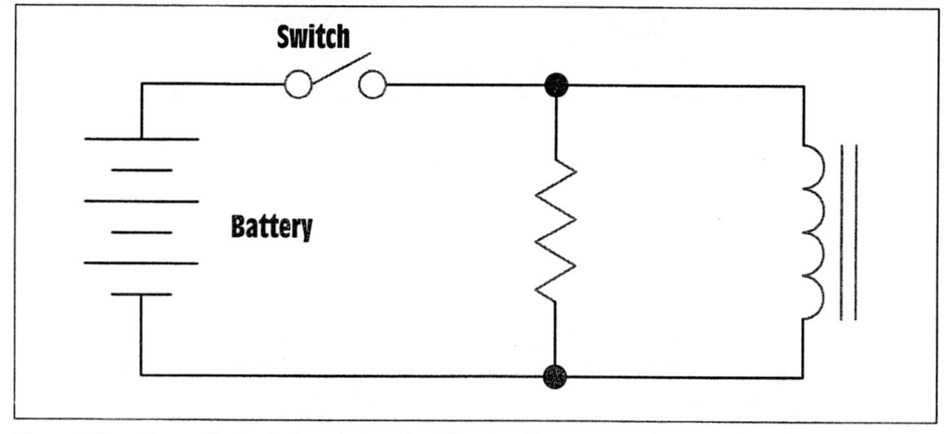

Figure 2-20 The resistor helps prevent voltage spikes caused by the inductor.

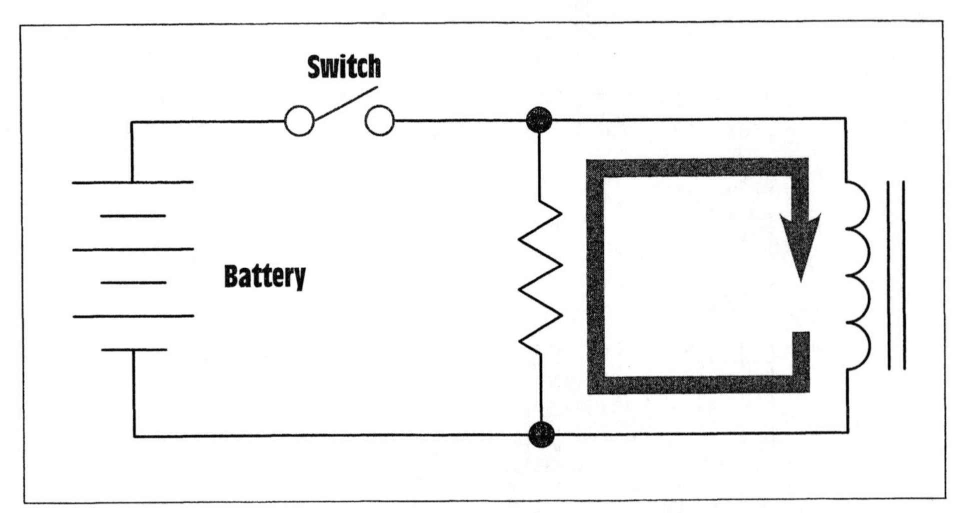

Figure 2-21 When the switch is opened, a series path is formed by the resistor and inductor.

and the total resistance in the circuit. In the circuit shown in *Figure 2-20*, assume that the inductor has a wire resistance of 6 Ω, and the resistor has a resistance of 100 Ω. Also assume that when the switch is closed a current of 2 A will flow through the inductor.

When the switch is opened, a series circuit exists composed of the resistor and inductor *(Figure 2-21)*. The maximum voltage developed in this circuit would be 212 V (2 A x 106 Ω = 212 V). If the circuit resistance were increased, the induced voltage would become greater. If the circuit resistance were decreased, the induced voltage would become less.

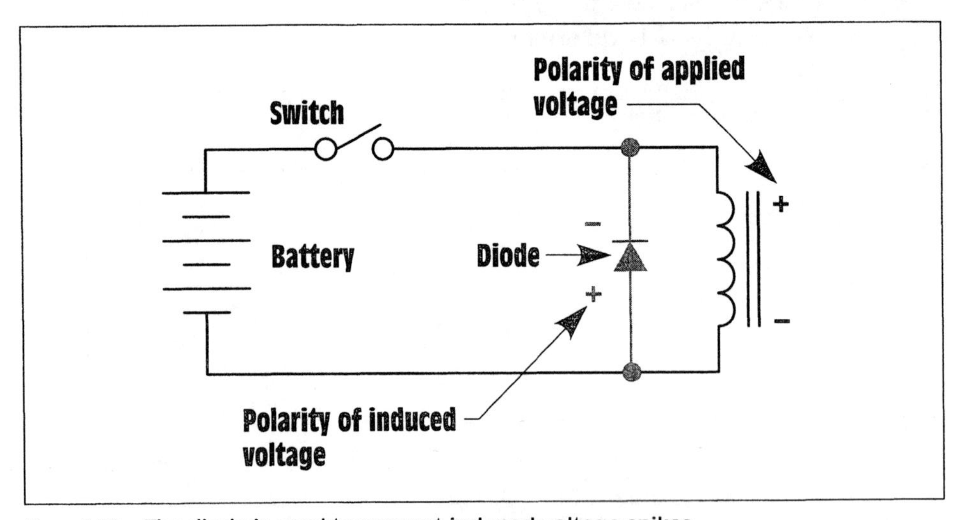

Figure 2-22 The diode is used to prevent induced voltage spikes.

Another device often used to prevent induced voltage spikes when the current flow through an inductor is stopped is the diode *(Figure 2-22)*. The diode is an electronic component that operates like an electrical check valve. The diode will permit current to flow through it in only one direction. The diode is connected in parallel with the inductor in such a manner that when voltage is applied to the circuit, the diode is reverse-biased and acts like an open switch. When the diode is reverse-biased no current will flow through it.

When the switch is opened, the induced voltage produced by the collapsing magnetic field will be opposite in polarity to the applied voltage. The diode then becomes forward-biased and acts like a closed switch. Current can now flow through the diode and complete a circuit back to the inductor. A silicon diode has a forward voltage drop of approximately 0.7 V regardless of the current flowing through it. Since the diode is connected in parallel with the inductor, and voltage drops of devices connected in parallel must be the same, the induced voltage is limited to approximately 0.7 V. The diode can be used to eliminate inductive voltage spikes in direct current circuits only; it cannot be used for this purpose in alternating current circuits.

A device that can be used for spike suppression in either direct or alternating-current circuits is the **metal oxide varistor (MOV).** The MOV is a bidirectional device, which means that it will conduct current in either direction, and can, therefore, be used in alternating-current circuits. The metal oxide varistor is an extremely fast-acting solid-state component that will exhibit a change of resistance when the voltage reaches a certain point. Assume that the MOV shown in *Figure 2-23* has a voltage rating of 140 V, and that the voltage applied to the circuit is 120 V. When the switch is closed and current flows through the circuit, a magnetic field will be established around the inductor *(Figure 2-24)*. As long as the voltage applied to the MOV is less than 140 V, it will exhibit an extremely high resistance, in the range of several hundred thousand ohms.

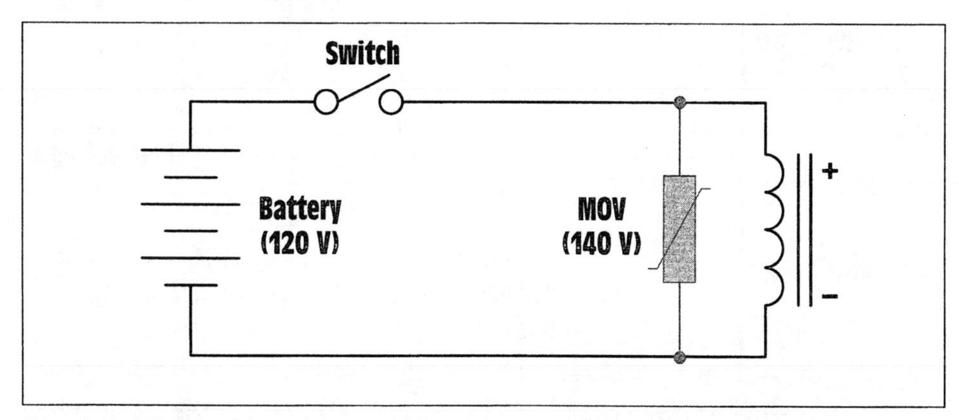

Figure 2-23 Metal oxide varistor used to suppress a voltage spike

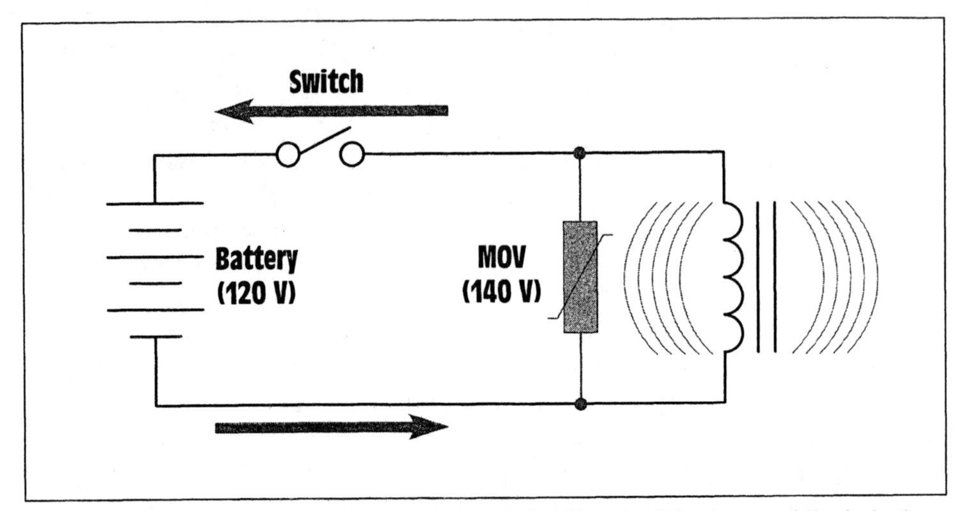

Figure 2-24 When the switch is closed, a magnetic field is established around the inductor.

When the switch is opened, current flow through the coil suddenly stops, and the magnetic field collapses. This sudden collapse of the magnetic field will cause an extremely high voltage to be induced in the coil. When this induced voltage reaches 140 V, however, the MOV will suddenly change from a high resistance to a low resistance, preventing the voltage from becoming greater than 140 V *(Figure 2-25)*.

Metal oxide varistors are extremely fast-acting. They can typically change resistance values in less than 20 ns (nanoseconds). They are often found

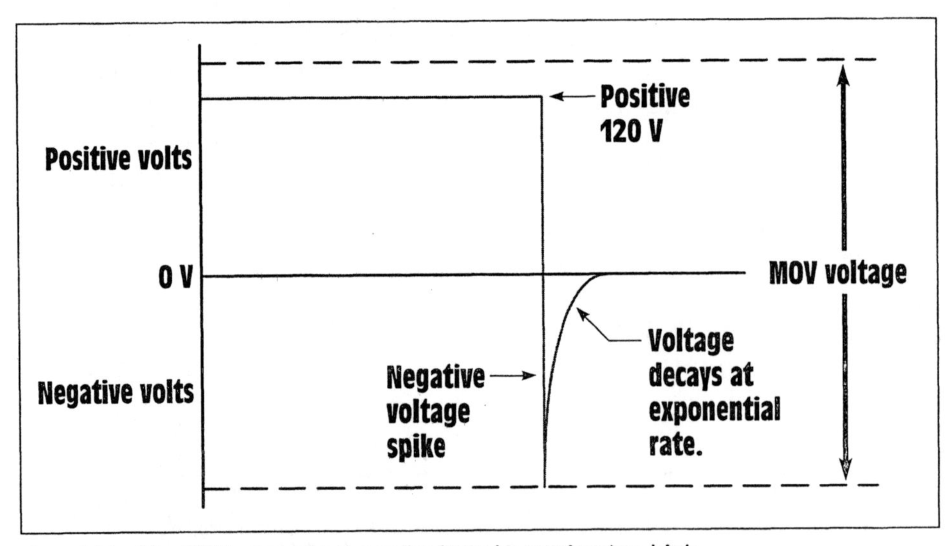

Figure 2-25 The MOV prevents the spike from becoming too high.

connected across the coils of relays and motor starters in control systems to prevent voltage spikes from being induced back into the line. They are also found in the surge protectors used to protect many home appliances such as televisions, stereos, and computers.

If nothing is connected in the circuit with the inductor when the switch opens, the induced voltage can become extremely high. In this instance, the resistance of the circuit is the air gap of the switch contacts, which is practically infinite. The inductor will attempt to produce any voltage necessary to prevent a change of current. Inductive voltage spikes can reach thousands of volts. This is the principle of operation of many high-voltage devices such as the ignition systems of many automobiles.

Another device that uses the collapsing magnetic field of an inductor to produce a high voltage is the electric-fence charger, shown in *Figure 2-26*. The switch is constructed in such a manner that it will pulse on and off. When the switch closes, current flows through the inductor, and a magnetic field is produced around the inductor. When the switch opens, the magnetic field collapses and induces a high voltage across the inductor. If anything or anyone standing on the ground touches the fence, a circuit is completed through the object or person and the ground. The coil is generally constructed of many turns of very small wire. This construction provides the coil with a high resistance and limits current flow when the field collapses.

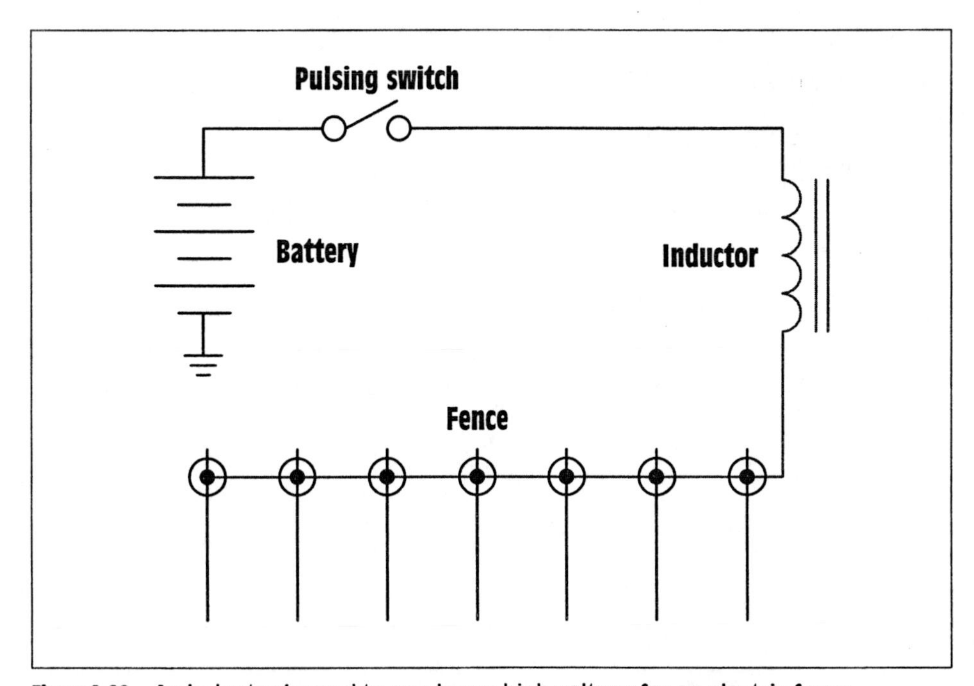

Figure 2-26 An inductor is used to produce a high voltage for an electric fence.

Summary

1. When current flows through a conductor, a magnetic field is created around the conductor.
2. When a conductor is cut by a magnetic field, a voltage is induced in the conductor.
3. The polarity of the induced voltage is determined by the polarity of the magnetic field in relation to the direction of motion.
4. Three factors that determine the amount of induced voltage are:
   a. the number of turns of wire,
   b. the strength of the magnetic field, and
   c. the speed of the cutting action.
5. One volt is induced in a conductor when magnetic lines of flux are cut at a rate of one weber per second.
6. Induced voltage is always opposite in polarity to the applied voltage.
7. Inductors oppose a change of current.
8. Current rises in an inductor at an exponential rate.
9. An exponential curve is divided into five time constants.
10. Each time constant is equal to 63.2% of some value.
11. Inductance is measured in units called henrys (H).
12. A coil has an inductance of 1 H when a current change of 1 A per second results in an induced voltage of 1 V.
13. Air-core inductors are inductors wound on cores of nonmagnetic material.
14. Iron-core inductors are wound on cores of magnetic material.
15. The amount of inductance an inductor will have is determined by the number of turns of wire and the physical construction of the coil.
16. Inductors can produce extremely high voltages when the current flowing through them is stopped.
17. Two devices used to help prevent large spike voltages are the resistor and diode.

Review Questions

1. What determines the polarity of magnetism when current flows through a conductor?
2. What determines the strength of the magnetic field when current flows through a conductor?
3. Name three factors that determine the amount of induced voltage in a coil.
4. How many lines of magnetic flux must be cut in 1 s to induce a voltage of 1 V?
5. What is the effect on induced voltage of adding more turns of wire to a coil?
6. Into how many time constants is an exponential curve divided?
7. Each time constant of an exponential curve is equal to what percentage of the whole?
8. An inductor has an inductance of 0.025 H and a wire resistance of 3 Ω. How long will it take the current to reach its full Ohm's law value?
9. Refer to the circuit shown in *Figure 2-20*. Assume that the inductor has a wire resistance of 0.2 Ω and the resistor has a value of 250 Ω. If a current of 3 A is flowing through the inductor what will be the maximum induced voltage when the switch is opened?
10. What electronic component is often used to prevent large voltage spikes from being produced when the current flow through an inductor is suddenly terminated?

3

Inductance in Alternating-Current Circuits

Objectives

After studying this unit, you should be able to

- Discuss the properties of inductance in an alternating current circuit
- Discuss inductive reactance
- Compute values of inductive reactance and inductance
- Discuss the relationship of voltage and current in a pure inductive circuit
- Be able to compute values for inductors connected in series or parallel
- Discuss reactive power (VARs)
- Determine the Q of a coil

This unit discusses the effects of inductance on alternating-current circuits. The unit explains how current is limited in an inductive circuit as well as the effect inductance has on the relationship of voltage and current.

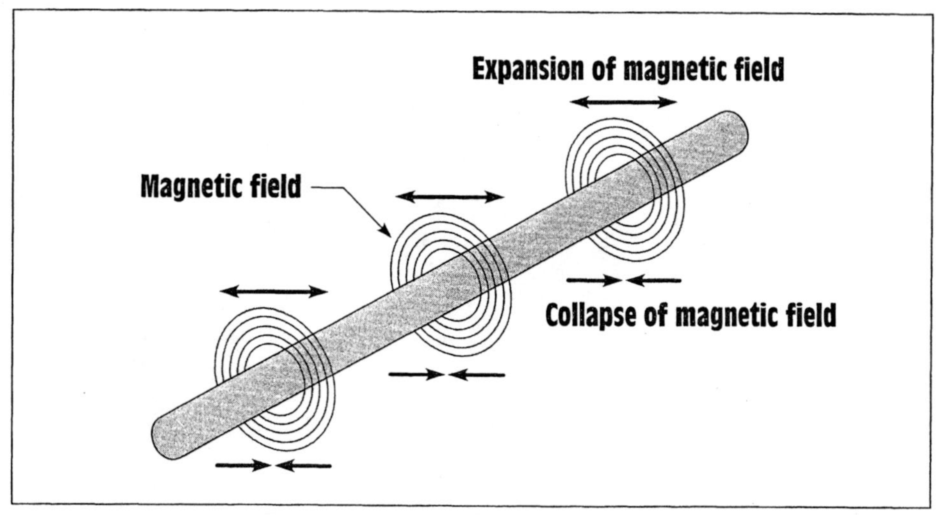

Figure 3-1 A continually changing magnetic field induces a voltage into any conductor.

3-1 Inductance

Inductance (L) is one of the primary types of loads in alternating-current circuits. Some amount of inductance is present in all alternating-current circuits because of the continually changing magnetic field *(Figure 3-1)*. The amount of inductance of a single conductor is extremely small, and in most instances it is not considered in circuit calculations. Circuits are generally considered to contain inductance when any type of load that contains a coil is used. For circuits that contain a coil, inductance *is* considered in circuit calculations. Loads such as motors, transformers, lighting ballast, and chokes all contain coils of wire.

In Unit 2, it was discussed that whenever current flows through a coil of wire a magnetic field is created around the wire *(Figure 3-2)*. If the amount of current decreases, the magnetic field will collapse *(Figure 3-3)*. Recall from Unit 2 several facts concerning inductance:

1. When magnetic lines of flux cut through a coil, a voltage is induced in the coil.
2. An induced voltage is always opposite in polarity to the applied voltage.
3. The amount of induced voltage is proportional to the rate of change of current.
4. An inductor opposes a change of current.

The inductors in *Figures 3-2* and *3-3* are connected to an alternating voltage. Therefore the magnetic field continually increases, decreases, and

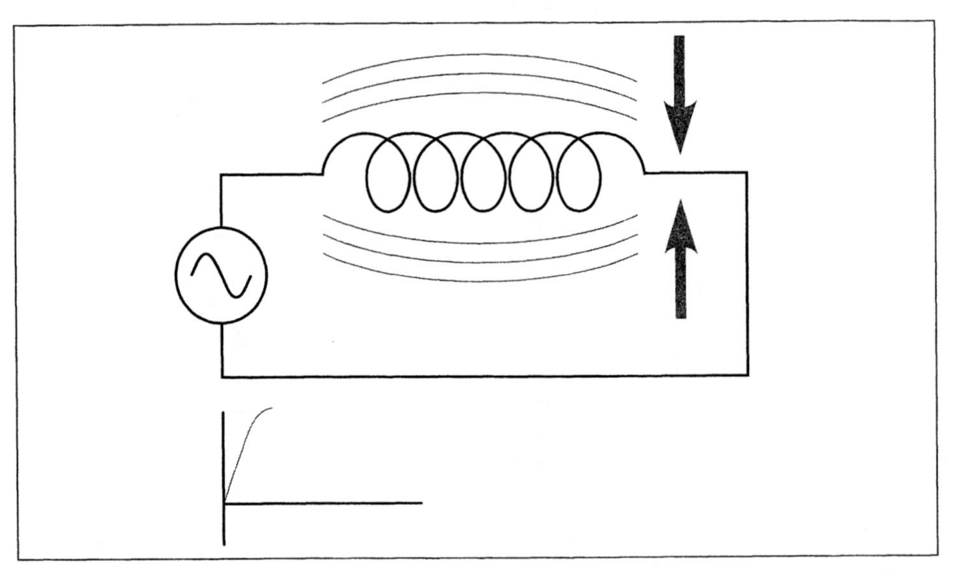

Figure 3-2 As current flow through a coil, a magnetic field is created around the coil.

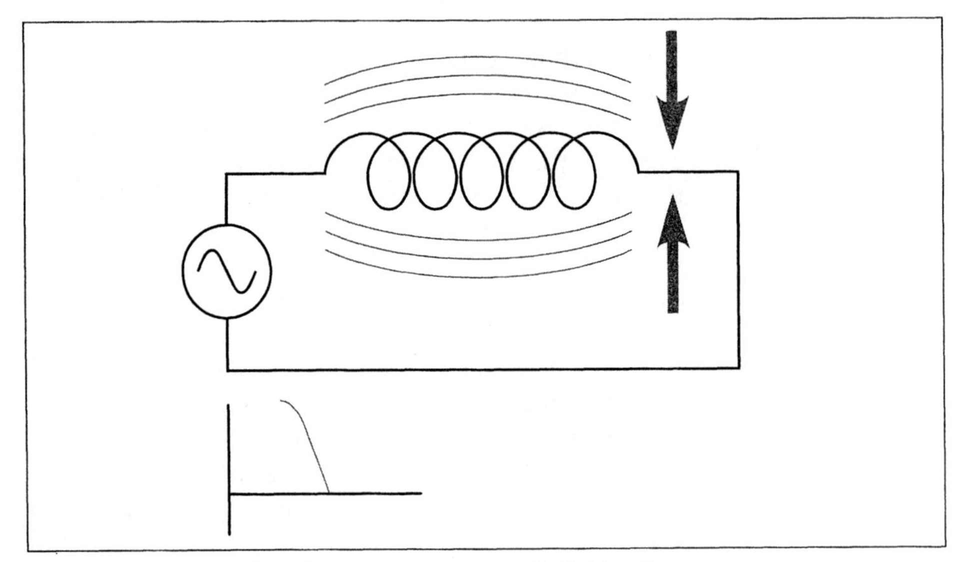

Figure 3-3 As current flow decreases, the magnetic field collapses.

reverses polarity. Since the magnetic field continually changes magnitude and direction, a voltage is continually being induced in the coil. This **induced voltage** is 180° out of phase with the applied voltage and is always in opposition to the applied voltage *(Figure 3-4)*. Since the induced voltage is always in opposition to the applied voltage, the applied voltage must

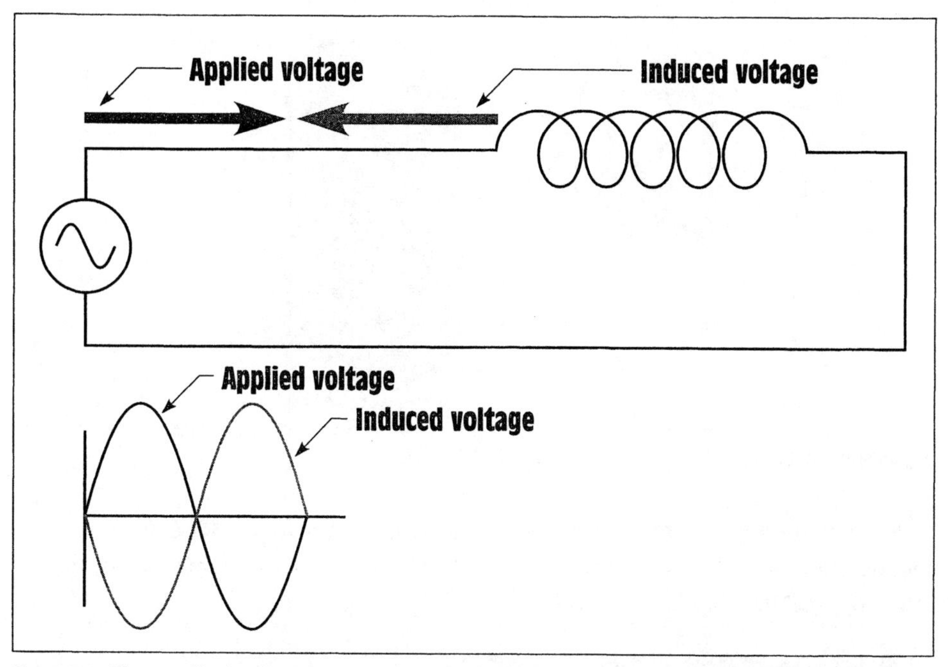

Figure 3-4 The applied voltage and induced voltage are 180° out of phase with each other.

overcome the induced voltage before current can flow through the circuit. For example, assume an inductor is connected to a 120-V AC line. Now assume that the inductor has an induced voltage of 116 V. Since an equal amount of applied voltage must be used to overcome the induced voltage, there will be only 4 V to push current through the wire resistance of the coil (120 – 116 = 4).

Computing the Induced Voltage

The amount of induced voltage in an inductor can be computed if the resistance of the wire in the coil and the amount of circuit current are known. For example, assume that an ohmmeter is used to measure the actual amount of resistance in a coil, and the coil is found to contain 6 Ω of wire resistance *(Figure 3-5)*. Now assume that the coil is connected to a 120-V AC circuit and an ammeter measures a current flow of 0.8 A *(Figure 3-6)*. Ohm's law can now be used to determine the amount of voltage necessary to push 0.8 A of current through 6 Ω of resistance.

$$E = I \times R$$

$$E = 0.8 \times 6$$

$$E = 4.8\ V$$

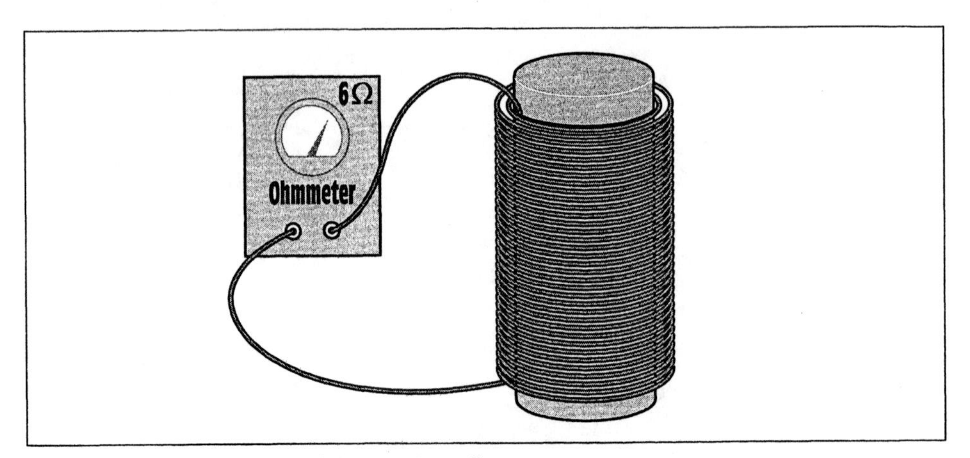

Figure 3-5 Measuring the resistance of a coil.

Since only 4.8 V is needed to push the current through the wire resistance of the inductor, the remainder of the 120 V is used to overcome the coil's induced voltage of 115.2 V (120 – 4.8 = 115.2).

3-2 Inductive Reactance

Notice that the induced voltage is able to limit the flow of current through the circuit in a manner similar to resistance. This induced voltage is *not* resistance, but it can limit the flow of current just as resistance does. This current-limiting property of the inductor is called **reactance** and is symbolized by the letter *X*. This reactance is caused by inductance, so it is called **inductive reactance** and is symbolized by $\mathbf{X_L}$, pronounced "X sub L." Inductive reactance is measured in ohms just as resistance is and can be

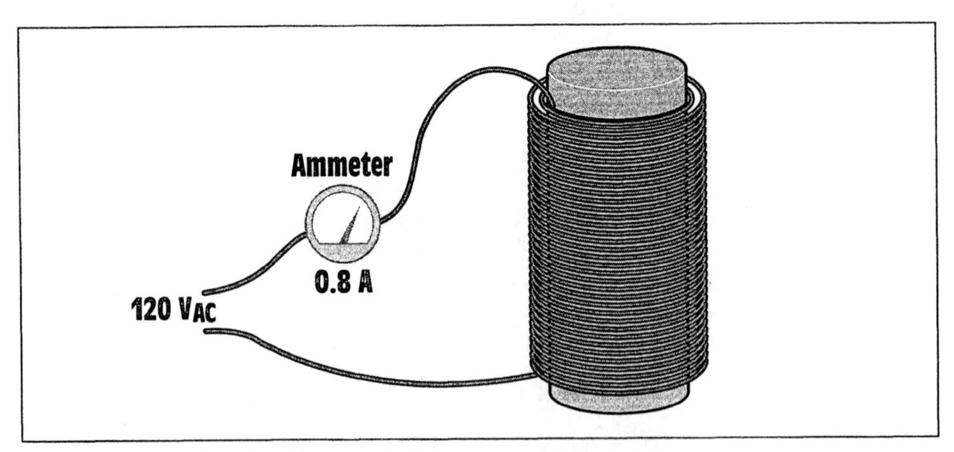

Figure 3-6 Measuring circuit current with an ammeter.

computed when the values of inductance and frequency are known. The following formula can be used to find inductive reactance.

$$X_L = 2\pi FL$$

where

X_L = inductive reactance

2 = a constant

π = 3.1416

F = frequency in hertz (Hz)

L = inductance in henrys (H)

Inductive reactance is an induced voltage and is, therefore, proportional to the three factors that determine induced voltage:

1. the **number** of turns of wire,
2. the **strength** of the magnetic field, and
3. the **speed** of the cutting action (relative motion between the inductor and the magnetic lines of flux).

The number of turns of wire and strength of the magnetic field are determined by the physical construction of the inductor. Factors such as the size of wire used, the number of turns, how close the turns are to each other, and the type of core material determine the amount of inductance (in henrys, H) of the coil *(Figure 3-7)*. The speed of the cutting action is proportional to the frequency (Hz). An increase of frequency will cause the magnetic lines of flux to cut the conductors at a faster rate, and thus will produce a higher induced voltage or more inductive reactance.

Example 1

The inductor shown in *Figure 3-8* has an inductance of 0.8 H and is connected to a 120-V, 60-Hz line. How much current will flow in this circuit if the wire resistance of the inductor is negligible?

Solution

The first step is to determine the amount of inductive reactance of the inductor.

$$X_L = 2\pi FL$$

$$X_L = 2 \times 3.1416 \times 60 \times 0.8$$

$$X_L = 301.6\ \Omega$$

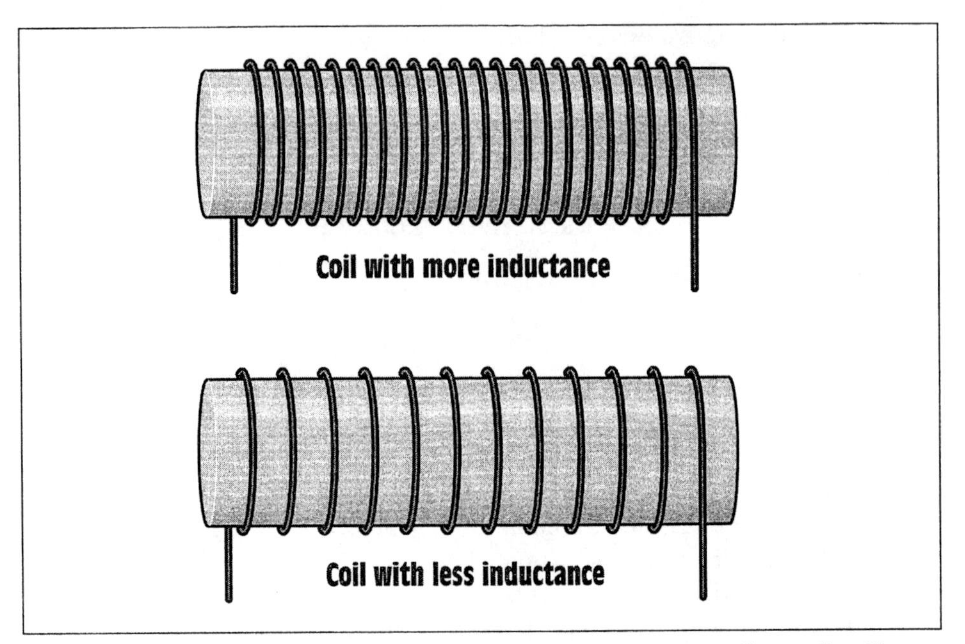

Figure 3-7 Coils with turns close together produce more inductance than coils with turns far apart.

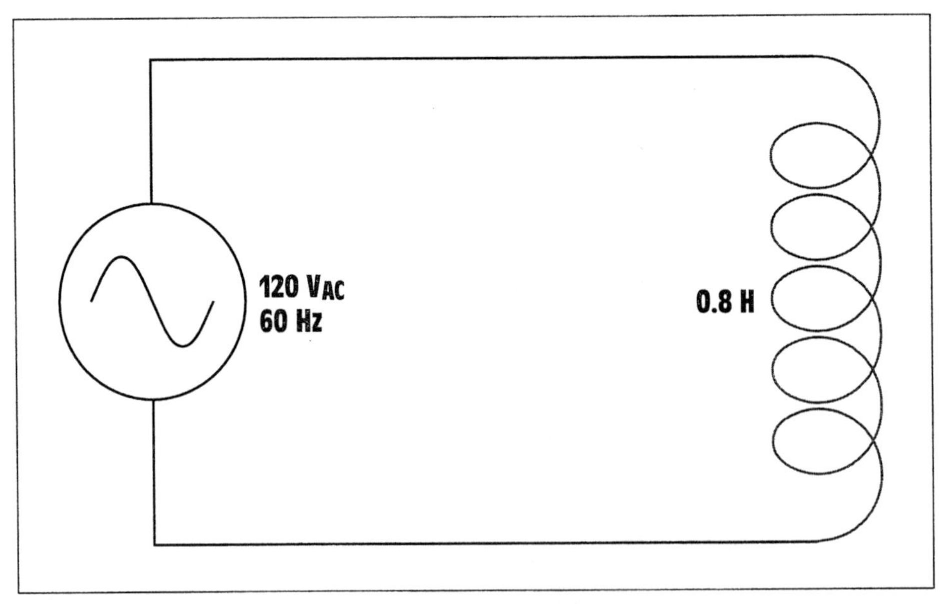

Figure 3-8 Circuit current is limited by inductive reactance.

Since inductive reactance is the current-limiting property of this circuit, it can be substituted for the value of R in an Ohm's law formula.

$$I = \frac{E}{X_L}$$

$$I = \frac{120}{301.6}$$

$$I = 0.398 \text{ A}$$

If the amount of inductive reactance is known, the inductance of the coil can be determined using the formula

$$L = \frac{X_L}{2\pi F}$$

Example 2

Assume an inductor with a negligible resistance is connected to a 36-V, 400-Hz line. If the circuit has a current flow of 0.2 A what is the inductance of the inductor?

Solution

The first step is to determine the inductive reactance of the circuit.

$$X_L = \frac{E}{I}$$

$$X_L = \frac{36}{0.2}$$

$$X_L = 180\ \Omega$$

Now that the inductive reactance of the inductor is known, the inductance can be determined.

$$L = \frac{X_L}{2\pi F}$$

$$L = \frac{180}{2 \times 3.1416 \times 400}$$

$$L = 0.0716 \text{ H}$$

Example 3

An inductor with negligible resistance is connected to a 480-V, 60-Hz line. An ammeter indicates a current flow of 24 A. How much current will flow in this circuit if the frequency is increased to 400 Hz?

Solution

The first step in solving this problem is to determine the amount of inductance of the coil. Since the resistance of the wire used to make the inductor is negligible, the current is limited by inductive reactance. The inductive reactance can be found by substituting X_L for R in an Ohm's law formula.

$$X_L = \frac{E}{I}$$

$$X_L = \frac{480}{24}$$

$$X_L = 20\ \Omega$$

Now that the inductive reactance is known, the inductance of the coil can be found using the formula

$$L = \frac{X_L}{2\pi F}$$

NOTE: When using a frequency of 60 Hz, $2 \times \pi \times 60 = 377$. Since 60 Hz is the major frequency used throughout the United States and Canada, 377 should be memorized for use when necessary.

$$L = \frac{20}{377}$$

$$L = 0.053\ H$$

Since the inductance of the coil is determined by its physical construction, it will not change when connected to a different frequency. Now that the inductance of the coil is known, the inductive reactance at 400 Hz can be computed.

$$X_L = 2\pi FL$$

$$X_L = 2 \times 3.1416 \times 400 \times 0.053$$

$$X_L = 133.2\ \Omega$$

The amount of current flow can now be found by substituting the value of inductive reactance for resistance in an Ohm's law formula.

$$I = \frac{E}{X_L}$$

$$I = \frac{480}{133.2}$$

$$I = 3.6 \text{ A}$$

3-3 Schematic Symbols

The schematic symbol used to represent an inductor depicts a coil of wire. Several symbols for inductors are shown in *Figure 3-9.* The symbols shown with the two parallel lines represent iron-core inductors, and the symbols without the parallel lines represent air-core inductors.

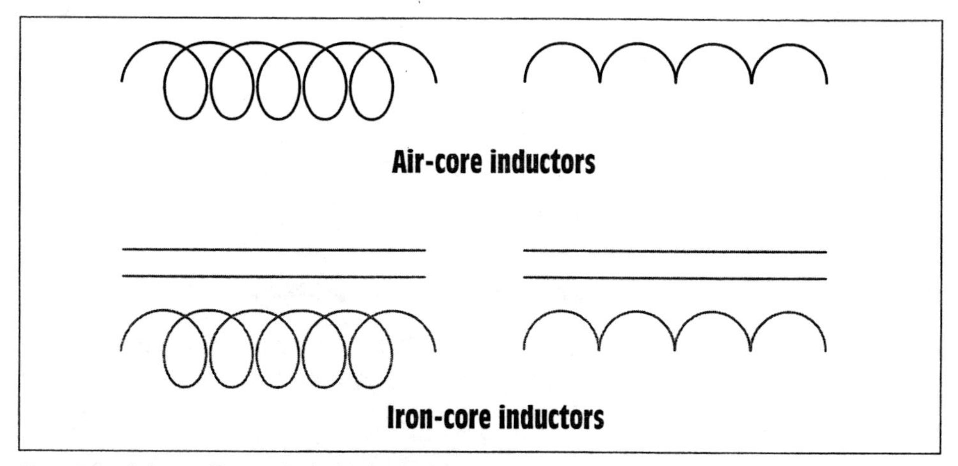

Figure 3-9 Schematic symbols for inductors.

3-4 Inductors Connected in Series

When inductors are connected in series, the total inductance of the circuit (L_T) equals the sum of the inductances of all the inductors.

$$L_T = L_1 + L_2 + L_3$$

The total inductive reactance (X_{LT}) of inductors connected in series equals the sum of the inductive reactances for all the inductors.

$$X_{LT} = X_{L1} + X_{L2} + X_{L3}$$

Example 4

Three inductors are connected in series. Inductor 1 has an inductance of 0.6 H, inductor 2 has an inductance of 0.4 H, and inductor 3 has an inductance of 0.5 H. What is the total inductance of the circuit?

Solution

$$L_T = 0.6 + 0.4 + 0.5$$

$$L_T = 1.5 \text{ H}$$

Example 5

Three inductors are connected in series. Inductor 1 has an inductive reactance of 180 Ω, inductor 2 has an inductive reactance of 240 Ω, and inductor 3 has an inductive reactance of 320 Ω. What is the total inductive reactance of the circuit?

Solution

$$X_{LT} = 180\ \Omega + 240\ \Omega + 320\ \Omega$$

$$X_{LT} = 740\ \Omega$$

3-5 Inductors Connected in Parallel

When inductors are connected in parallel, the total inductance can be found in a similar manner to finding the total resistance of a parallel circuit. The reciprocal of the total inductance is equal to the sum of the reciprocals of all the inductors.

$$\frac{1}{L_T} = \frac{1}{L_1} + \frac{1}{L_2} + \frac{1}{L_3}$$

or

$$L_T = \frac{1}{\frac{1}{L_1} + \frac{1}{L_2} + \frac{1}{L_3}}$$

Another formula that can be used to find the total inductance of parallel inductors is the product over sum formula.

$$L_T = \frac{L_1 \times L_2}{L_1 + L_2}$$

If the values of all the inductors are the same, total inductance can be found by dividing the inductance of one inductor by the total number of inductors.

$$L_T = \frac{L}{N}$$

Similar formulas can be used to find the total inductive reactance of inductors connected in parallel.

$$\frac{1}{X_{LT}} = \frac{1}{X_{L1}} + \frac{1}{X_{L2}} + \frac{1}{X_{L3}}$$

or

$$X_{LT} = \frac{1}{\frac{1}{X_{L1}} + \frac{1}{X_{L2}} + \frac{1}{X_{L3}}}$$

or

$$X_{LT} = \frac{X_{L1} \times X_{L2}}{X_{L1} + X_{L2}}$$

or

$$X_{LT} = \frac{X_L}{N}$$

Example 6

Three inductors are connected in parallel. Inductor 1 has an inductance of 2.5 H, inductor 2 has an inductance of 1.8 H, and inductor 3 has an inductance of 1.2 H. What is the total inductance of this circuit?

Solution

$$L_T = \frac{1}{\frac{1}{2.5} + \frac{1}{1.8} + \frac{1}{1.2}}$$

$$L_T = \frac{1}{1.789}$$

$$L_T = 0.559 \text{ H}$$

3-6 Voltage and Current Relationships in an Inductive Circuit

When current flows through a pure resistive circuit, the current and voltage are in phase with each other. **In a pure inductive circuit the current lags the voltage by 90°.** At first this may seem to be an impossible condition until the relationship of applied voltage and induced voltage is considered. How the current and applied voltage can become 90° out of phase with each other can best be explained by comparing the relationship of the current and induced voltage *(Figure 3-10)*. Recall that the induced voltage is proportional to the rate of change of the current (speed of cutting action). At the beginning of the wave form, the current is shown at its maximum value in the negative direction. At this time, the current is not changing, so induced voltage is zero. As the current begins to decrease in value, the magnetic field produced by the flow of current decreases or collapses and begins to induce a voltage into the coil as it cuts through the conductors *(Figure 3-3)*.

The greatest rate of current change occurs when the current passes from negative, through zero, and begins to increase in the positive direction *(Figure 3-11)*. Since the current is changing at the greatest rate, the induced voltage is maximum. As current approaches its peak value in the positive direction, the rate of change decreases, causing a decrease in the induced voltage. The induced voltage will again be zero when the current reaches its peak value and the magnetic field stops expanding.

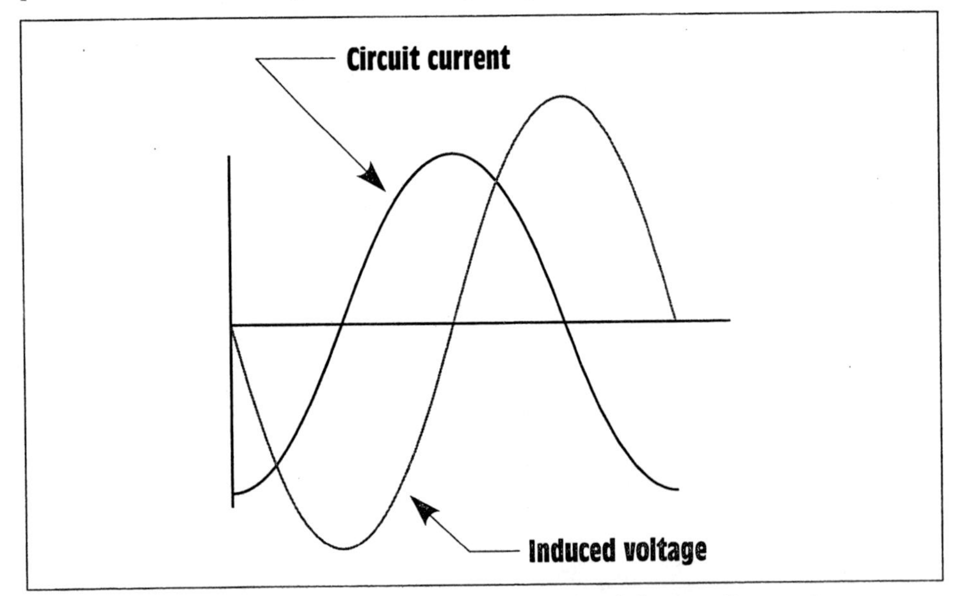

Figure 3-10 Induced voltage is proportional to the rate of change of current.

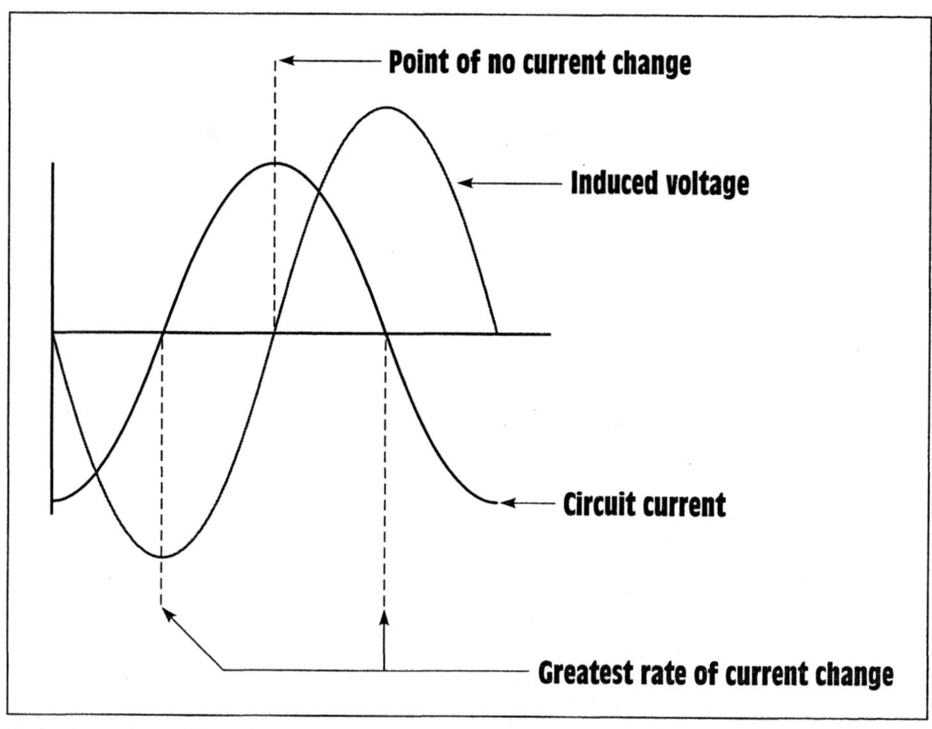

Figure 3-11 No voltage is induced when the current does not change.

It can be seen that the current flowing through the inductor is leading the induced voltage by 90°. Since the induced voltage is 180° out of phase with the applied voltage, the current will lag the applied voltage by 90° *(Figure 3-12)*.

3-7 Power in an Inductive Circuit

In a pure resistive circuit, the true power, or watts, is equal to the product of the voltage and current. In a pure inductive circuit, however, no true power, or watts, is produced. Recall that voltage and current must both be either positive or negative before true power can be produced. Since the voltage and current are 90° out of phase with each other in a pure inductive circuit, the current and voltage will be at different polarities 50% of the time and at the same polarity 50% of the time. During the period of time that the current and voltage have the same polarity, power is being given to the circuit in the form of creating a magnetic field. When the current and voltage are opposite in polarity, power is being given back to the circuit as the magnetic field collapses and induces a voltage back into the circuit. Since power is stored in the form of a magnetic field and then given back, no power is used by the inductor. Any power used in an inductor is caused by losses

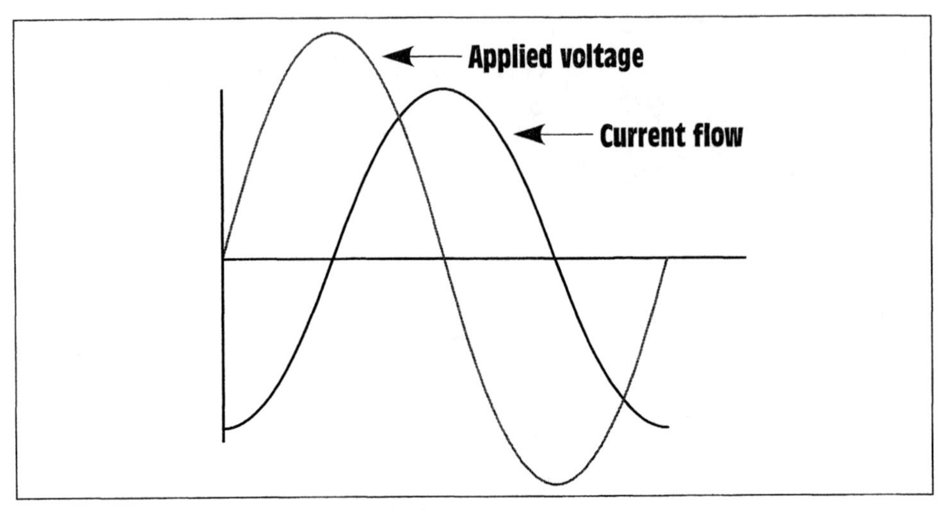

Figure 3-12 The current lags the applied voltage by 90°.

such as the resistance of the wire used to construct the inductor, generally referred to as I^2R losses, eddy current losses, and hysteresis losses.

The current and voltage wave form in *Figure 3-13* has been divided into four sections: A, B, C, and D. During the first time period, indicated by A, the current is negative and the voltage is positive. During this period, energy is being given to the circuit as the magnetic field collapses. During the second time period, section B, both the voltage and current are positive. Power is being used to produce the magnetic field. In the third time period, C, the

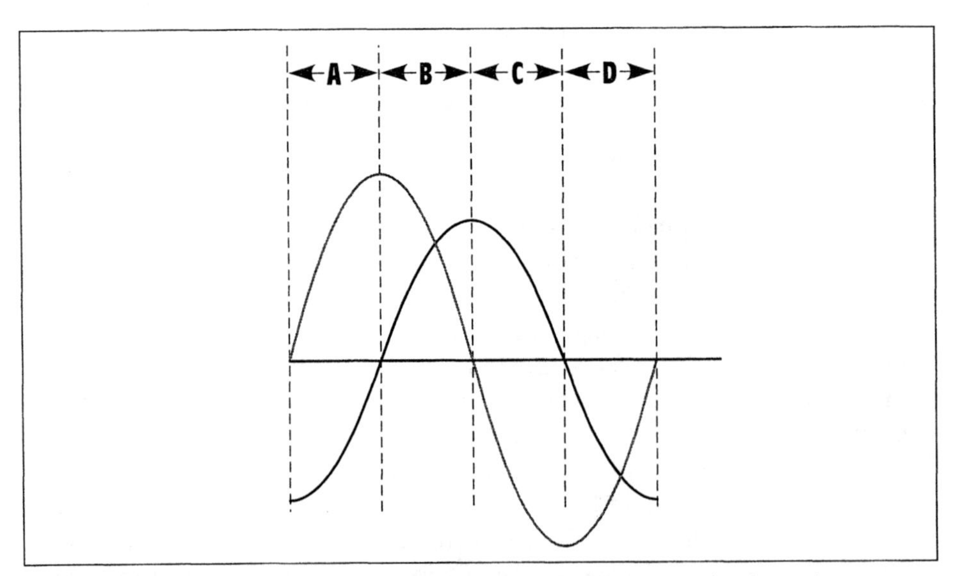

Figure 3-13 Voltage and current relationships during different parts of a cycle.

current is positive and the voltage is negative. Power is again being given back to the circuit as the field collapses. During the fourth time period, D, both the voltage and current are negative. Power is again being used to produce the magnetic field. If the amount of power used to produce the magnetic field is subtracted from the power given back, the result will be zero.

3-8 Reactive Power

Although essentially no true power is being used, except by previously mentioned losses, an electrical measurement called **VARs** is used to measure the **reactive power** in a pure inductive circuit. **VARs** is an abbreviation for volt-amps-reactive. VARs can be computed in the same way as watts except that inductive values are substituted for resistive values in the formulas. VARs is equal to the amount of current flowing through an inductive circuit times the voltage applied to the inductive part of the circuit. Several formulas for computing VARs are:

$$VARs = E_L \times I_L$$

$$VARs = \frac{E_L^2}{X_L}$$

$$VARs = I_L^2 \times X_L$$

where

E_L = voltage applied to an inductor

I_L = current flow through an inductor

X_L = inductive reactance

3-9 Q of an Inductor

So far in this unit, it has been generally assumed that an inductor has no resistance and that inductive reactance is the only current-limiting factor. In reality, that is not true. Since inductors are actually coils of wire they all contain some amount of internal resistance. Inductors actually appear to be a coil connected in series with some amount of resistance *(Figure 3-14)*. The amount of resistance compared with the inductive reactance determines the Q of the coil. The letter ***Q*** stands for **quality**. Inductors that have a higher ratio of inductive reactance to resistance are considered to be inductors of higher quality. An inductor constructed with a large wire will have a low wire resistance and, therefore, a higher Q *(Figure 3-15)*. Inductors con-

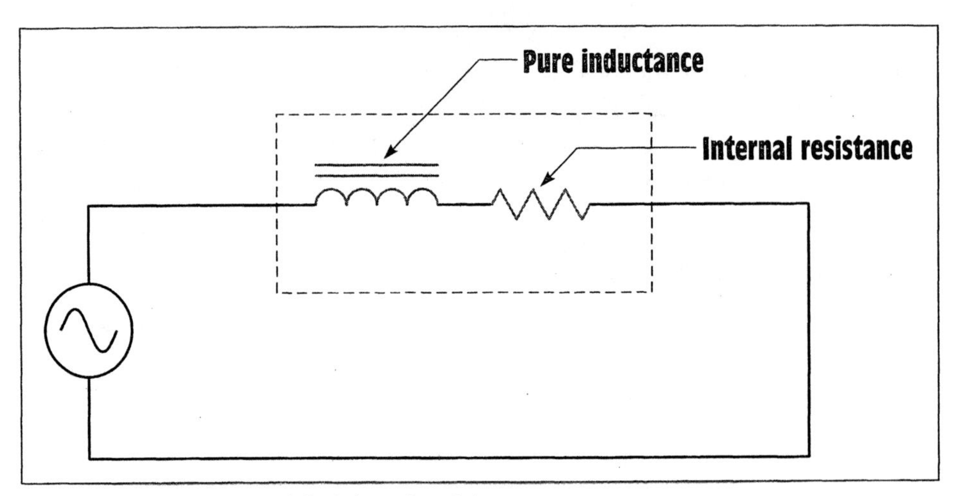

Figure 3-14 Inductors contain internal resistance.

structed with many turns of small wire have a much higher resistance, and, therefore, a lower Q. To determine the Q of an inductor, divide the inductive reactance by the resistance.

$$Q = \frac{X_L}{R}$$

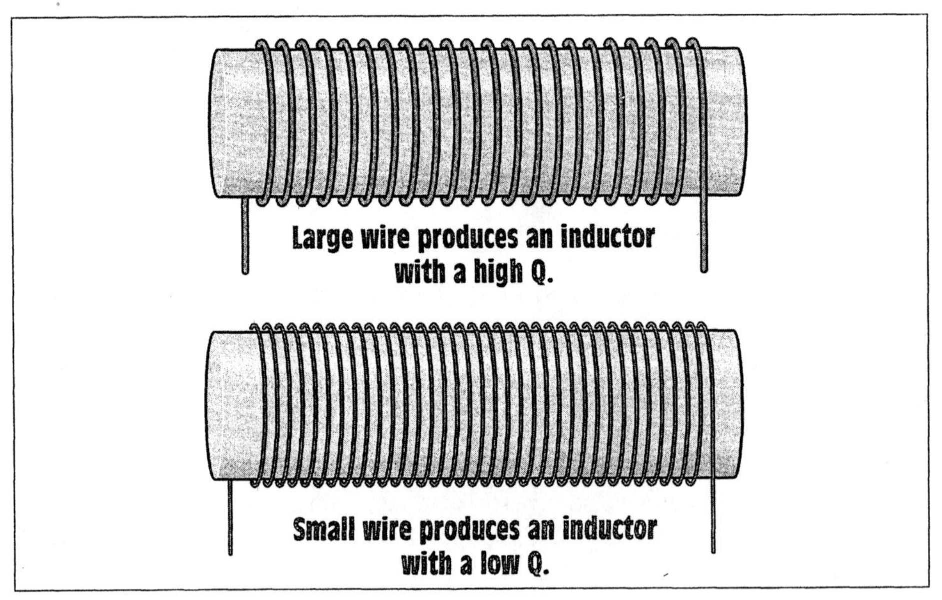

Figure 3-15 The Q of an inductor is a ratio of inductive reactance as compared to resistance. The letter "Q" stands for quality.

Although inductors have some amount of resistance, inductors that have a Q of 10 or greater are generally considered to be pure inductors. Once the ratio of inductive reactance becomes 10 times as great as resistance, the amount of resistance is considered negligible. For example, assume an inductor has an inductive reactance of 100 Ω and a wire resistance of 10 Ω. The inductive reactive component in the circuit is 90° out of phase with the resistive component. This relationship produces a right triangle *(Figure 3-16)*. The total current-limiting effect of the inductor is a combination of the inductive reactance and resistance. This total current-limiting effect is called **impedance** and is symbolized by the letter **Z**. The impedance of the circuit is represented by the hypotenuse of the right triangle formed by the inductive reactance and the resistance. To compute the value of impedance for the coil, the inductive reactance and resistance must be added. Since these two components form the legs of a right triangle and the impedance forms the hypotenuse, the value of impedance can be computed using the Pythagorean theorem: the sum of the squares of the sides of a right triangle is equal to the square of the hypotenuse.

$$Z = \sqrt{R^2 + X_L^2}$$

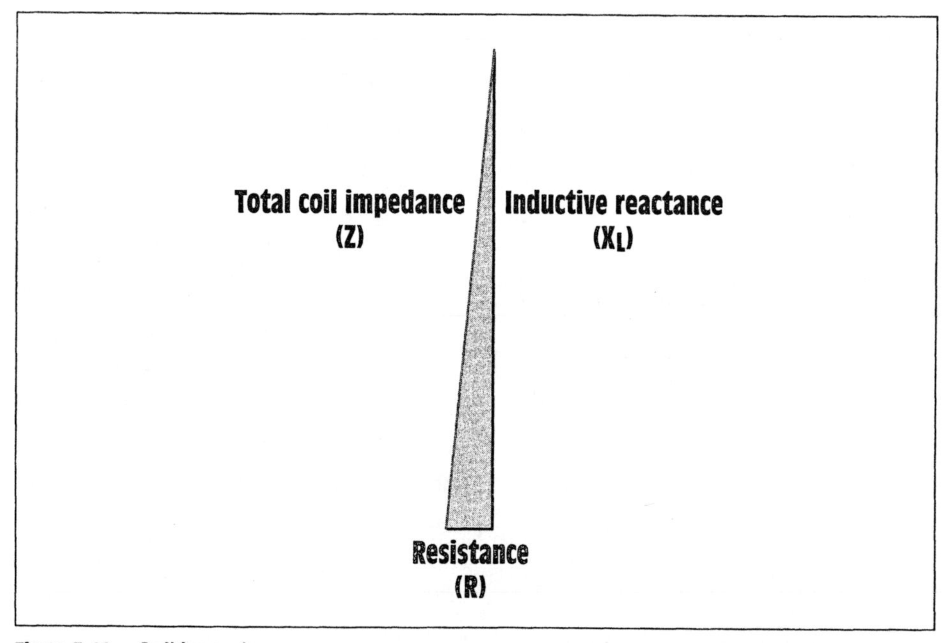

Figure 3-16 Coil impedance.

$$Z = \sqrt{10^2 + 100^2}$$

$$Z = \sqrt{10{,}100}$$

$$Z = 100.5\ \Omega$$

Notice that the value of total impedance for the inductor is only 0.5 Ω greater than the value of inductive reactance.

Summary

1. Induced voltage is proportional to the rate of change of current.
2. Induced voltage is always opposite in polarity to the applied voltage.
3. Inductive reactance is a countervoltage that limits the flow of current, as does resistance.
4. Inductive reactance is measured in ohms.
5. Inductive reactance is proportional to the inductance of the coil and the frequency of the line.
6. Inductive reactance is symbolized by X_L.
7. Inductance is measured in henrys (H) and is symbolized by the letter *L*.
8. When inductors are connected in series the total inductance is equal to the sum of all the inductors.
9. When inductors are connected in parallel the reciprocal of the total inductance is equal to the sum of the reciprocals of all the inductors.
10. The current lags the applied voltage by 90° in a pure inductive circuit.
11. All inductors contain some amount of resistance.
12. The Q of an inductor is the ratio of the inductive reactance to the resistance.
13. Inductors with a Q of 10 are generally considered to be "pure" inductors.
14. Pure inductive circuits contain no true power or watts.
15. Reactive power is measured in VARs.
16. VARs is an abbreviation for volt-amps-reactive.

Review Questions

1. How many degrees are the current and voltage out of phase with each other in a pure resistive circuit?
2. How many degrees are the current and voltage out of phase with each other in a pure inductive circuit?
3. To what is inductive reactance proportional?
4. Four inductors, each having an inductance of 0.6 H, are connected in series. What is the total inductance of the circuit?
5. Three inductors are connected in parallel. Inductor 1 has an inductance of 0.06 H; inductor 2 has an inductance of 0.05 H; and inductor 3 has an inductance of 0.1 H. What is the total inductance of this circuit?
6. If the three inductors in question 5 were connected in series, what would be the inductive reactance of the circuit? Assume the inductors are connected to a 60-Hz line.
7. An inductor is connected to a 240-V, 1000-Hz line. The circuit current is 0.6 A. What is the inductance of the inductor?
8. An inductor with an inductance of 3.6 H is connected to a 480-V, 60-Hz line. How much current will flow in this circuit?
9. If the frequency in question 8 is reduced to 50 Hz how much current will flow in the circuit?
10. An inductor has an inductive reactance of 250 Ω when connected to a 60-Hz line. What will be the inductive reactance if the inductor is connected to a 400-Hz line?

Practice Problems

Inductive Circuits

Fill in all the missing values. Refer to the formulas given below.

$$X_L = 2\pi FL$$

$$L = \frac{X_L}{2\pi F}$$

$$F = \frac{X_L}{2\pi L}$$

Inductance (H)	Frequency (Hz)	Induct. Rct. (Ω)
1.2	60	
0.085		213.628
	1000	4712.389
0.65	600	
3.6		678.584
	25	411.459
0.5	60	
0.85		6408.849
	20	201.062
0.45	400	
4.8		2412.743
	1000	40.841

4

Single-Phase Isolation Transformers

Objectives

After studying this unit, you should be able to

- Discuss the different types of transformers
- Calculate values of voltage, current, and turns for single-phase transformers using formulas
- Calculate values of voltage, current, and turns for single-phase transformers using the turns ratio
- Connect a transformer and test the voltage output of different windings
- Discuss polarity markings on a schematic diagram
- Test a transformer to determine the proper polarity marks

Transformers are one of the most common devices found in the electrical field. They range in size from occupying a space less than one cubic inch to requiring rail cars to move them after they have been broken into sections. Their rating can range from mVA (milli-volt-amps) to GVA (giga-volt-amps).

A **transformer** is a magnetically operated machine that can change values of voltage, current, and impedance without a change of frequency. Transformers are the most efficient machines known. Their efficiencies commonly range from 90% to 99% at full load. Transformers can be divided into several classifications such as:

1. Isolation
2. Auto
3. Current

A basic law concerning transformers is that **all values of a transformer are proportional to its turns ratio**. This does not mean that the exact number of turns of wire on each winding must be known to determine different values of voltage and current for a transformer. What must be known is the *ratio* of turns. For example, assume a transformer has two windings. One winding, the primary, has 1000 turns of wire, and the other, the secondary, has 250 turns of wire *(Figure 4-1)*. The **turns ratio** of this transformer is 4 to 1, or 4:1 (1000/250 = 4). This indicates there are four turns of wire on the primary for every one turn of wire on the secondary.

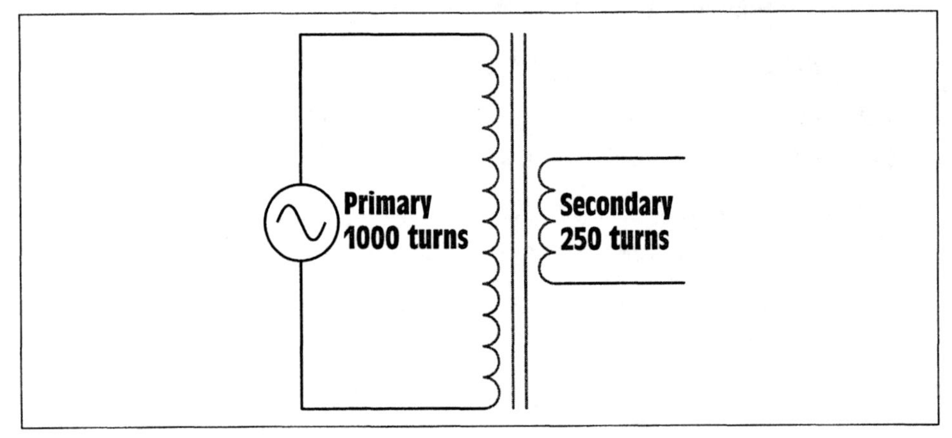

Figure 4-1 All values of a transformer are proportional to its turns ratio.

Transformer Formulas

There are different formulas that can be used to find the values of voltage and current for a transformer. The following is a list of standard formulas, where:

N_P = number of turns in the primary

N_S = number of turns in the secondary

E_P = voltage of the primary

E_S = voltage of the secondary

I_P = current in the primary

I_S = current in the secondary

$$\frac{E_P}{E_S} = \frac{N_P}{N_S}$$

$$\frac{E_P}{E_S} = \frac{I_S}{I_P}$$

$$\frac{N_P}{N_S} = \frac{I_S}{I_P}$$

or

$$E_P \times N_S = E_S \times N_P$$

$$E_P \times I_P = E_S \times I_S$$

$$N_P \times I_P = N_S \times I_S$$

The **primary winding** of a transformer is the power-input winding. It is the winding that is connected to the incoming power supply. The **secondary winding** is the load winding, or output winding. It is the side of the transformer that is connected to the driven load *(Figure 4-2)*. Any winding of a transformer can be used as a primary or secondary wiring provided its voltage or current rating is not exceeded. Transformers can also be operated at a lower voltage than their rating indicates, but they cannot be connected to a higher voltage. Assume the transformer shown in *Figure 4-2*, for example, has a primary voltage rating of 480 volts and the secondary has a voltage rating of 240 volts. Now assume that the primary winding is connected to a 120-volt source. No damage would occur to the transformer but the secondary winding would produce only 20 volts.

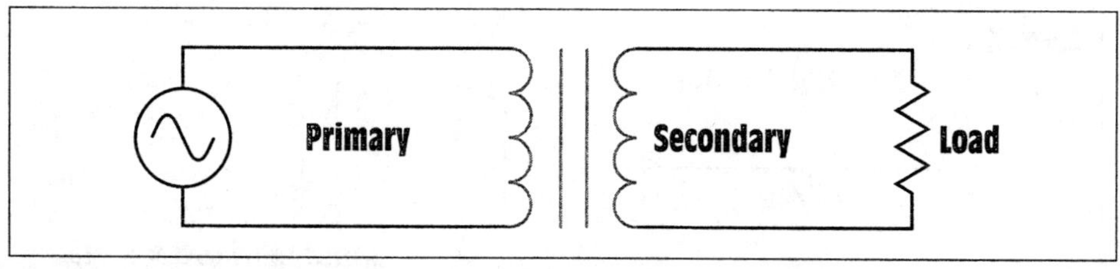

Figure 4-2 Isolation transformer.

4-2 Isolation Transformers

The transformers shown in *Figures 4-1* and *4-2* are **isolation transformers**. This means that the secondary winding is physically and electrically isolated from the primary winding. There is no electrical connection between the primary and secondary winding. This transformer is magnetically coupled, not electrically coupled. This "line isolation" is often a very desirable characteristic. Since there is no electrical connection between the load and power supply, the transformer becomes a filter between the two. The isolation transformer will greatly reduce any voltage spikes that originate on the supply side before they are transferred to the load side. Some isolation transformers are built with a turns ratio of 1:1. A transformer of this type will have the same input and output voltage and is used for the purpose of isolation only.

The reason that the isolation transformer can greatly reduce any voltage spikes before they reach the secondary is because of the rise time of current through an inductor. Recall from Unit 2 it was discussed that the current in an inductor rises at an exponential rate *(Figure 4-3)*. As the current increases in value, the expanding magnetic field cuts through the conductors of the coil and induces a voltage that is opposed to the applied voltage. The amount of induced voltage is proportional to the rate of change of current. This simply means that the faster current attempts to increase, the greater the opposition to that increase will be. Spike voltages and currents are generally of very short duration, which means that they increase in value very rapidly *(Figure 4-4)*. This rapid change of value causes the opposition to the change to increase just as rapidly. By the time the spike has been transferred to the

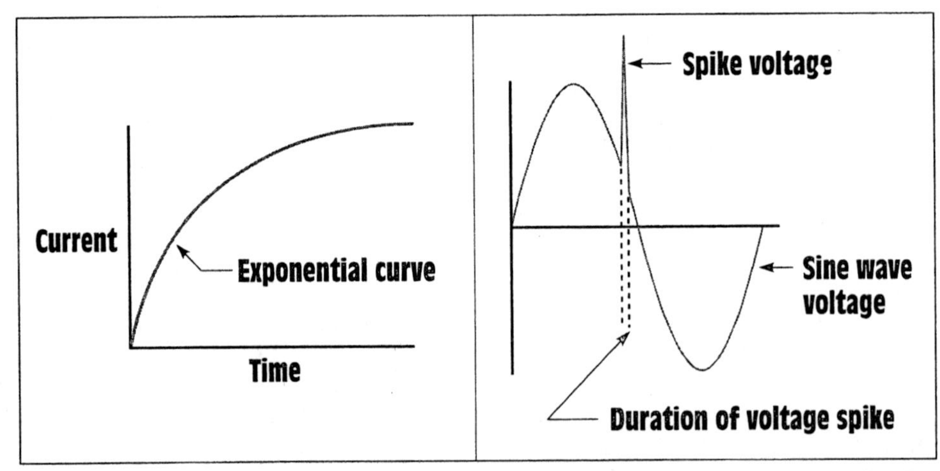

Figure 4-3 The current through an inductor rises at an exponential rate.

Figure 4-4 Voltage spikes are generally of very short duration.

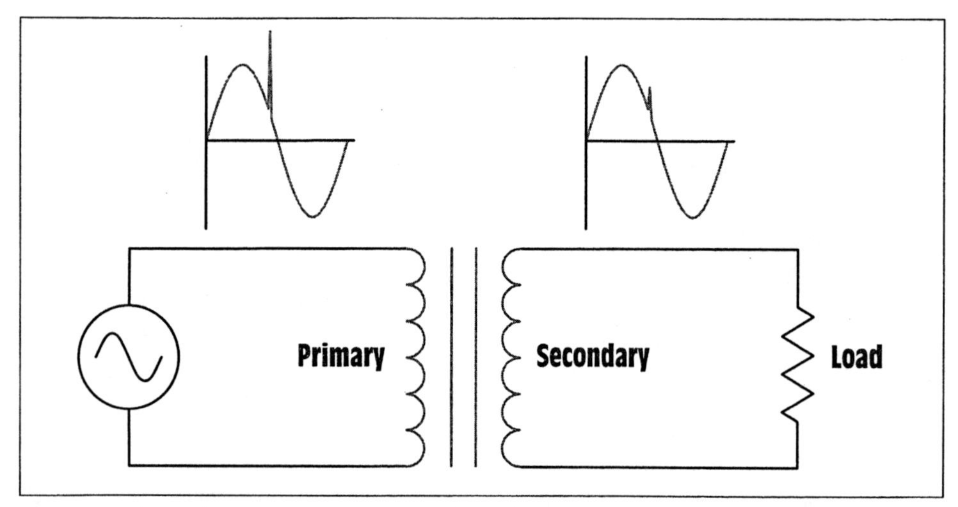

Figure 4-5 The isolation transformer greatly reduces the voltage spike.

secondary winding of the transformer it has been eliminated or greatly reduced *(Figure 4-5)*.

Another purpose of isolation transformers is to remove some piece of electrical equipment from ground. It is sometimes desirable that a piece of electrical equipment not be connected directly to ground. This is often done as a safety precaution to eliminate the hazard of an accidental contact between a person at ground potential and the undergrounded conductor. If the equipment case should come into contact with the ungrounded conductor, the isolation transformer would prevent a circuit from being completed to ground by someone touching the equipment case. Many alternating-current circuits have one side connected to ground. A familiar example of this is the common 120-volt circuit with a grounded neutral conductor *(Figure 4-6)*.

Transformer Construction

The basic construction of an isolation transformer is shown in *Figure 4-7*. A metal core is used to provide good magnetic coupling between the two windings. The core is generally made of laminations stacked together. Laminating the core helps reduce power losses due to eddy current induction. This illustration shows the basic design of electrically separated winding.

Transformer Core Types

There are several different types of cores used in the construction of transformers. Most cores are made from thin steel punchings laminated to form a solid metal core. The core for a 600 mVA (mega-volt-amp) three-phase trans-

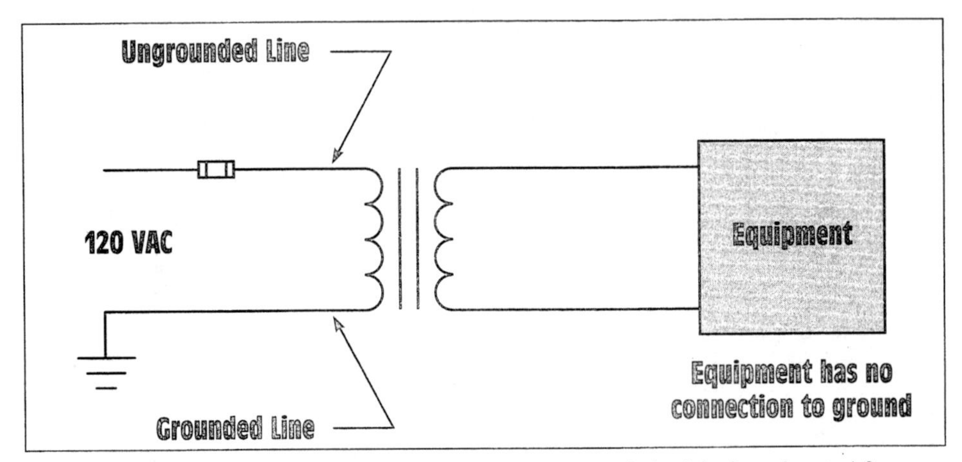

Figure 4-6 Isolation transformer used to remove a piece of electrical equipment from ground.

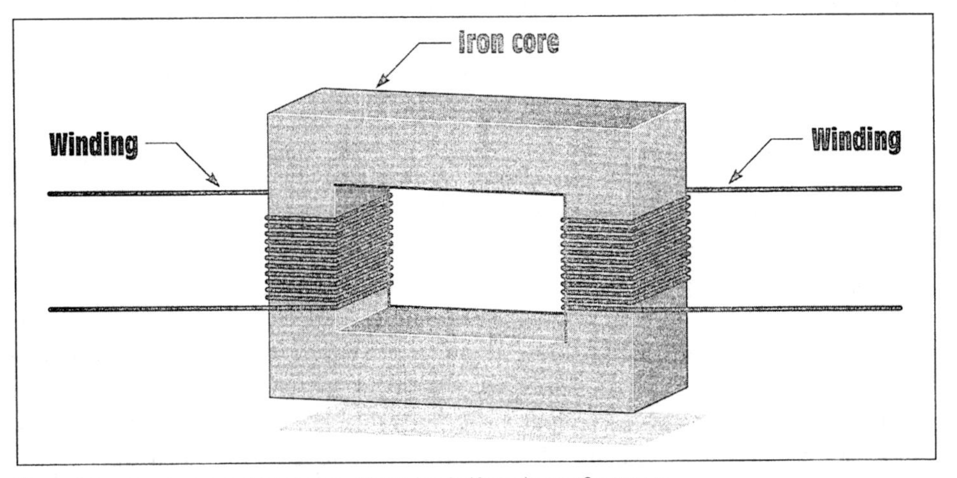

Figure 4-7 Basic construction of an isolation transformer.

former is shown in *Figure 4-8*. Laminated cores are preferred because a thin layer of oxide forms on the surface of each lamination which acts as an insulator to reduce the formation of eddy currents inside the core material. The amount of core material needed for a particular transformer is determined by the power rating of the transformer. The amount of core material must be sufficient to prevent saturation at full load. The type and shape of the core generally determines the amount of magnetic coupling between the windings and to some extent the efficiency of the transformer.

The transformer illustrated in *Figure 4-9* is known as a *core*-type transformer. The windings are placed around each end of the core material. The metal core provides a good magnetic path between the two windings.

Figure 4-8 Core of a 600-MVA three-phase transformer *(Courtesy of Houston Lighting and Power.)*

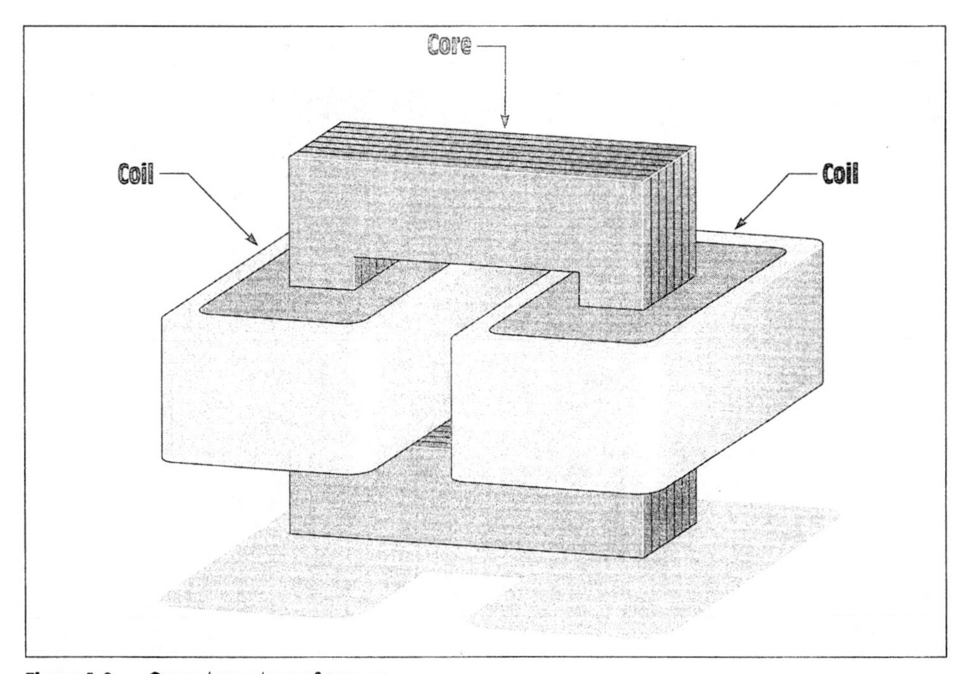

Figure 4-9 Core-type transformer.

The *shell*-type transformer is constructed in a similar manner as the core-type, except that the shell-type has a metal core piece through the middle of the window *(Figure 4-10)*. The primary and secondary windings are wound around the center core piece, with the low-voltage winding being closest to the metal core. This arrangement permits the transformer to be surrounded by the core which provides excellent magnetic coupling. When the transformer is in operation, all the magnetic flux must pass through the center core piece. It then divides through the two outer core pieces. Shell-type cores are sometimes referred to as *E-I* cores because the steel punchings used to construct the core are in the shape of an E and an I *(Figure 4-11)*.

The H-type core shown in *Figure 4-12* is similar to the shell-type in that it has an iron core through its center around which the primary and secondary windings are wound. The H core, however, surrounds the windings on four sides instead of two. This extra metal helps reduce stray leakage flux and improve the efficiency of the transformer. The H-type core is often found on high-voltage distribution transformers.

The *tape wound* or *torroid* core, *(Figure 4-13)* is constructed by winding tightly one long continuous silicon-steel tape into a spiral. The tape may or may not be housed in a plastic container depending on the application. This type core does not require laminated steel punchings. Since the core is one continuous length of metal, flux leakage is kept to a minimum. The tape-wound core is one of the most efficient core designs available.

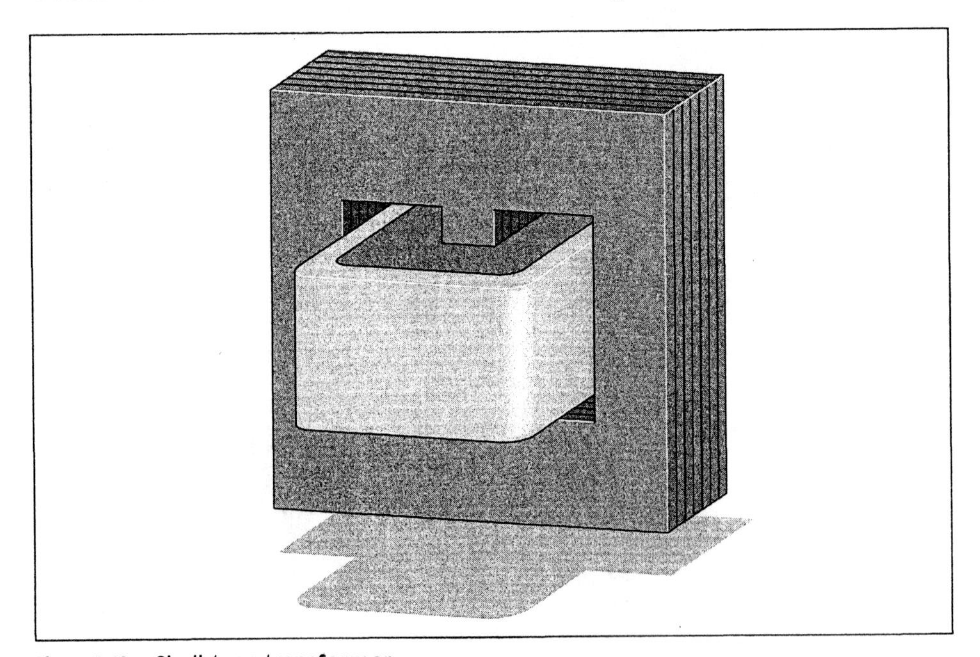

Figure 4-10 Shell-type transformer.

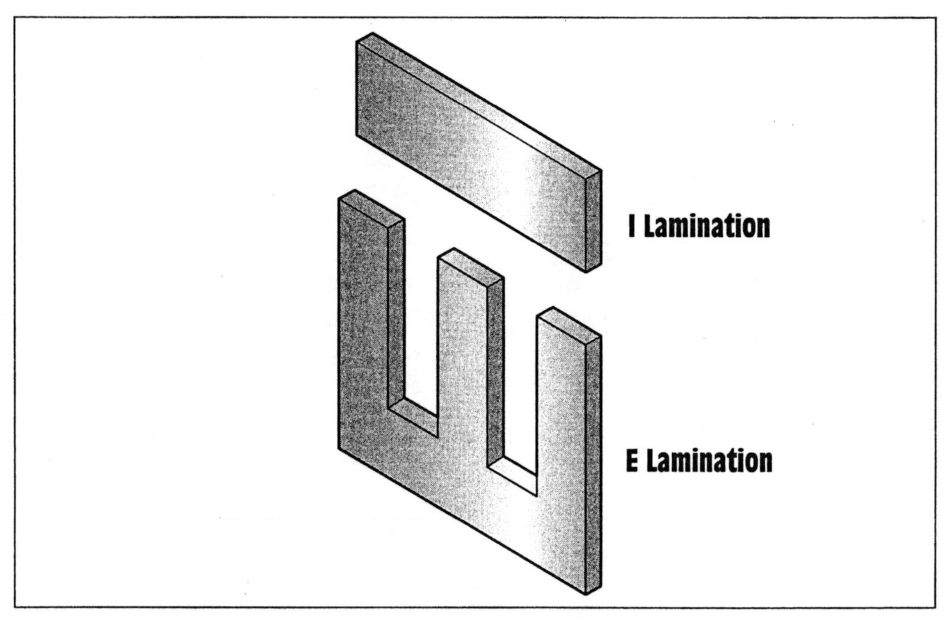

Figure 4-11 Shell-type cores are made of E and I laminations.

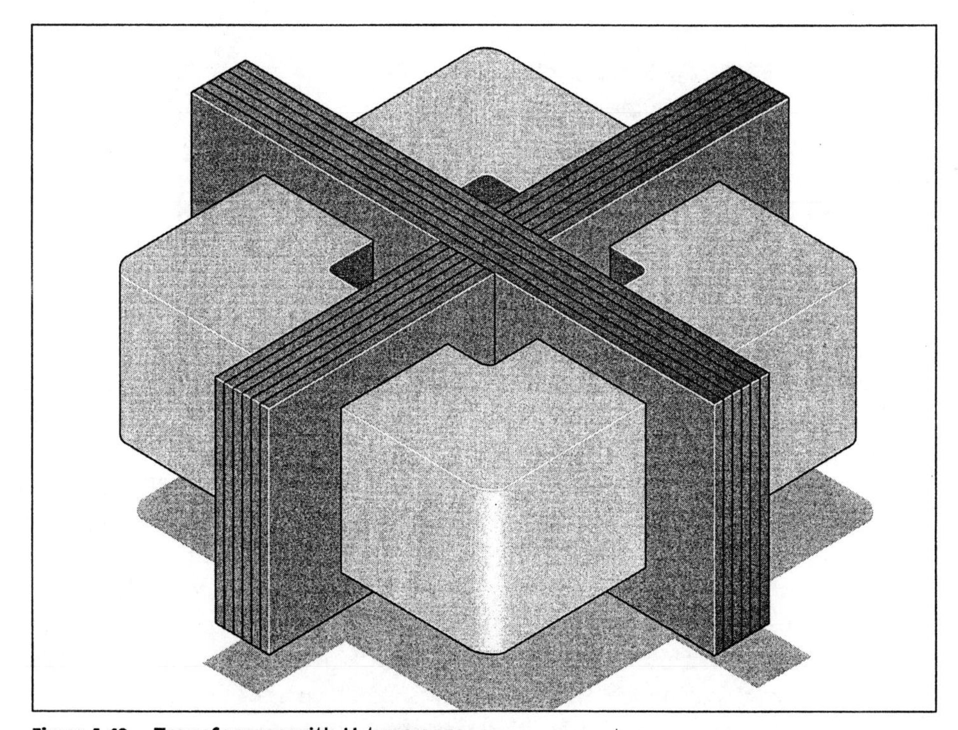

Figure 4-12 Transformer with H-type core.

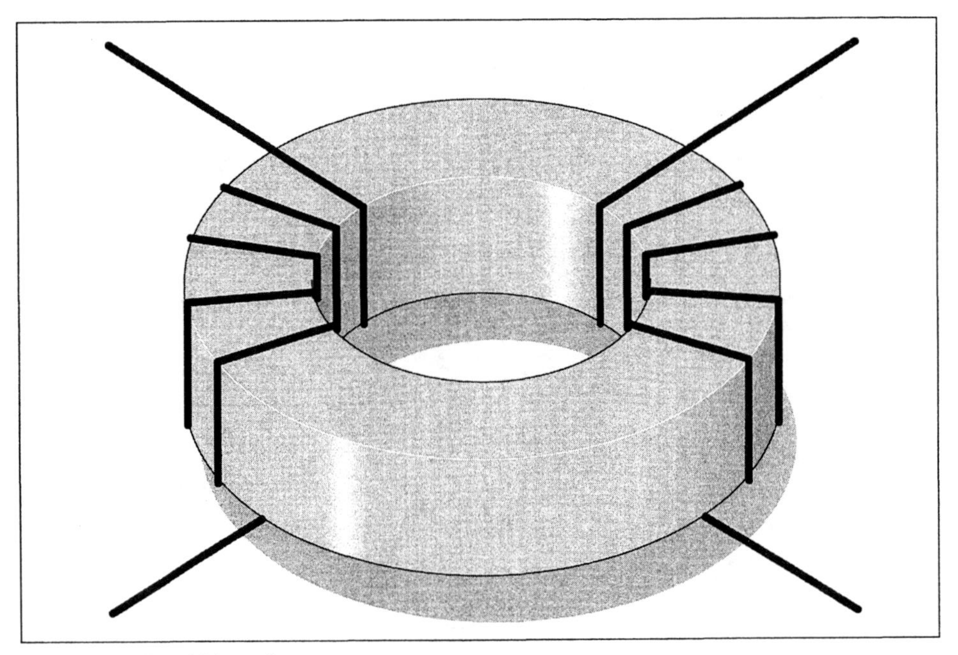

Figure 4-13 Toroid transformer.

Basic Operating Principles

In *Figure 4-14*, one winding of the transformer has been connected to an alternating-current supply, and the other winding has been connected to a load. As current increases from zero to its peak positive point, a magnetic field expands outward around the coil. When the current decreases from its peak positive point toward zero, the magnetic field collapses. When the current increases toward its negative peak, the magnetic field again expands, but with an opposite polarity of that previously. The field again collapses when the current decreases from its negative peak toward zero. This continually expanding and collapsing magnetic field cuts the windings of the primary and induces a voltage into it. This induced voltage opposes the applied voltage and limits the current flow of the primary. When a coil induces a voltage into itself, it is known as *self-induction*. It is this induced voltage, inductive reactance, that limits the flow of current in the primary winding. If the resistance of the primary winding is measured with an ohmmeter, it will indicate only the resistance of the wire used to construct the winding and will not give an indication of the actual current-limiting effect of the winding. Most transformers with a large kVA rating will appear to be almost a short circuit when measured with an ohmmeter. When connected to power, however, the actual no-load current is generally relatively small.

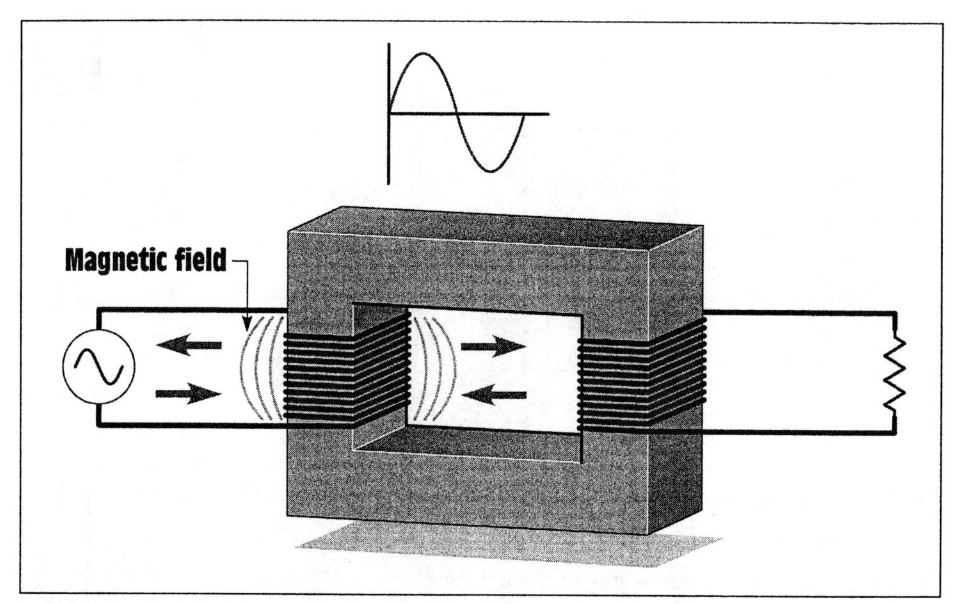

Figure 4-14 Magnetic field produced by alternating current.

Excitation Current

There will always be some amount of current flow in the primary of any voltage transformer even if there is no load connected to the secondary. This is called the **excitation current** of the transformer. The excitation current is the amount of current required to magnetize the core of the transformer. The excitation current remains constant from no load to full load. As a general rule, the excitation current is such a small part of the full-load current it is often omitted when making calculations.

Mutual Induction

Since the secondary windings of an isolation transformer are wound on the same core as the primary, the magnetic field produced by the primary winding cuts the windings of the secondary also *(Figure 4-15)*. This continually changing magnetic field induces a voltage into the secondary winding. The ability of one coil to induce a voltage into another coil is called *mutual induction*. The amount of voltage induced in the secondary is determined by the ratio of the number of turns of wire in the secondary to those in the primary. For example, assume the primary has 240 turns of wire and is connected to 120 VAC. This gives the transformer a **volts-per-turn ratio** of 0.5 (120 V/240 turns = 0.5-volt per turn). Now assume the secondary winding contains 100 turns of wire. Since the transformer has a volts-per-turn ratio of 0.5, the secondary voltage will be 50 V (100 x 0.5 = 50).

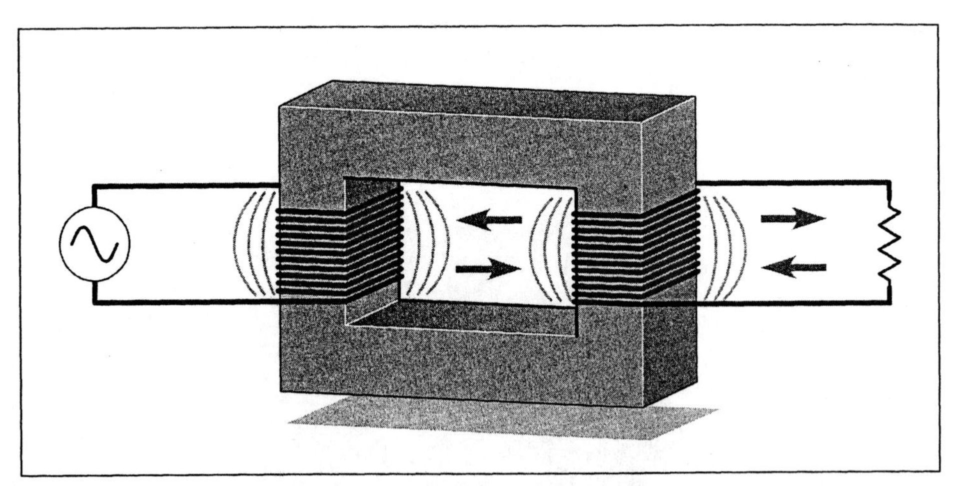

Figure 4-15 The magnetic field of the primary induces a voltage into the secondary.

Transformer Calculations

In the following examples, values of voltage, current, and turns for different transformers will be computed.

Example 1

Assume that the isolation transformer shown in *Figure 4-2* has 240 turns of wire on the primary and 60 turns of wire on the secondary. This is a ratio of 4:1 (240/60 = 4). Now assume that 120 V is connected to the primary winding. What is the voltage of the secondary winding?

$$\frac{E_P}{E_S} = \frac{N_P}{N_S}$$

$$\frac{120}{E_S} = \frac{240}{60}$$

$$E_S = 30 \text{ V}$$

The transformer in this example is known as a **step-down transformer** because it has a lower secondary voltage than primary voltage.

Now assume that the load connected to the secondary winding has an impedance of 5 Ω. The next problem is to calculate the current flow in the secondary and primary windings. The current flow of the secondary can be computed using Ohm's law since the voltage and impedance are known.

$$I = \frac{E}{Z}$$

$$I = \frac{30}{5}$$

$$I = 6 \text{ amps}$$

Now that the amount of current flow in the secondary is known, the primary current can be computed using the formula

$$\frac{E_P}{E_S} = \frac{I_S}{I_P}$$

$$\frac{120}{30} = \frac{6}{I_P}$$

$$120\ I_P = 180$$

$$I_P = 1.5\text{ A}$$

Notice that the primary voltage is higher than the secondary voltage, but the primary current is much less than the secondary current. **A good rule for any type of transformer is that power in must equal power out.** If the primary voltage and current are multiplied together, it should equal the product of the voltage and current of the secondary.

Primary	Secondary
120 X 1.5 = 180 voltamps	30 x 6 = 180 voltamps

Example 2

In the next example, assume that the primary winding contains 240 turns of wire and the secondary contains 1200 turns of wire. This is a turns ratio of 1:5 (1200/240 = 5). Now assume that 120 V is connected to the primary winding. Compute the voltage output of the secondary winding.

$$\frac{E_P}{E_S} = \frac{N_P}{N_S}$$

$$\frac{120}{E_S} = \frac{240}{1200}$$

$$240\ E_S = 144{,}000$$

$$E_S = 600\text{ V}$$

Notice that the secondary voltage of this transformer is higher than the primary voltage. This type of transformer is known as a **step-up transformer**.

Now assume that the load connected to the secondary has an impedance of 2400 Ω. Find the amount of current flow in the primary and secondary windings. The current flow in the secondary winding can be computed using Ohm's law.

$$I = \frac{E}{Z}$$

$$I = \frac{600}{2400}$$

$$I = 0.25\ A$$

Now that the amount of current flow in the secondary is known, the primary current can be computed using the formula

$$\frac{E_P}{E_S} = \frac{I_S}{I_P}$$

$$\frac{120}{600} = \frac{0.25}{I_P}$$

$$120\ I_P = 150$$

$$I_P = 1.25\ A$$

Notice that the amount of power input equals the amount of power output.

Primary	Secondary
120 x 1.25 = 150 VA	600 x 0.25 = 150 VA

Calculating Transformer Values Using the Turns Ratio

As illustrated in the previous examples, transformer values of voltage, current, and turns can be computed using formulas. It is also possible to compute these same values using the turns ratio. There are several ways in which turns ratios can be expressed. One method is to use a whole number value such as 13:5 or 6:21. The first ratio indicates that one winding has 13 turns of wire for every 5 turns of wire in the other winding. The second ratio indicates that there are 6 turns of wire in one winding for every 21 turns in the other.

A second method is to use the number 1 as a base. When using this method, the number 1 is always assigned to the winding with the lowest voltage rating. The ratio is found by dividing the higher voltage by the lower voltage. The number on the left side of the ratio represents the primary winding and the number on the right of the ratio represents the secondary winding. For example, assume a transformer has a primary rated at 240 volts and a secondary rated at 96 volts, (*Figure 4-16*). The turns ratio can be computed by dividing the higher voltage by the lower voltage.

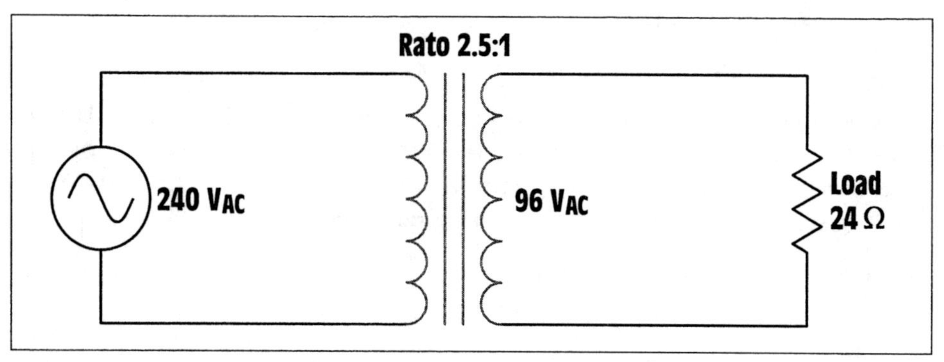

Figure 4-16 Computing transformer values using the turns ratio.

$$\text{Ratio} = \frac{240}{96}$$

$$\text{Ratio} = 2.5{:}1$$

Notice in this example that the primary winding has the higher voltage rating and the secondary has the lower. Therefore, the 2.5 is placed on the left and the base unit, 1, is placed on the right. This ratio indicates that there are 2.5 turns of wire in the primary winding for every 1 turn of wire in the secondary.

Now assume that a resistance of 24 Ω is connected to the secondary winding. The amount of secondary current can be found using Ohm's law.

$$I_S = \frac{96}{24}$$

$$I_S = 4 \text{ amps}$$

The primary current can be found using the turns ratio. Recall that the volt-amps of the primary must equal the volt-amps of the secondary. Since the primary voltage is greater, the primary current will have to be less than the secondary current. Therefore, the secondary current will be divided by the turns ratio.

$$I_P = \frac{I_S}{\text{turns ratio}}$$

$$I_P = \frac{4}{2.5}$$

$$I_P = 1.6 \text{ A}$$

To check the answer, find the volt-amps of the primary and secondary.

Primary	Secondary
240 X 1.6 = 384	96 X 4 = 384

Now assume that the secondary winding contains 150 turns of wire. The primary turns can be found by using the turns ratio also. Since the primary voltage is higher than the secondary voltage, the primary must have more turns of wire, the secondary turns will be multiplied by the turns ratio.

$$N_P = N_S \text{ x turns ratio}$$

$$N_P = 150 \text{ x } 2.5$$

$$N_P = 375 \text{ turns}$$

In the next example, assume a transformer has a primary voltage of 120 volts and a secondary voltage of 500 volts. The secondary has a load impedance of 1200 Ω. The secondary contains 800 turns of wire *(Figure 4-17)*.

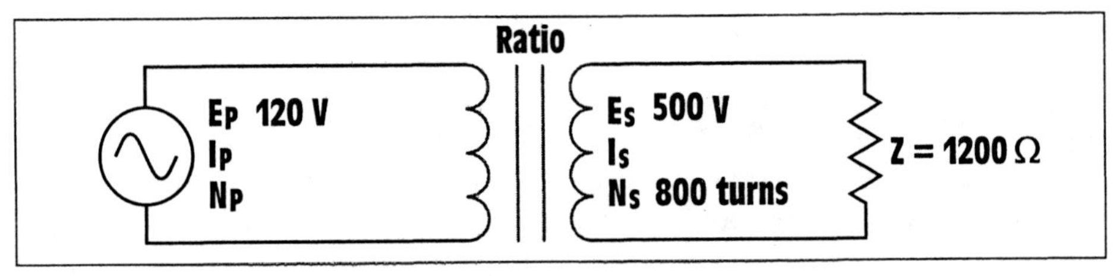

Figure 4-17 Calculating transformer values.

The turns ratio can be found by dividing the higher voltage by the lower voltage.

$$\text{ratio} = \frac{500}{120}$$

$$\text{ratio} = 1:4.17$$

The secondary current can be found using Ohm's law.

$$I_S = \frac{500}{1200}$$

$$I_S = 0.417 \text{ amps}$$

In this example, the primary voltage is lower than the secondary voltage. Therefore, the primary current must be higher. To find the primary current, multiply the secondary current by the turns ratio.

$$I_P = I_S \text{ x turns ratio}$$

$$I_P = 0.417 \times 4.17$$

$$I_P = 1.74 \text{ amps}$$

To check this answer, compute the volt-amps of both windings.

Primary	Secondary
120 X 1.74 = 208.8	500 X 0.417 = 208.5

The slight difference in answers is caused by rounding off values.

Since the primary voltage is less than the secondary voltage, the turns of wire in the primary will be less also. The primary turns will be found by dividing the turns of wire in the secondary by the turns ratio.

$$N_P = \frac{N_S}{\text{turns ratio}}$$

$$N_P = \frac{800}{4.17}$$

$$N_P = 192 \text{ turns}$$

Figure 4-18 shows the transformer with all completed values.

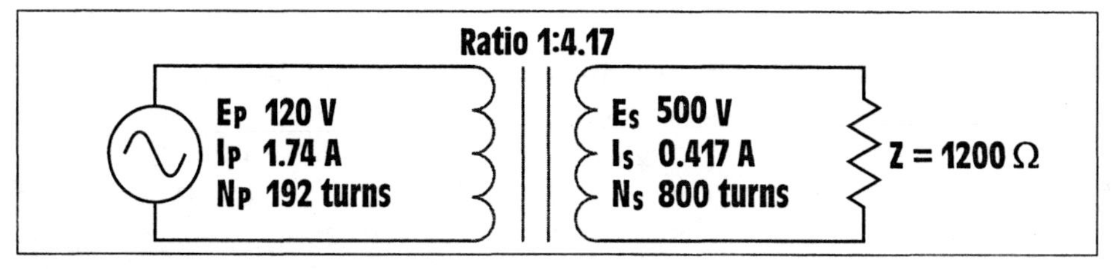

Figure 4-18 Transformer with completed values.

Multiple-Tapped Windings

It is not uncommon for isolation transformers to be designed with windings that have more than one set of lead wires connected to the primary or secondary. The transformer shown in *Figure 4-19* contains a secondary winding rated at 24 V. The primary winding contains several taps, however. One of the primary lead wires is labeled C and is the common for the other leads. The other leads are labeled 120, 208, and 240. This transformer is designed in such a manner that it can be connected to different primary voltages without changing the value of the secondary voltage. In this example, it is assumed that the secondary winding has a total of 120 turns of wire. To maintain the proper turns ratio, the primary would have 600 turns of wire

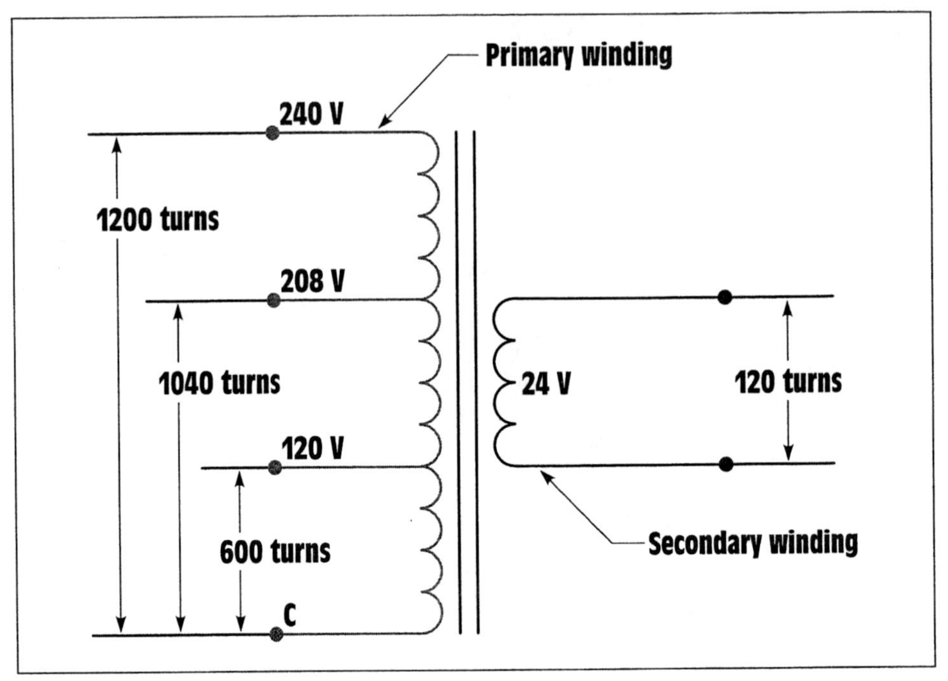

Figure 4-19 Transformer with multiple-tapped primary winding.

between C and 120, 1040 turns between C and 208, and 1200 turns between C and 240.

The transformer shown in *Figure 4-20* contains a single primary winding. The secondary winding, however, has been tapped at several points. One of the secondary lead wires is labeled C and is common to the other lead wires. When rated voltage is applied to the primary, voltages of 12, 24, and 48 V

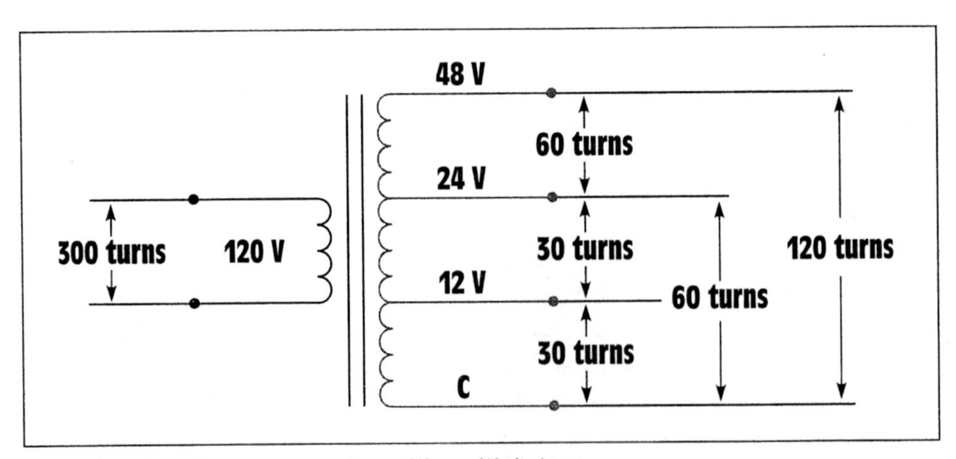

Figure 4-20 Transformer secondary with multiple taps.

can be obtained at the secondary. It should also be noted that this arrangement of taps permits the transformer to be used as a center-tapped transformer for two of the voltages. If a load is placed across the lead wires labeled C and 24, the lead wire labeled 12 becomes a center tap. If a load is placed across the C and 48 lead wires, the 24 lead wire becomes a center tap.

In this example, it is assumed that the primary winding has 300 turns of wire. In order to produce the proper turns ratio it would require 30 turns of wire between C and 12, 60 turns of wire between C and 24, and 120 turns of wire between C and 48.

The transformer shown in *Figure 4-21* is similar to the transformer in *Figure 4-20.* The transformer in *Figure 4-21*, however, has multiple secondary windings instead of a single secondary winding with multiple taps. The advantage of the transformer in *Figure 4-21* is that the secondary windings are electrically isolated from each other. These secondary windings can be either step-up or step-down depending on the application of the transformer.

Computing Values for Isolation Transformers with Multiple Secondaries

When computing the values of a transformer with multiple secondary windings, each secondary must be treated as a different transformer. For

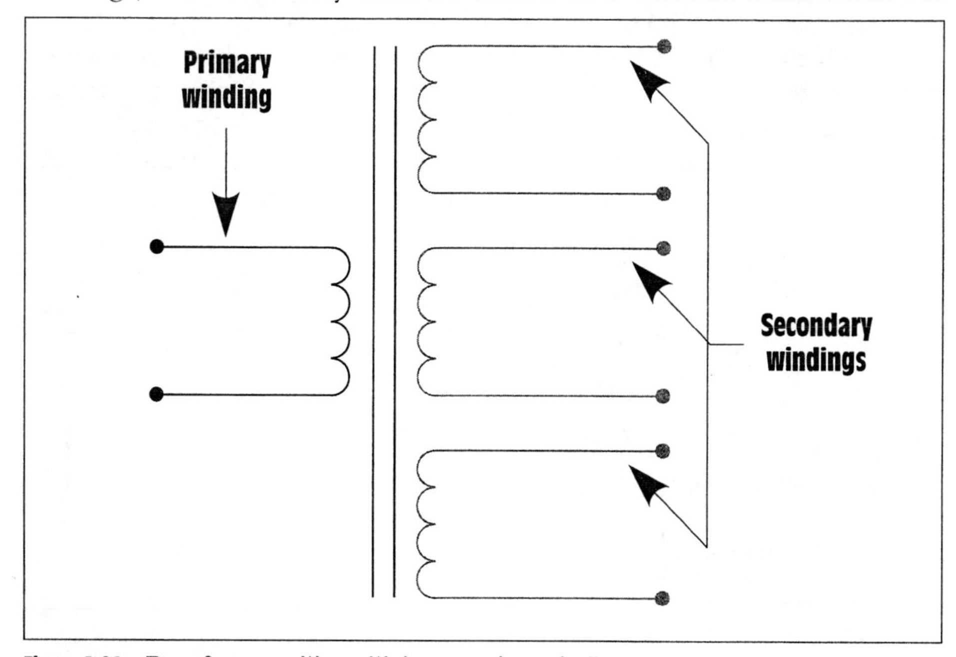

Figure 4-21 Transformer with multiple secondary windings.

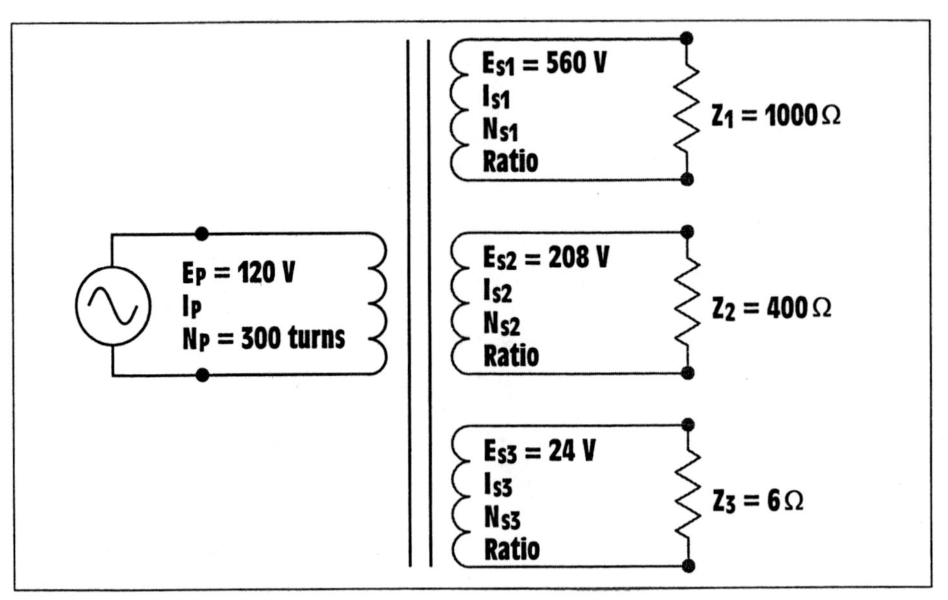

Figure 4-22 Computing values for a transformer with multiple secondary windings.

example, the transformer in *Figure 4-22* contains one primary winding and three secondary windings. The primary is connected to 120 VAC and contains 300 turns of wire. One secondary has an output voltage of 560 V and a load impedance of 1000 Ω. The second secondary has an output voltage of 208 V and a load impedance of 400 Ω, and the third secondary has an output voltage of 24 V and a load impedance of 6 Ω. The current, turns of wire, and ratio for each secondary and the current of the primary will be found.

The first step will be to compute the turns ratio of the first secondary. The turns ratio can be found by dividing the smaller voltage into the larger.

$$\text{ratio} = \frac{E_{S1}}{E_P}$$

$$\text{ratio} = \frac{560}{120}$$

$$\text{ratio} = 1{:}4.67$$

The current flow in the first secondary can be computed using Ohm's law.

$$I_{S1} = \frac{560}{1000}$$

$$I_{S1} = 0.56 \text{ amps}$$

The number of turns of wire in the first secondary winding will be found using the turns ratio. Since this secondary has a higher voltage than the primary, it must have more turns of wire. The number of primary turns will be multiplied by the turns ratio.

$$N_{S1} = N_P \times \text{turns ratio}$$

$$N_{S1} = 300 \times 4.67$$

$$N_{S1} = 1401 \text{ turns}$$

The amount of primary current needed to supply this secondary winding can be found using the turns ratio also. Since the primary has less voltage, it will require more current. The primary current can be determined by multiplying the secondary current by the turns ratio.

$$I_{P(\text{First Secondary})} = I_{S1} \times \text{turns ratio}$$

$$I_{P(\text{First Secondary})} = 0.56 \times 4.67$$

$$I_{P(\text{First Secondary})} = 2.61 \text{ amps}$$

The turns ratio of the second secondary winding will be found by dividing the higher voltage by the lower.

$$\text{ratio} = \frac{208}{120}$$

$$\text{ratio} = 1{:}1.73$$

The amount of current flow in this secondary can be determined using Ohm's law.

$$I_{S2} = \frac{208}{400}$$

$$I_{S2} = 0.52 \text{ amps}$$

Since the voltage of this secondary is greater than the primary, it will have more turns of wire than the primary. The turns of this secondary will be found using the turns ratio.

$$N_{S2} = N_P \times \text{turns ratio}$$

$$N_{S2} = 300 \times 1.73$$

$$N_{S2} = 519 \text{ turns}$$

The voltage of the primary is less than this secondary. The primary will, therefore, require a greater amount of current. The amount of current required to operate this secondary will be computed by multiplying the secondary current by the turns ratio.

$$I_{P(Second\ Secondary)} = I_{S2} \times \text{turns ratio}$$

$$I_{P(Second\ Secondary)} = 0.52 \times 1.732$$

$$I_{P(Second\ Secondary)} = 0.9 \text{ amps}$$

The turns ratio of the third secondary winding will be computed in the same way as the other two. The larger voltage will be divided by the smaller.

$$\text{ratio} = \frac{120}{24}$$

$$\text{ratio} = 5{:}1$$

The primary current will be found using Ohm's law.

$$I_{S3} = \frac{24}{6}$$

$$I_{S3} = 4 \text{ amps}$$

The output voltage of the third secondary is less than the primary. The number of turns of wire will, therefore, be less than the primary turns. To find the number of secondary turns, divide the primary turns by the turns ratio.

$$N_{S3} = \frac{N_P}{\text{turns ratio}}$$

$$N_{S3} = \frac{300}{5}$$

$$N_{S3} = 60 \text{ turns}$$

The primary has a higher voltage than this secondary. The primary current will, therefore, be less by the amount of the turns ratio.

$$I_{P(Third\ Secondary)} = \frac{I_{S3}}{\text{turns ratio}}$$

$$I_{P(TThird\ Secondary)} = \frac{4}{5}$$

$$I_{P(Third\ Secondary)} = 0.8 \text{ amps}$$

The primary must supply current to each of the three secondary windings. Therefore, the total amount of primary current will be the sum of the currents required to supply each secondary.

$$I_{P(Total)} = I_{P1} + I_{P2} + I_{P3}$$

$$I_{P(Total)} = 2.61 + 0.9 + 0.8$$

$$I_{P(Total)} = 4.31 \text{ amps}$$

The transformer with all computed values is shown in *Figure 4-23*.

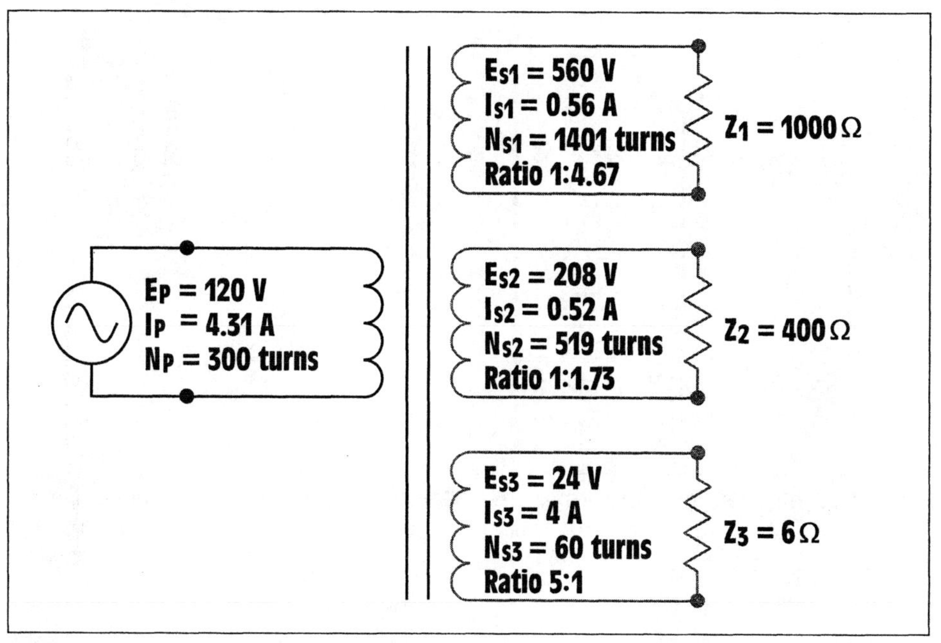

Figure 4-23 The transformer with all computed values.

Distribution Transformers

A very common type of isolation transformer is the **distribution transformer**, *Figure 4-24*. This transformer is used to supply power to most homes and many businesses. In this example, it is assumed that the primary is connected to a 7200-volt line. The secondary is 240 volts with a center tap. The center tap is grounded and becomes the **neutral conductor**. If voltage is measured across the entire secondary, a voltage of 240 volts will be seen. If voltage is measured from either line to the center tap, half of the secondary voltage, or 120 volts, will be seen *(Figure 4-25)*. The reason is that the voltages between the two secondary lines are 180° out of phase with each other. If a vector diagram is drawn to illustrate this condition, it will be seen that the grounded neutral conductor is connected to the axis point of the two voltage vectors *(Figure 4-26)*. Loads that are intended to operate on

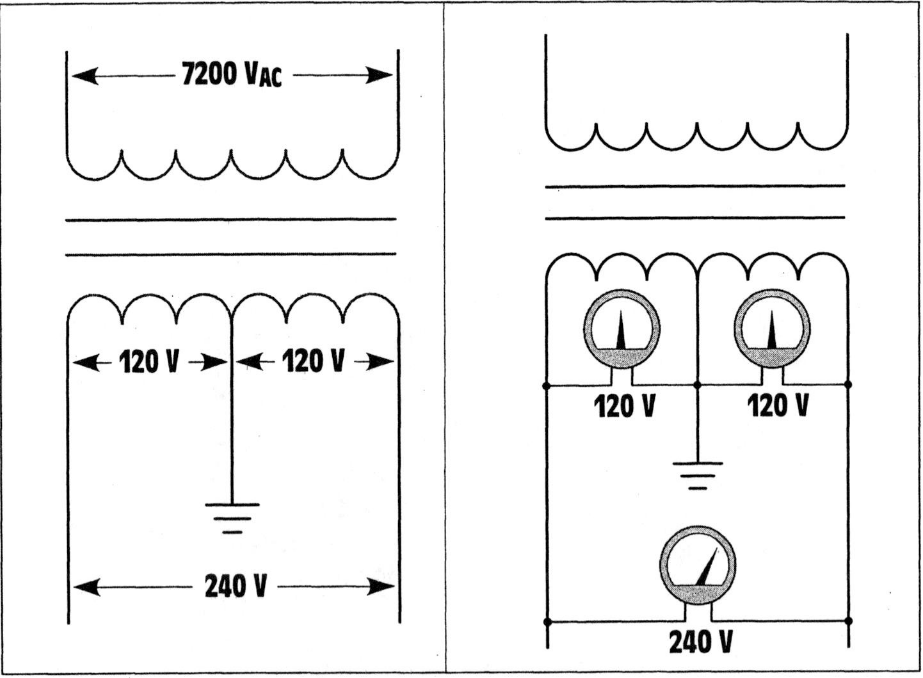

Figure 4-24 Distribution transformer.

Figure 4-25 The voltage from either line to neutral is 120 volts. The voltage across the entire secondary winding is 240 volts.

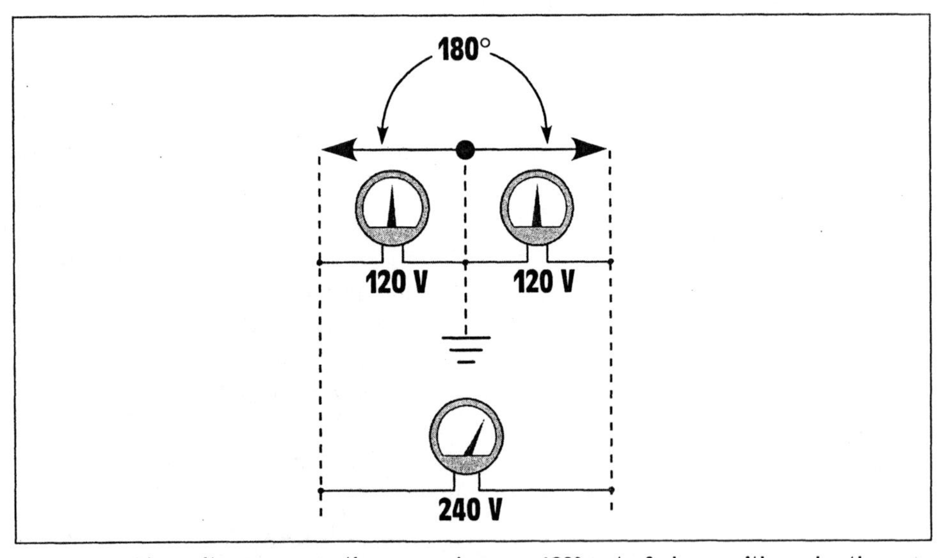

Figure 4-26 The voltages across the secondary are 180° out of phase with each other.

240 volts, such as water heaters, electric-resistance heating units, and central air conditioners are connected directly across the lines of the secondary *(Figure 4-27)*.

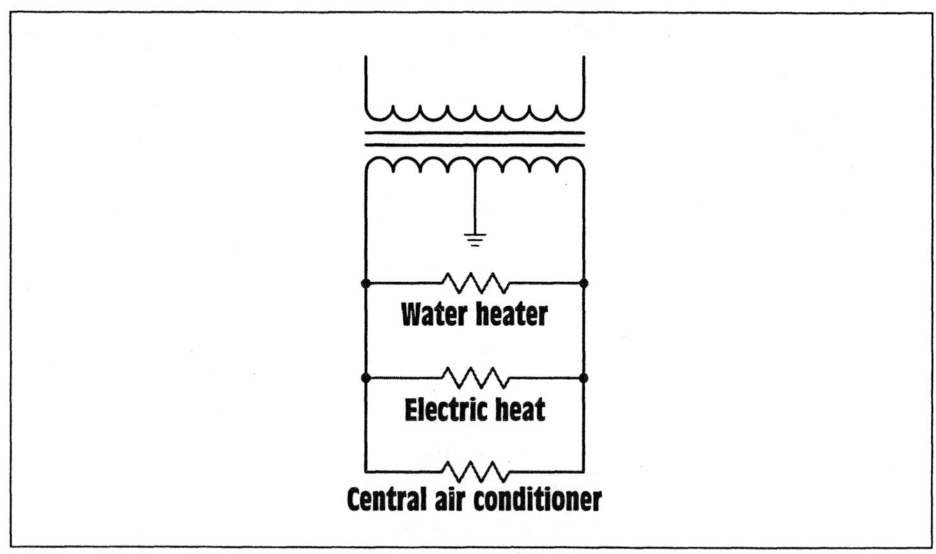

Figure 4-27 240-volt loads connect directly across the secondary winding.

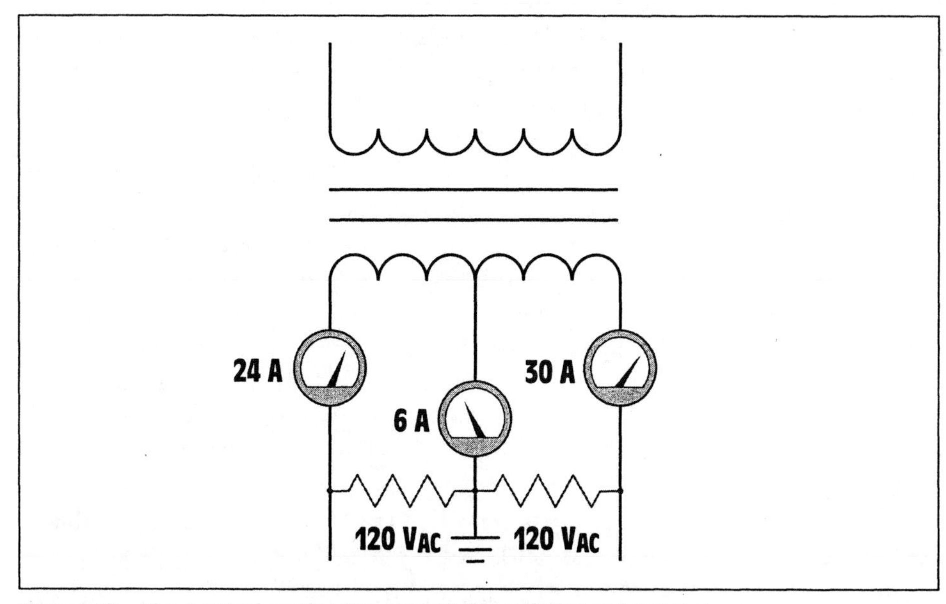

Figure 4-28 The neutral carries the sum of the unbalanced current.

Loads intended to operate on 120 volts connect from the center tap, or neutral, to one of the secondary lines. The function of the neutral is to carry the difference in current between the two secondary lines and maintain a balanced voltage. In the example shown in Figure *4-28*, it is assumed that one of the secondary lines has a current flow of 30 amperes and the other has a current flow of 24 amperes. The neutral will conduct the sum of the unbalanced load. In this example, the neutral current will be 6 A (30 – 24 = 6). Common kVA ratings for single-phase distribution transformers are given in *table 4-1*.

3	5	10	15	25
37.5	50	75	100	167
250	333	500	833	1250
1667	2500	3333	5000	6667
8333	10,000	12,500	16,667	20,000
25,000	33,333			

Table 4-1 Common kVA Values for Single-Phase Transformers

Control Transformers

Another common type of isolation transformer found throughout industry is the **control transformer** *(Figure 4-29)*. The control transformer is used to reduce the line voltage to the value needed to operate control circuits. The most common type of control transformer contains two primary windings and one secondary. The primary windings are generally rated at 240 volts each, and the secondary is rated at 120 volts. This provides a 2:1 turns ratio between each of the primary windings and the secondary. For example, assume that each of the primary windings contains 200 turns of wire. The secondary will contain 100 turns of wire.

One of the primary windings is labeled H_1 and H_2. The other is labeled H_3 and H_4. The secondary winding is labeled X_1 and X_2. If the primary of the transformer is to be connected to 240 volts, the two primary windings will be connected in parallel by connecting H_1 and H_3 together, and H_2 and H_4 together, *(Figure 4-30)*. When the primary windings are connected in parallel, the same voltage is applied across both windings. This has the same effect as using one primary winding with a total of 200 turns of wire. A turns ratio of 2:1 is maintained and the secondary voltage will be 120 volts.

Figure 4-29 Control transformer with fuse protection added to the secondary winding. *(Courtesy of Hevi-Duty Electric.)*

If the transformer is to be connected to 480 volts, the two primary windings will be connected in series by connecting H_2 and H_3 together *(Figure 4-31)*. The incoming power is connected to H_1 and H_4. Series-connecting the primary windings has the effect of increasing the number of turns in the primary to 400. This produces a turns ratio of 4:1. When 480 volts is connected to the primary the secondary voltage will remain at 120.

The primary leads of a control transformer are generally cross-connected as shown in *Figure 4-32*. This is done so that metal links can be used to connect the primary for 240 or 480 volt operation. If the primary is to be connected for 240 volt operation, the metal links will be connected under screws as shown in *Figure 4-33*. Notice that leads H_1 and H_3 are connected

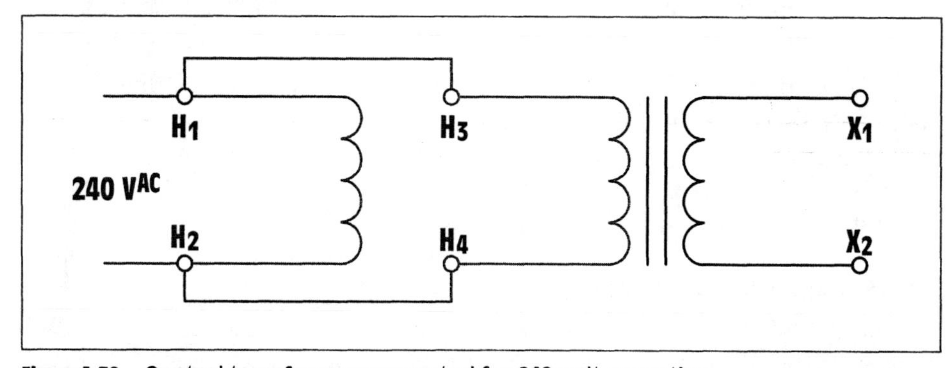

Figure 4-30 Control transformer connected for 240-volt operation.

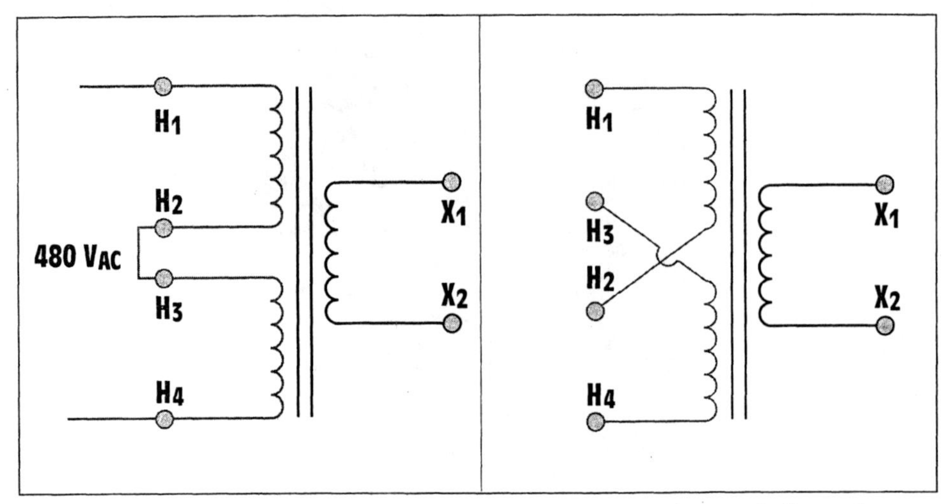

Figure 4-31 Control transformer connected for 480-volt operation.

Figure 4-32 The primary windings of a control transformer are crossed.

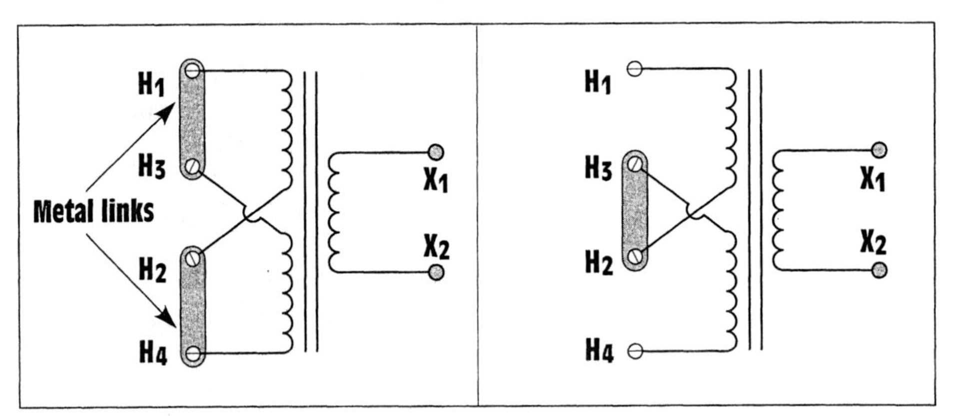

Figure 4-33 Metal links connect transformer for 240-volt operation.

Figure 4-34 Control transformer connected for 480-volt operation.

together and leads H_2 and H_4 are connected together. Compare this connection with the connection shown in *Figure 4-30*.

If the transformer is to be connected for 480 volt operation, terminals H_2 and H_3 are connected as shown in *Figure 4-34*. Compare this connection with the connection shown in *Figure 4-31*.

Potential Transformer

The potential transformer is basically an isolation transformer with a high-voltage rating on the primary winding intended for use in metering circuits.

They generally have low power ratings, such as 100 to 500 VA, and have a standard secondary voltage rating of 120 volts. Potential transformers are commonly used to operate meters that require a voltage input such as voltmeters, wattmeters, varmeters, and so forth, *(Figure 4-35)*. Since the secondary voltage will be proportional to the voltage applied to the primary, they can be used to accurately measure high voltages without the necessity of trying to insulate a meter for operation at a high voltage. Potential transformers intended for connection to voltages of 25,000 volts or less can be designed for indoor and outdoor usage. For voltage above 25,000 volts they are generally designed for outdoor connection only.

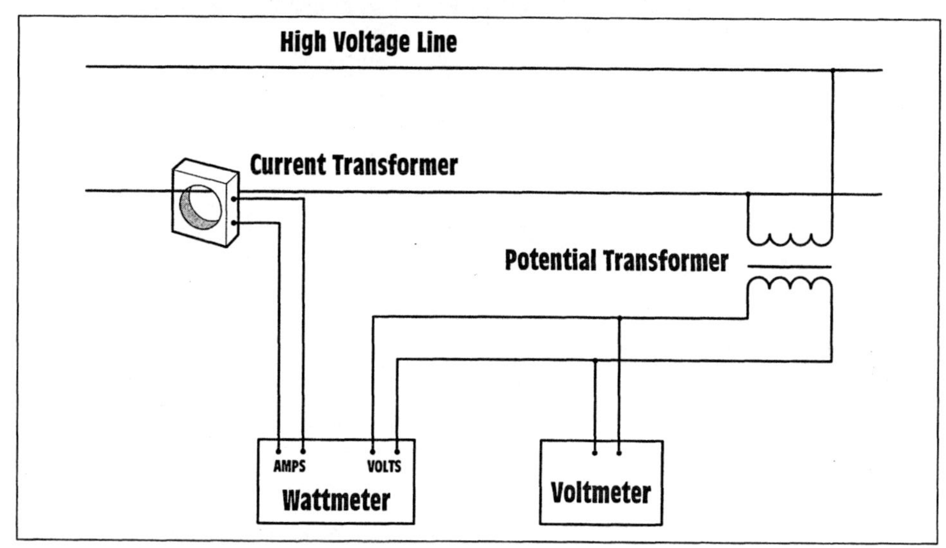

Figure 4-35 The potential transformer is used for metering purposes.

Transformer Inrush Current

Although transformers and reactors are both inductive devices, there is a great difference in their operating characteristics. Reactors (chokes) are often connected in series with the primary winding of large kVA transformers to prevent inrush current from becoming excessive when a circuit is first turned on *(Figure 4-36)*. Transformers can produce extremely high inrush currents when power is first applied to the primary winding. The type of core used when constructing inductors and transformers is primarily responsible for this difference in characteristics.

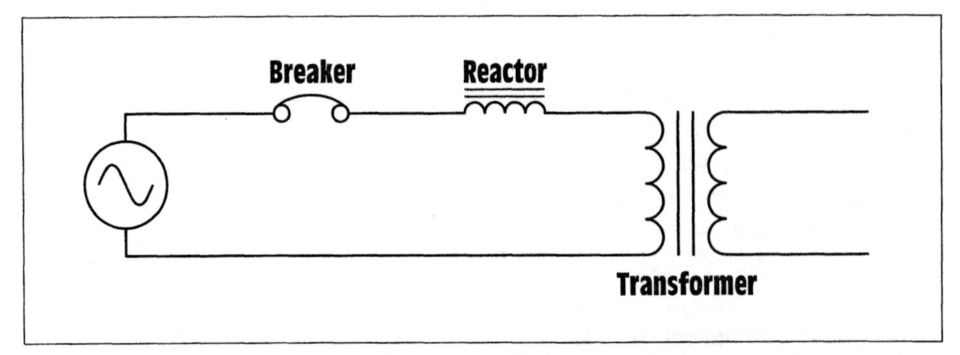

Figure 4-36 Reactors are connected in series with large kVA transformers to prevent excessive inrush current.

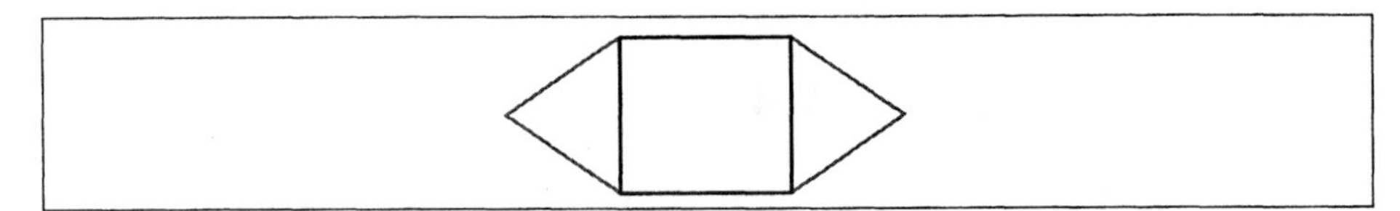

Figure 4-37 Magnetic domain in neutral position.

Magnetic Domains

Magnetic materials contain tiny magnetic structures in their molecular material known as *magnetic domains* . These domains can be affected by outside sources of magnetism. *Figure 4-37* illustrates a magnetic domain that has not been polarized by an outside magnetic source.

Now assume that the north pole of a magnet is placed toward the top of the material that contains the magnetic domains *(Figure 4-38)*. Notice that the structure of the domain has changed to realign the molecules in the direction of the outside magnetic field. If the polarity of the magnetic pole is changed *(Figure 4-39),* the molecular structure of the domain will change to realign itself with the new magnetic lines of flux. This external influence can be produced by an electromagnet as well as a permanent magnet.

In certain types of cores, the molecular structure of the domain will snap back to its neutral position when the magnetizing force is removed. This type of core is used in the construction of reactors or chokes *(Figure 4-40)*. A core of this type is constructed by separating sections of the steel laminations with an air gap. This air gap breaks the magnetic path through the core material and is responsible for the domains returning to their neutral position once the magnetizing force is removed.

The core construction of a transformer, however, does not contain an air gap. The steel laminations are connected together in such a manner as to

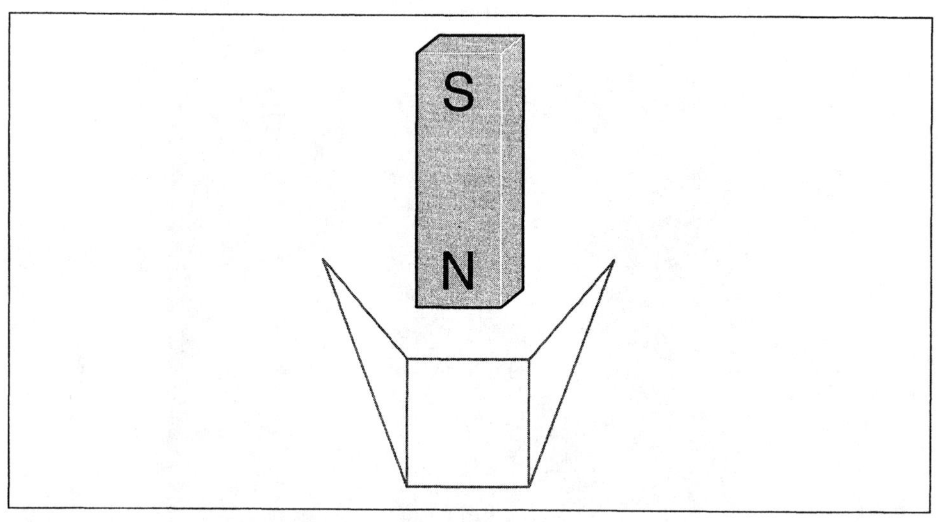

Figure 4-38 Domain influenced by a north magnetic field.

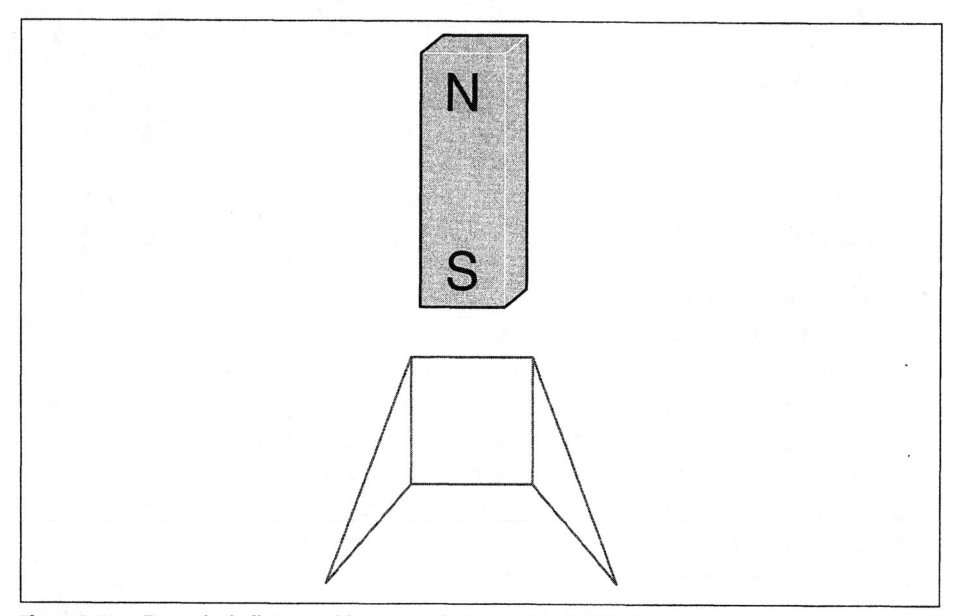

Figure 4-39 Domain influenced by a south magnetic field.

produce a very low-reluctance path for the magnetic lines of flux. In this type of core, the domains remain in their set position once the magnetizing force has been removed. This type of core "remembers" where it was last set. This was the principle of operation of the core memory of early computers. It is also the reason that transformers can have extremely high inrush

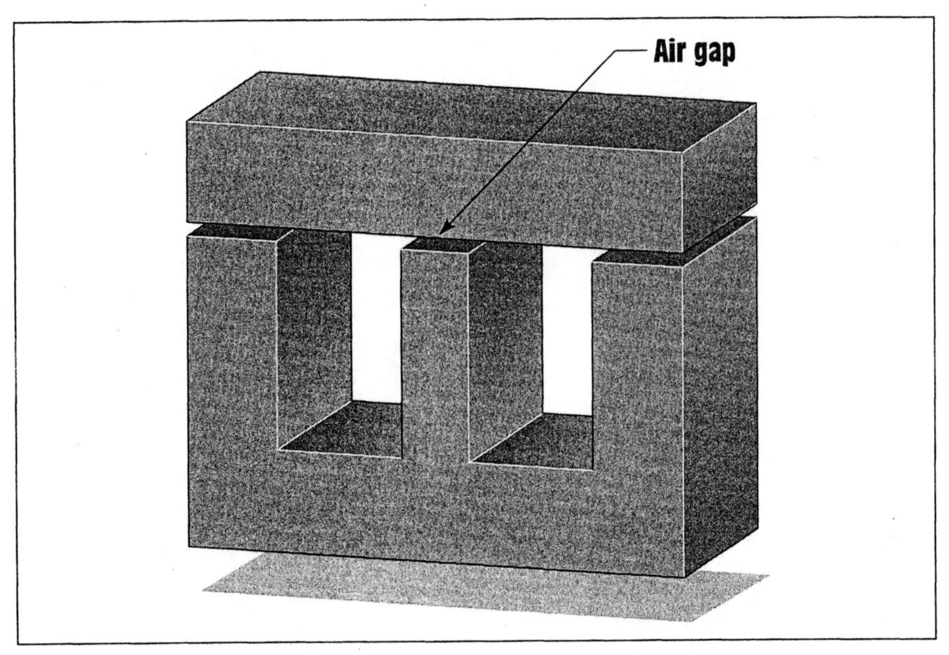

Figure 4-40 Core of an inductor.

currents when they are first connected to the power line.

The amount of inrush current in the primary of a transformer is limited by three factors which are:

1. amount of applied voltage
2. resistance of the wire in the primary winding
3. flux change of the magnetic field in the core.

The amount of flux change determines the amount of inductive reactance produced in the primary winding when power is applied. *Figure 4-41* illustrates a simple isolation-type transformer. The alternating current applied to the primary winding produces a magnetic field around the winding. As the current changes in magnitude and direction, the magnetic lines of flux change also. Since the lines of flux in the core are continually changing polarity, the magnetic domains in the core material are changing also. As stated previously, the magnetic domains in the core of a transformer remember their last set position. For this reason, the point on the wave form at which current is disconnected from the primary winding can have a great bearing on the amount of inrush current when the transformer is reconnected to power. For example, assume the power supplying the primary winding is disconnected at the zero crossing point *(Figure 4-42)*. In this instance, the magnetic domains would be set at the neutral point. When

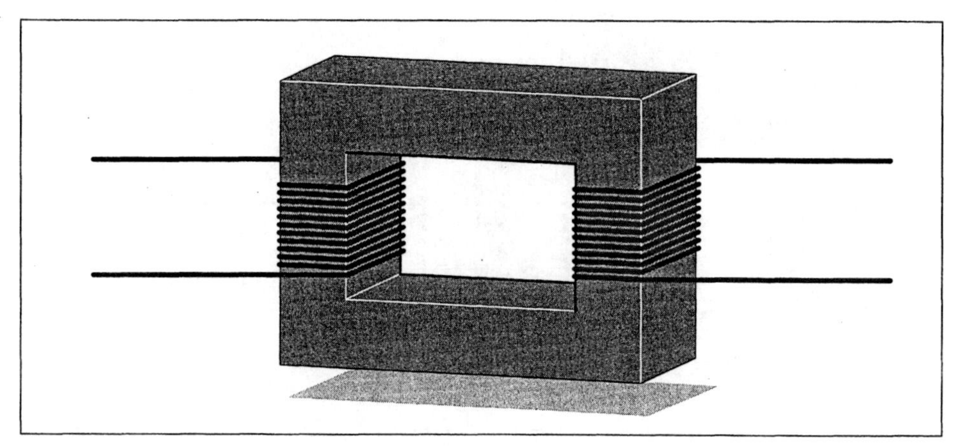

Figure 4-41 Isolation transformer.

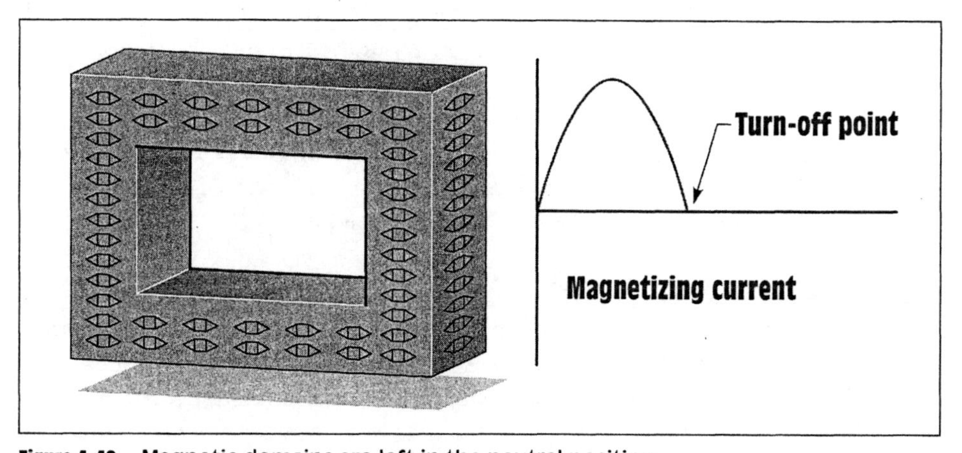

Figure 4-42 Magnetic domains are left in the neutral position.

power is restored to the primary winding, the core material can be magnetized by either magnetic polarity. This permits a change of flux, which is the dominant current-limiting factor. In this instance, the amount of inrush current would be relatively low.

If the power supplying current to the primary winding is interrupted at the peak point of the positive or negative half-cycle, however, the domains in the core material will be set at that position. *Figure 4-43* illustrates this condition. It is assumed that the current was stopped as it reached its peak positive point. If the power is reconnected to the primary winding during the positive half-cycle, only a very small amount of flux change can take place. Since the core material is saturated in the positive direction, the primary winding of the transformer is essentially an air-core inductor, which

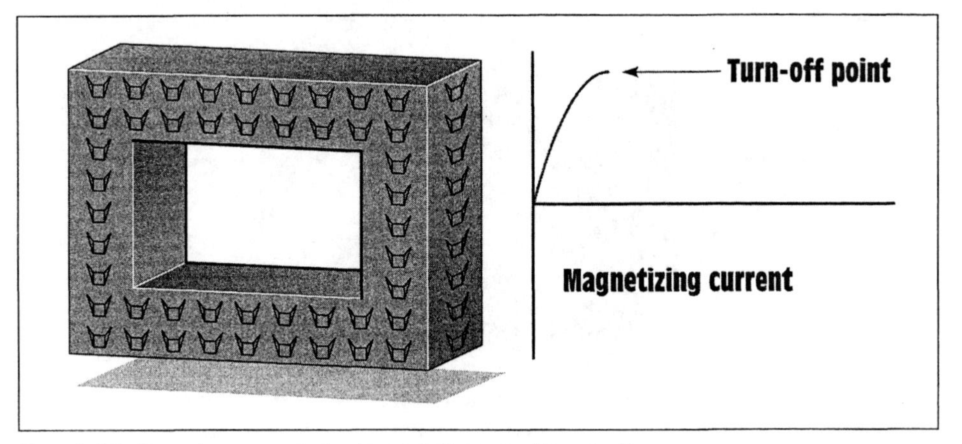

Figure 4-43 Domains are set at one end of magnetic polarity.

greatly decreases the inductive characteristics of the winding. The inrush current in this situation would be limited by the resistance of the winding and a very small amount of inductive reactance.

This characteristic of transformers can be demonstrated with a clamp-on ammeter that has a "peak hold" capability. If the ammeter is connected to one of the primary leads and power is switched on and off several times, the amount of inrush current will vary over a wide range.

Transformer Polarities

To understand what is meant by transformer polarity, the voltage produced across a winding must be considered during some point in time. In a 60-Hz AC circuit, the voltage changes polarity 120 times per second. When discussing transformer polarity, it is necessary to consider the relationship between the different windings at the same point in time. It will, therefore, be assumed that this point in time is when the peak positive voltage is being produced across the winding.

Polarity Markings on Schematics

When a transformer is shown on a schematic diagram it is common practice to indicate the polarity of the transformer windings by placing a dot beside one end of each winding as shown in *Figure 4-44*. These dots signify that the polarity is the same at that point in time for each winding. For example, assume the voltage applied to the primary winding is at its peak positive value at the terminal indicated by the dot. The voltage at the dotted lead of the secondary will be at its peak positive value at the same time.

This same type of polarity notation is used for transformers that have more

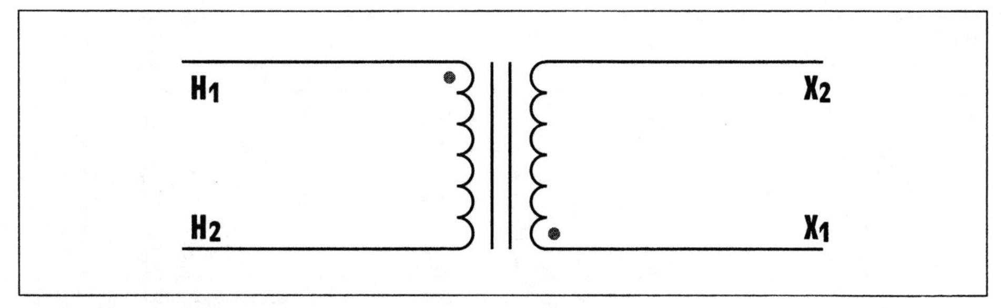

Figure 4-44 Transformer polarity dots.

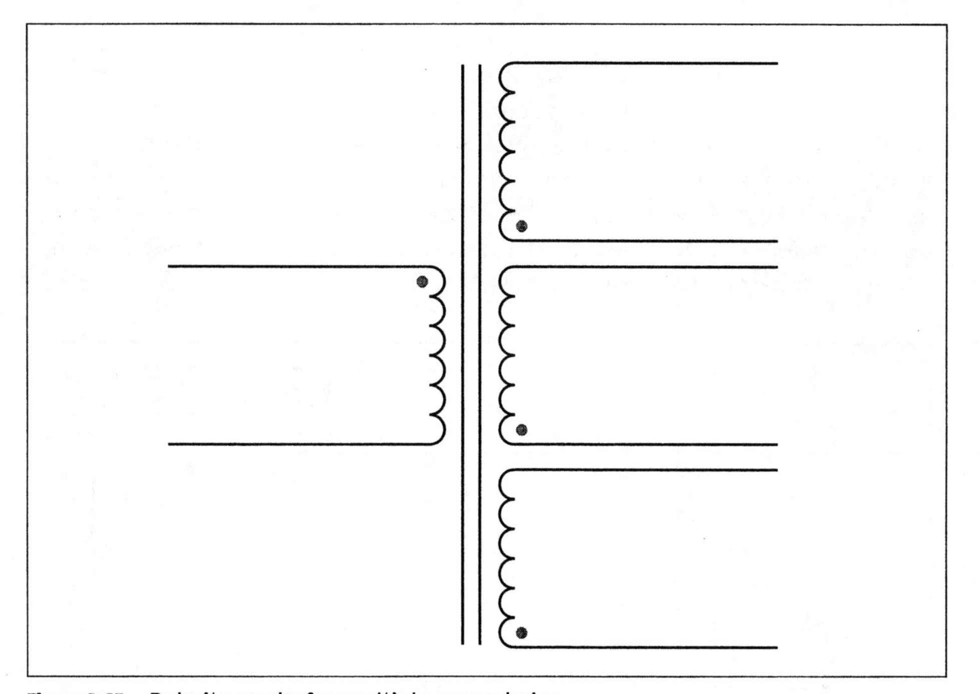

Figure 4-45 Polarity marks for multiple secondaries.

than one primary or secondary winding. An example of a transformer with a multisecondary is shown in *Figure 4-45*.

Additive and Subtractive Polarities

The polarity of transformer windings can be determined by connecting one lead of the primary to one lead of the secondary and testing for an increase or decrease in voltage. This is often referred to as a *buck* or *boost* connection, *(Figure 4-46)*. The transformer shown in the example has a primary voltage rating of 120 volts and a secondary voltage rating of 24 volts.

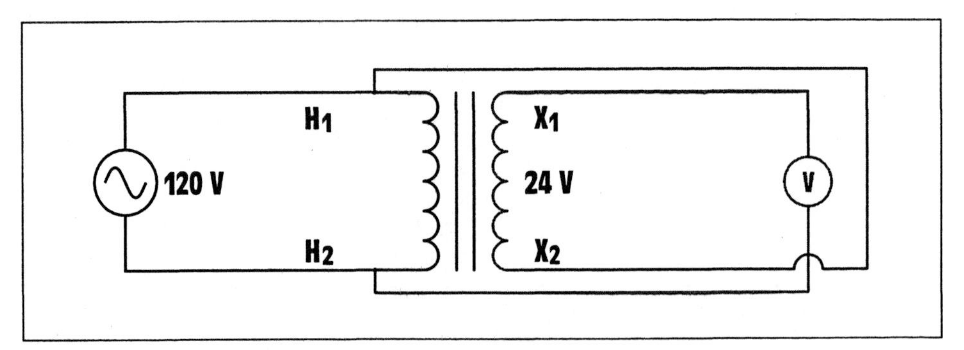

Figure 4-46 Connecting the secondary and primary windings.

This same circuit has been redrawn in *Figure 4-47* to show the connection more clearly. Notice that the secondary winding has been connected in series with the primary winding. When 120 volts is applied to the primary winding, the voltmeter connected across the secondary will indicate either the *sum* of the two voltages or the *difference* between the two voltages. If this voltmeter indicates 144 volts (120 + 24 = 144) the windings are connected additive (boost), and polarity dots can be placed as shown in *Figure*

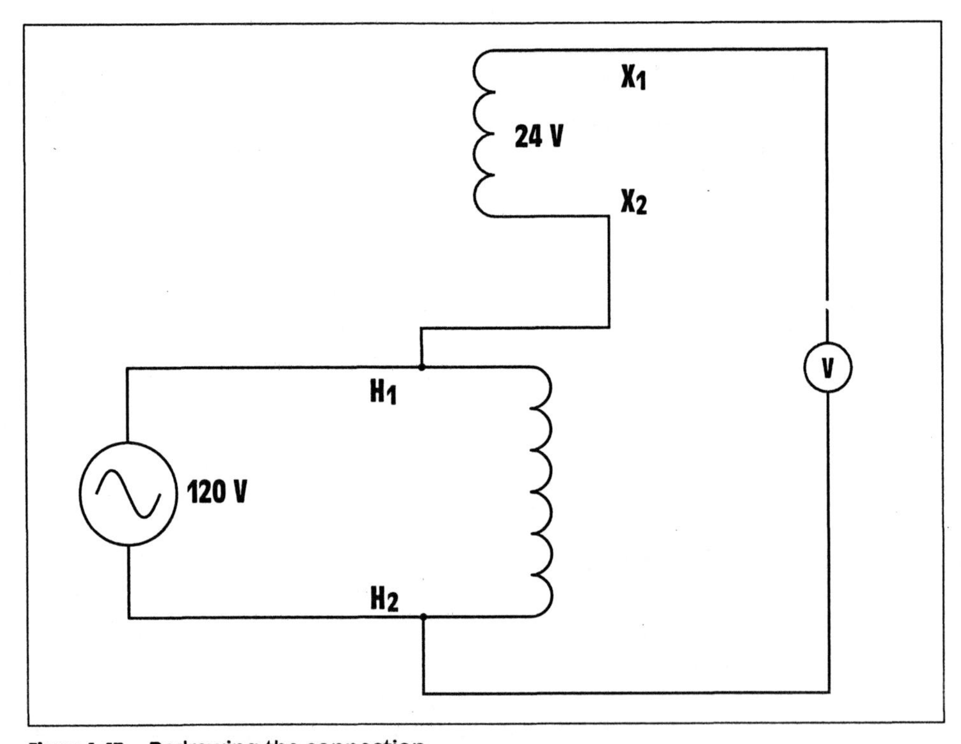

Figure 4-47 Redrawing the connection.

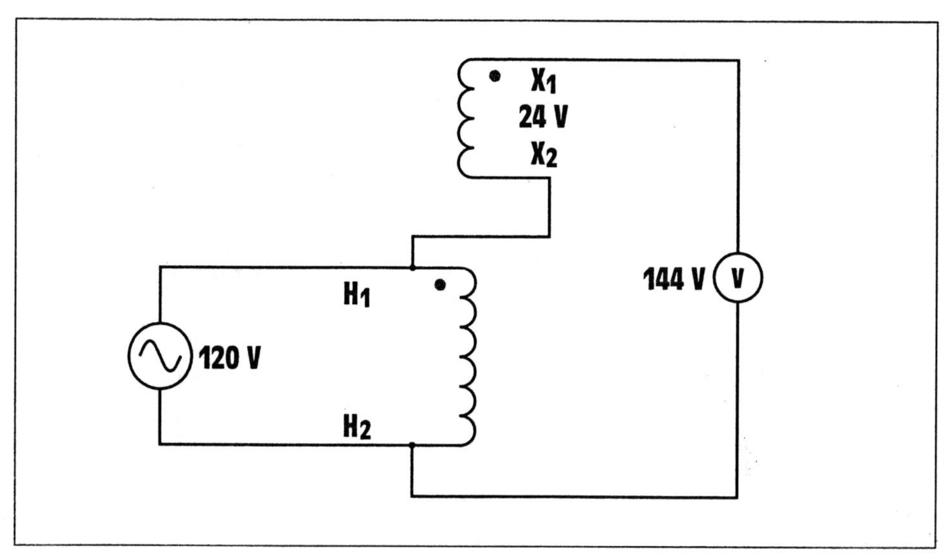

Figure 4-48 Placing polarity dots to indicate additive polarity.

4-48. Notice in this connection that the secondary voltage is added to the primary voltage.

If the voltmeter connected to the secondary winding indicates a voltage of 96 volts (120 – 24 = 96) the windings are connected subtractive (buck), and polarity dots are placed as shown in *Figure 4-49.*

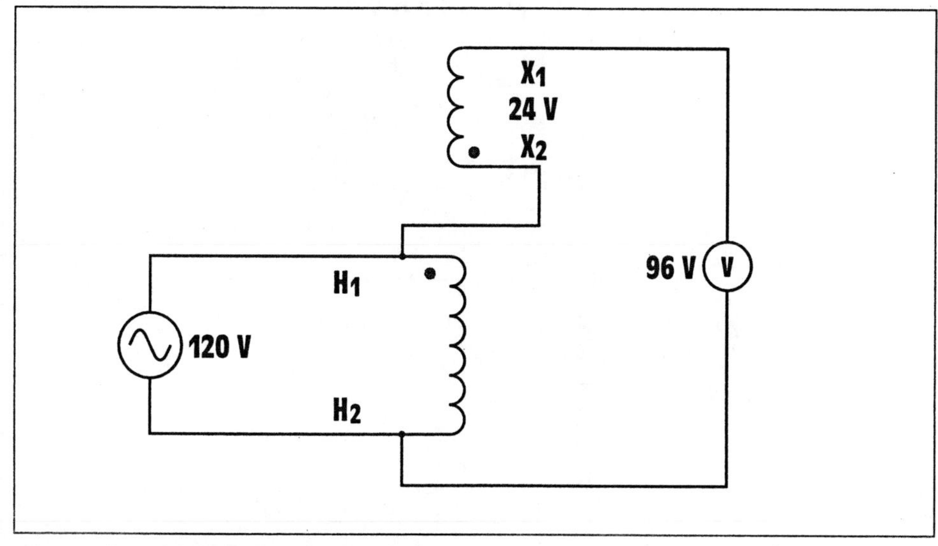

Figure 4-49 Polarity dots indicate subtractive polarity.

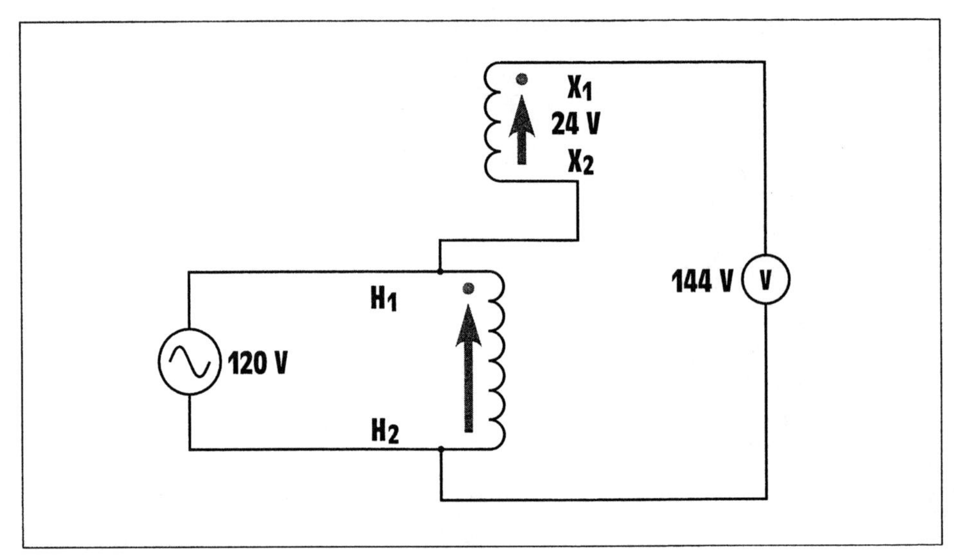

Figure 4-50 Arrows help indicate the placement of the polarity dots.

Using Arrows to Place Dots

To help in the understanding of additive and subtractive polarity, arrows can be used to indicate a direction of greater than or less than values. In *Figure 4-50,* arrows have been added to indicate the direction in which the dot is to be placed. In this example, the transformer is connected additive, or boost, and both of the arrows point in the same direction. Notice that the arrow points to the dot. In *Figure 4-51* it is seen that values of the two arrows add to produce 144 volts.

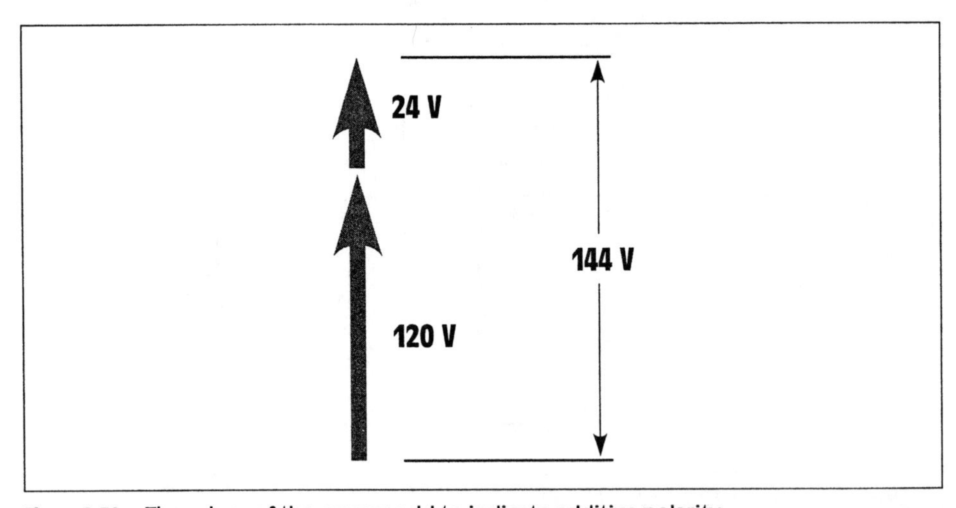

Figure 4-51 The values of the arrows add to indicate additive polarity.

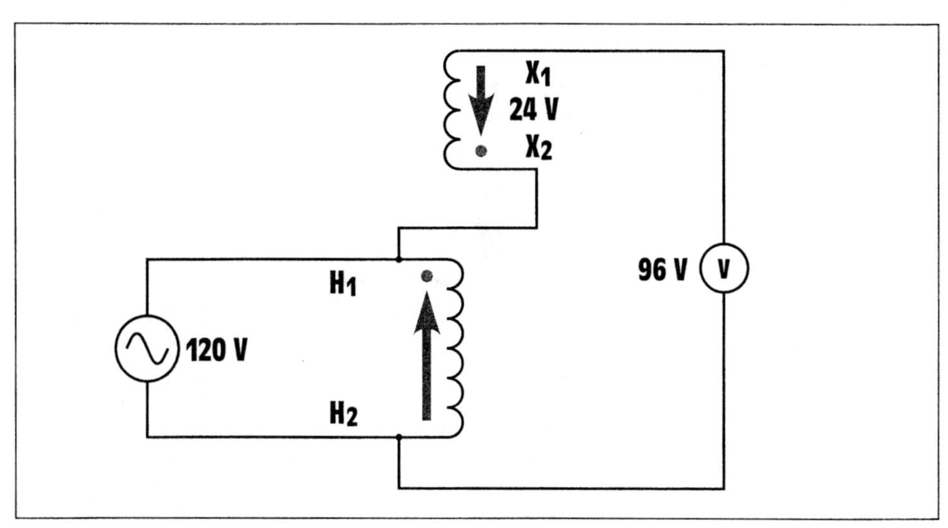

Figure 4-52 The arrows help indicate subtractive polarity.

In *Figure 4-52*, arrows have been added to a subtractive, or buck, connection. In this instance, the arrows point in opposite directions, and the voltage of one tries to cancel the voltage of the other. The result is that the smaller value is eliminated, and the larger value is reduced as shown in *Figure 4-53*.

Voltage and Current Relationships in a Transformer

When the primary of a transformer is connected to power but there is no load connected to the secondary, current is limited by the inductive reactance of the primary. At this time, the transformer is essentially an inductor, and the excitation current is lagging the applied voltage by 90° *(Figure 4-54)*. The primary current induces a voltage in the secondary. This induced voltage is proportional to the rate of change of current. The secondary voltage will be maximum during the periods that the primary current is changing the most (0°, 180°, and 360°), and it will be zero when the primary current is not changing (90° and 270°). If the primary current and secondary voltage are plotted, it will be seen that the secondary voltage lags the primary current by 90° *(Figure 4-55)*. Since the secondary voltage lags the primary current by 90° and the applied voltage leads the primary current by 90°, the secondary voltage is 180° out of phase with the applied voltage and in phase with the counter induced in the primary.

Adding Load to the Secondary

When a load is connected to the secondary, current begins to flow. Because the transformer is an inductive device, the secondary current lags

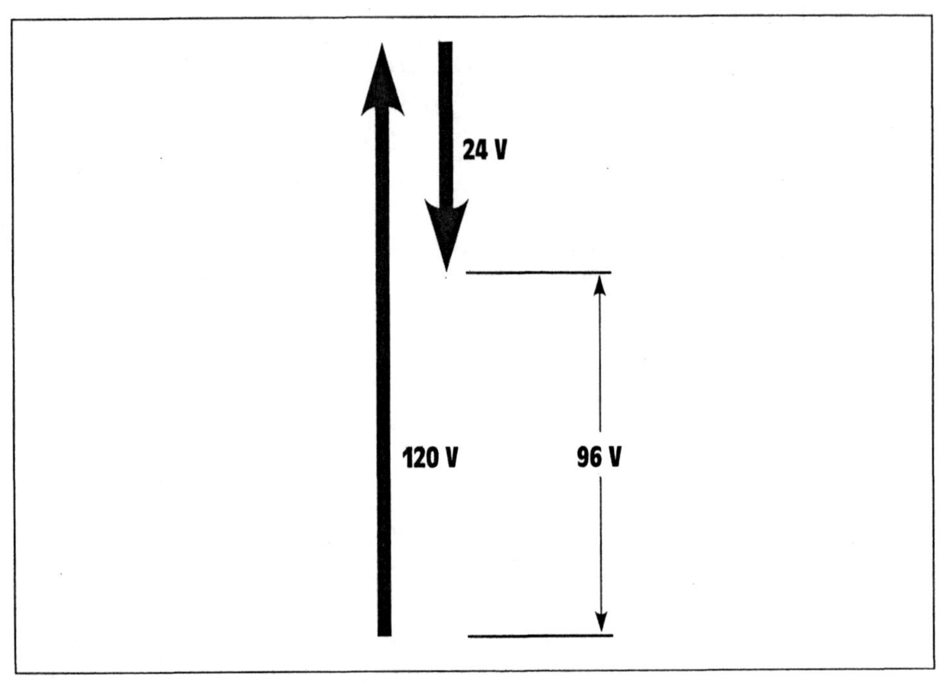

Figure 4-53 The values of the arrows subtract.

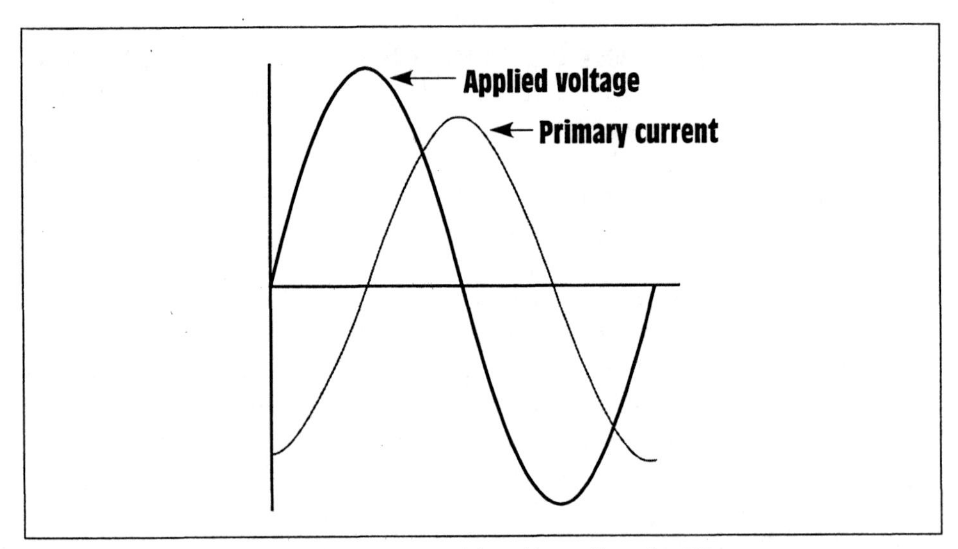

Figure 4-54 At no load, the primary current lags the voltage by 90°.

the secondary voltage by 90°. Since the secondary voltage lags the primary current by 90°, the secondary current is 180° out of phase with the primary current *(Figure 4-56)*.

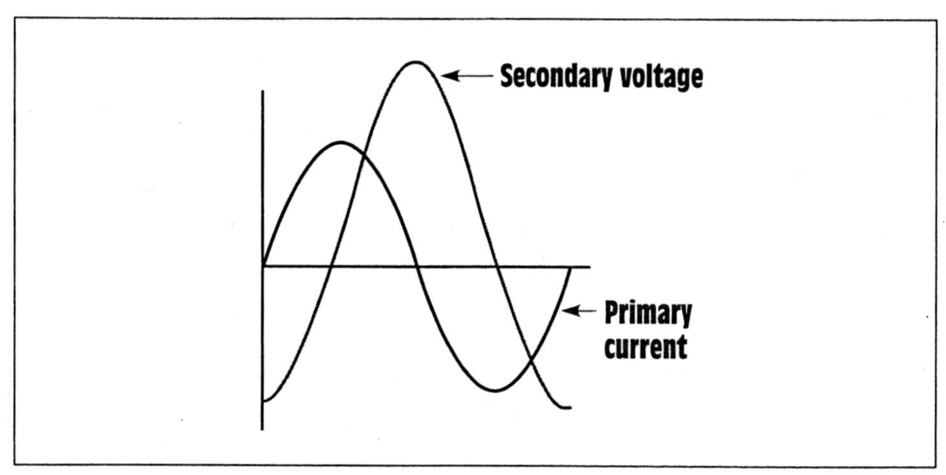

Figure 4-55 The secondary voltage lags the primary current by 90°.

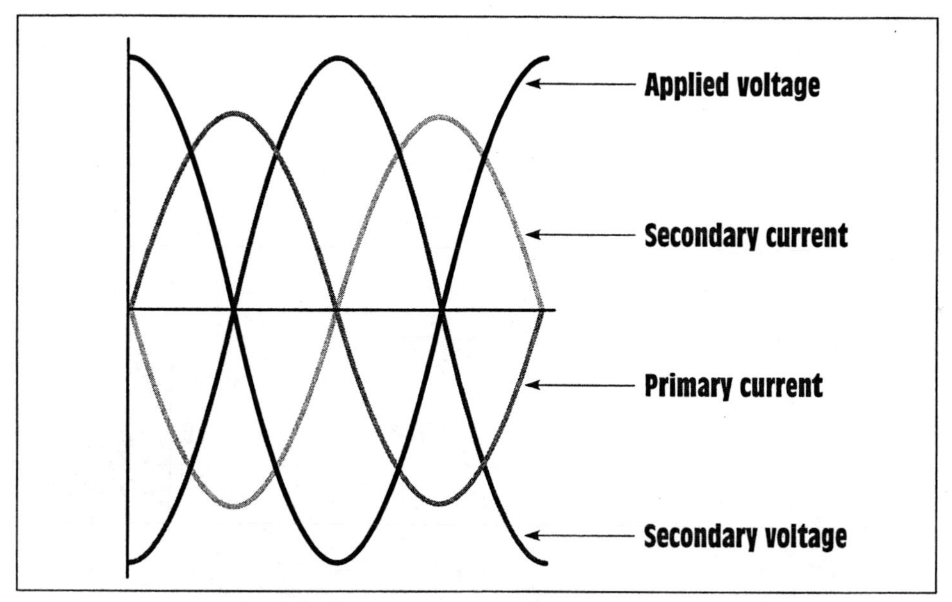

Figure 4-56 Voltage and current relationships of the primary and secondary windings

The current of the secondary induces a counter voltage in the secondary windings that is in opposition to the counter voltage induced in the primary. The counter voltage of the secondary weakens the counter voltage of the primary and permits more primary current to flow. As secondary current increases, primary current increases proportionally.

Since the secondary current causes a decrease in the counter voltage produced in the primary, the current of the primary is limited less by inductive reactance and more by the resistance of the windings as load is added to the secondary. If a wattmeter were connected to the primary, you would see that the true power would increase as load was added to the secondary.

Testing the Transformer

There are several tests that can be made to determine the condition of the transformer. A simple test for grounds, shorts, or opens can be made with an ohmmeter *(Figure 4-57)*. Ohmmeter A is connected to one lead of the primary and one lead of the secondary. This test checks for shorted windings between the primary and secondary. The ohmmeter should indicate infinity. If there is more than one primary or secondary winding, all isolated windings should be tested for shorts. Ohmmeter B illustrates testing the windings for grounds. One lead of the ohmmeter is connected to the case of the transformer, and the other is connected to the winding. All windings should be tested for grounds, and the ohmmeter should indicate infinity for each winding. Ohmmeter C illustrates testing the windings for continuity. The wire resistance of the winding should be indicated by the ohmmeter. Each winding should be tested for continuity.

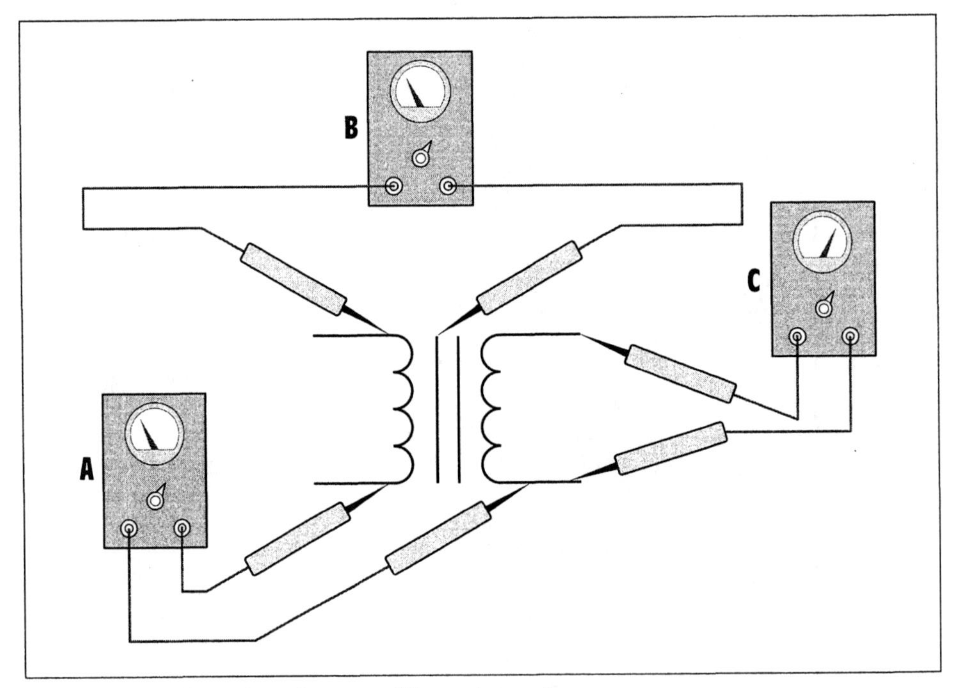

Figure 4-57 Testing a transformer with an ohmmeter.

If the transformer appears to be in good condition after the ohmmeter test, it should then be tested for shorts and grounds with a megohmmeter. A MEGGER® will reveal problems of insulation breakdown that an ohmmeter will not. Large oil-filled transformers should have the condition of the dielectric oil tested at periodic intervals. This involves taking a sample of the oil and performing tests for dielectric strength and contamination.

Transformer Ratings

Most transformers contain a nameplate that lists information concerning the transformer. The information listed is generally determined by the size, type, and manufacturer. Almost all nameplates will list the primary voltage, secondary voltage, and kVA rating. Transformers are rated in kilo-volt-amps and not kilowatts because the true power is determined by the power factor of the load. Other information that may or may not be listed is frequency, temperature rise in C°, percent impedance (%Z), type of insulating oil, gallons of insulating oil, serial number, type number, model number, and whether the transformer is single-phase or three-phase.

Determining Maximum Current

Notice the nameplate does not list the current rating of the windings. Since power input must equal power output, the current rating for a winding can be determined by dividing the kVA rating by the winding voltage. For example, assume a transformer has a kVA rating of 0.5 kVA, a primary voltage of 480 volts, and a secondary voltage of 120 volts. To determine the maximum current that can be supplied by the secondary, divide the kVA rating by the secondary voltage.

$$I_S = \frac{kVA}{E_S}$$

$$I_S = \frac{500}{120}$$

$$I_S = 4.16 \text{ amps}$$

The primary current can be computed in the same way.

$$I_P = \frac{kVA}{E_P}$$

$$I_P = \frac{500}{480}$$

$$I_P = 1.04 \text{ amps}$$

Transformers with multiple secondary windings will generally have the current rating listed with the voltage rating.

Transformer Impedance

Transformer impedance is determined by the physical construction of the transformer. Factors such as the amount and type of core material, wire size used to construct the windings, number of turns, and the degree of magnetic coupling between the windings greatly affect the transformer's impedance. Impedance is expressed as a percent and is measured by connecting a short circuit across the low-voltage winding of the transformer and then connecting a variable-voltage source to the high-voltage winding *(Figure 4-58)*. The variable voltage is then increased until rated current flows in the low-voltage winding. The transformer impedance is determined by calculating the percentage of variable voltage compared to the rated voltage of the high-voltage winding.

Example

Assume that the transformer shown in *Figure 4-58* is a 2400/480 volt 15-kVA transformer. To determine the impedance of the transformer, first compute the full-load current rating of the secondary winding.

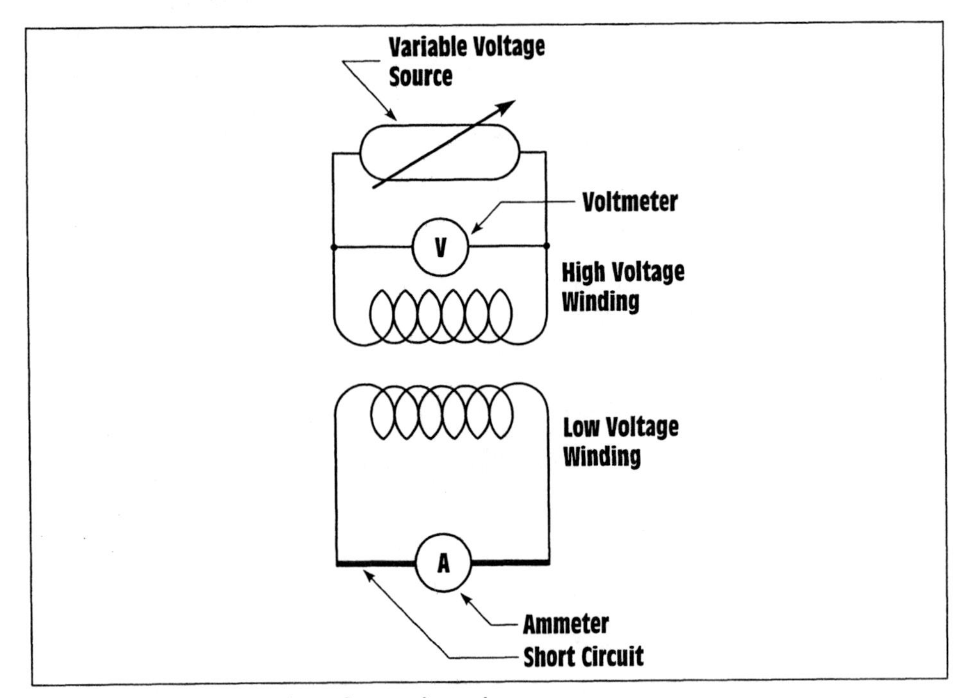

Figure 4-58 Determining transformer impedance.

$$I = \frac{VA}{E}$$

$$I = \frac{15,000}{480}$$

$$I = 31.25A$$

Next, increase the source voltage connected to the high-voltage winding until a current of 31.25 amperes flows in the low-voltage winding. For the purpose of this example, assume that voltage value is 138 volts. Finally, determine the percentage of applied voltage as compared to the rated voltage.

$$\%Z = \frac{\text{Source Voltage}}{\text{Rated Voltage}} \times 100$$

$$\%Z = \frac{138}{2400} \times 100$$

$$\%Z = 0.0575 \times 100$$

$$\%Z = 5.75$$

The impedance of this transformer is 5.75%.

Transformer impedance is a major factor in determining the amount of voltage drop a transformer will exhibit between no load and full load and in determining the amount of current flow in a short-circuit condition. Short-circuit current can be computed using the formula:

$$\text{(Single-Phase) } I_{SC} = \frac{VA}{E \times \%Z}$$

$$\text{(Three-Phase) } I_{SC} = \frac{VA}{E \times \sqrt{3} \times \%Z}$$

Since one of the formulas for determining current in a single-phase circuit is:

$$I = \frac{VA}{E}$$

And one of the formulas for determining current in a three-phase circuit is:

$$I = \frac{VA}{E \times \sqrt{3}}$$

The preceding formulas for determining short-circuit current can be modified to show that the short-circuit current can be computed by dividing the rated secondary current by the %Z.

$$I_{SC} = \frac{I_{Rated}}{\%Z}$$

Transformer Markings

The *American National Standards Institute* (ANSI) sets standardization rules concerning the way in which the terminals of transformers are to be marked. According to ANSI rules, the high-voltage leads of a transformer are to be marked $H_1 - H_2$, and so forth, and the low-voltage leads are marked $X_1 - X_2$, and so forth. The order is to be such that if the H_1 and X_1 terminals are connected together, and voltage is applied to the primary of the transformer, the voltage measured between the highest numbered H lead and the highest numbered X lead should be less than the voltage of the high-voltage winding, *(Figure 4-59)*. When the leads are marked in this manner, the transformer is considered to be subtractive polarity when the H_1 and X_1 leads are adjacent to each other. If the H_1 and X_1 leads are located diagonally, the transformer is considered to be additive polarity, *(Figure 4-60)*.

Standard markings can become very important when it is necessary to connect transformers together. It is sometimes necessary to use single-phase transformers to form a three-phase bank. (This will be discussed fully later in this text.) Another instance is when transformers are to be connected in series or parallel. The secondary windings of transformers are sometimes connected in series to increase the output voltage or to form a center-tap connection. When this is done, the primary windings are connected in paral-

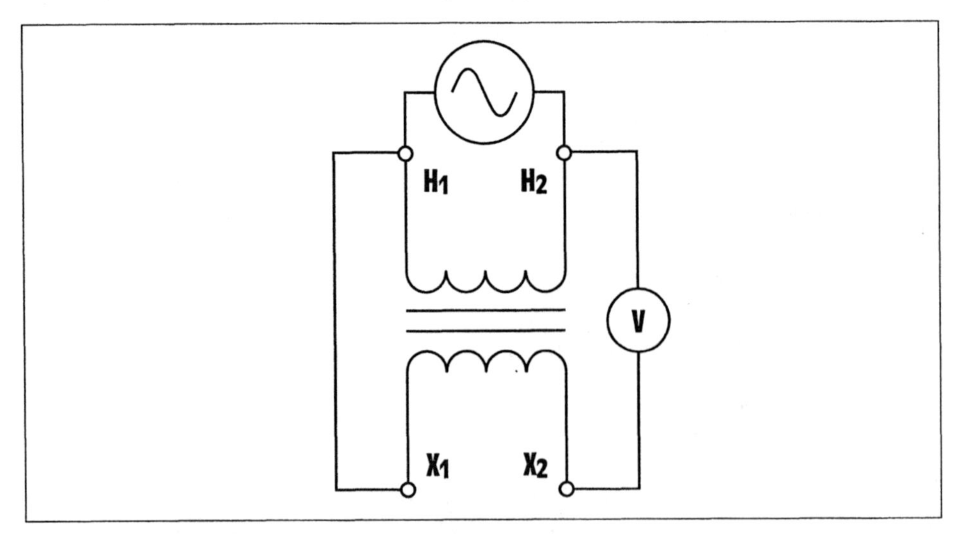

Figure 4-59 Determining transformer markings.

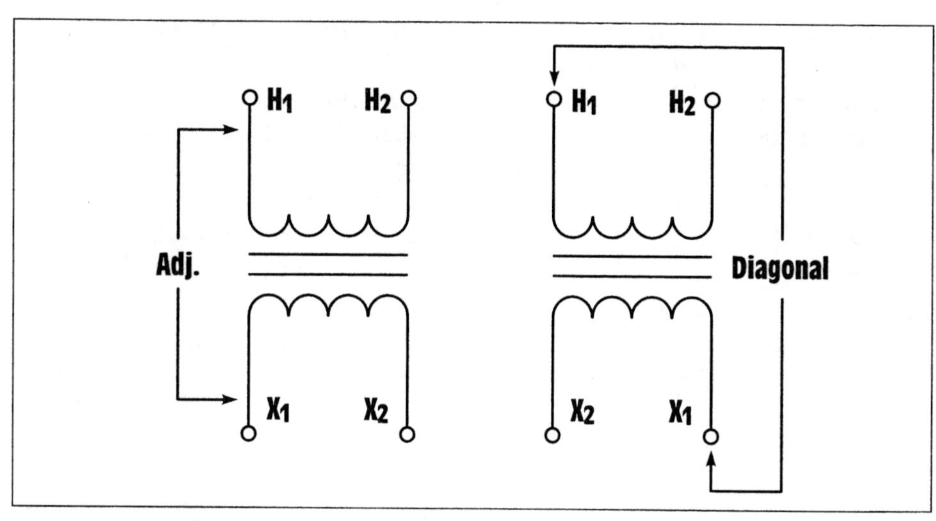

Figure 4-60 Additive and subtractive polarity.

lel to the power source, *(Figure 4-61)*. Series connection of the secondary windings does not present a problem with balance because the same current must flow through both windings.

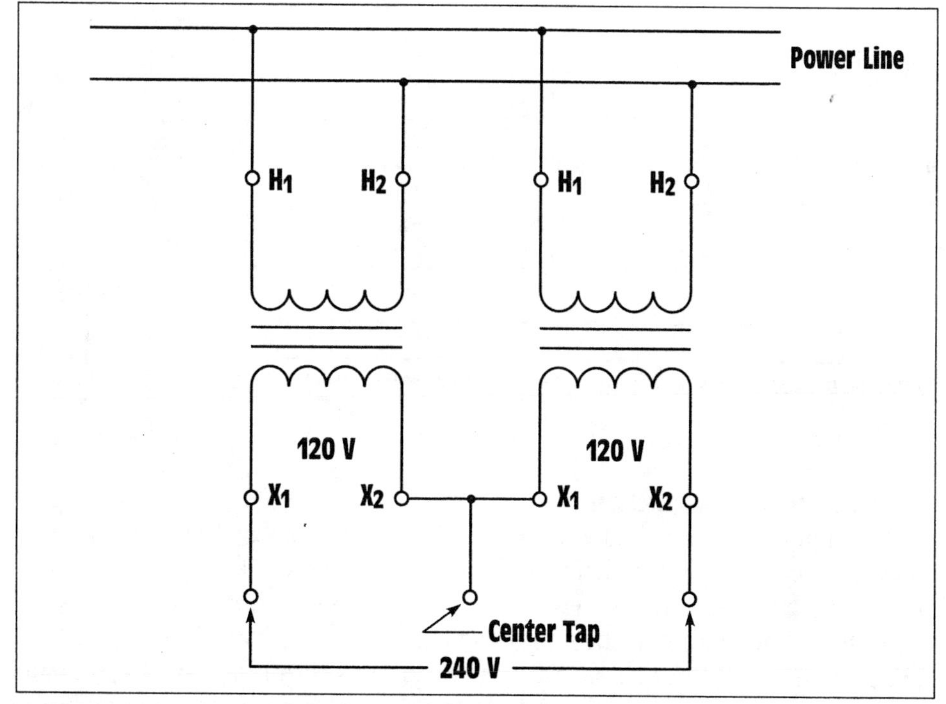

Figure 4-61 Secondary windings connected in series.

Connecting the secondary windings of transformers in parallel is generally avoided, but is sometimes necessary. When this connection is made, the primary windings are connected in parallel to the power source, *(Figure 4-62).* When transformers are connected in parallel, their characteristics must be the same or problems of imbalance can occur, causing one transformer to begin supplying current to the other. This causes one transformer to produce excessive current and the other to produce almost no current. Transformers intended for parallel connection should have the same voltage ratings, kVA ratings, and impedance.

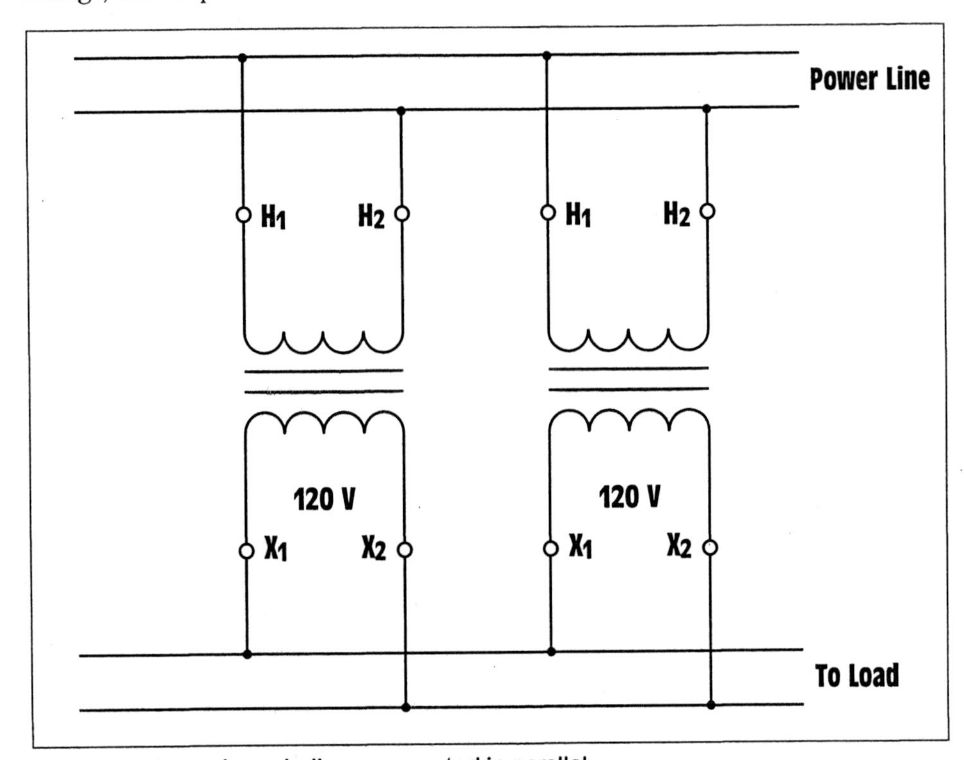

Figure 4-62 Secondary windings connected in parallel.

Transformer Losses

Although transformers are probably the most efficient machines known, they are not perfect. A transformer operating at 90% efficiency has a power loss of 10%. Some of these losses are I^2R losses, eddy current losses, hysteresis losses, and magnetic flux leakage. Most of these losses result in heat production. Recall that I^2R is one of the formulas for finding power or watts. In the case of a transformer, it describes the power loss associated with heat

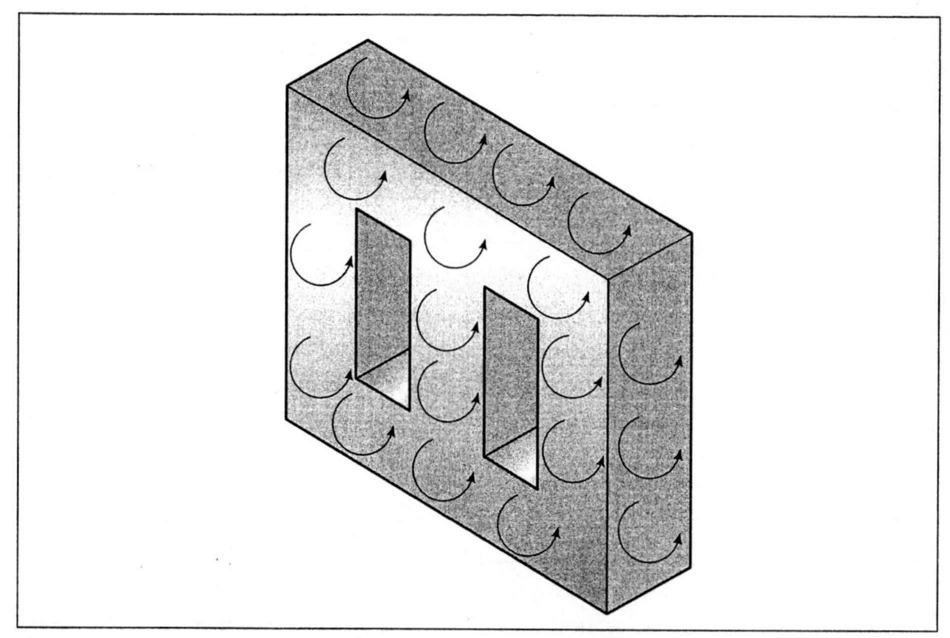

Figure 4-63 Eddy currents circulate inside the core material.

due to the resistance of the wire in both primary and secondary windings.

Eddy currents are currents that are induced into the metal-core material by the changing magnetic field as alternating current produces a changing flux. Eddy currents are so named because they circulate inside the metal in a manner similar to the swirling eddies in a river, *(Figure 4-63)*. These swirling currents produce heat which is a power loss. Transformers are constructed with laminated cores to help reduce eddy currents. The surface of each lamination forms a layer of iron oxide which acts as an insulator to help prevent the formation of eddy currents.

Hysteresis losses are due to molecular friction. As discussed previously, the reversal of the direction of current flow causes the molecules of iron in the core to realign each time the current changes direction. The molecules of iron are continually rubbing against each other as they realign magnetically. The friction of the molecules rubbing together causes heat which is a power loss. Hysteresis loss is proportional to frequency. The higher the frequency, the greater the loss. A special steel called *silicon steel* is often used in transformer cores to help reduce hysteresis loss. Power loss due to hysteresis and eddy currents is often called core losses.

Magnetic flux leakage does not produce heat, but does constitute a power loss. Flux leakage is caused by magnetic lines of flux radiating away from the transformer and not cutting the secondary windings. Flux leakage can be reduced by better core designs.

Transformer Voltage Regulation

The voltage regulation of a transformer is expressed as a percentage based on the difference between the secondary voltage at no load compared to full load. To determine regulation, load is added to the secondary until rated secondary voltage is applied across the load. The load is then removed, which will cause the secondary voltage to increase. The voltage regulation is the percentage of voltage increase compared to the rated voltage. The voltage regulation is proportional to the impedance of the transformer. Transformers with a lower-percent impedance will have better voltage regulation characteristics.

Constant-Current Transformers

A very special type of isolation transformer is the *constant-current transformer*, or *current regulator*. Constant-current transformers are designed to deliver a constant current, generally 6.6 amperes, under varying load conditions. They are most often used to provide the power for series-connected street lights. Street lights are often connected in series instead of parallel because of the savings in wire. Series-connected lights require one conductor run from lamp to lamp instead of two conductors, *(Figure 4-64)*. When series-connected lights are installed, a device must be used to provide a connection if one of the lamps should fail. Some lights use a reactor coil

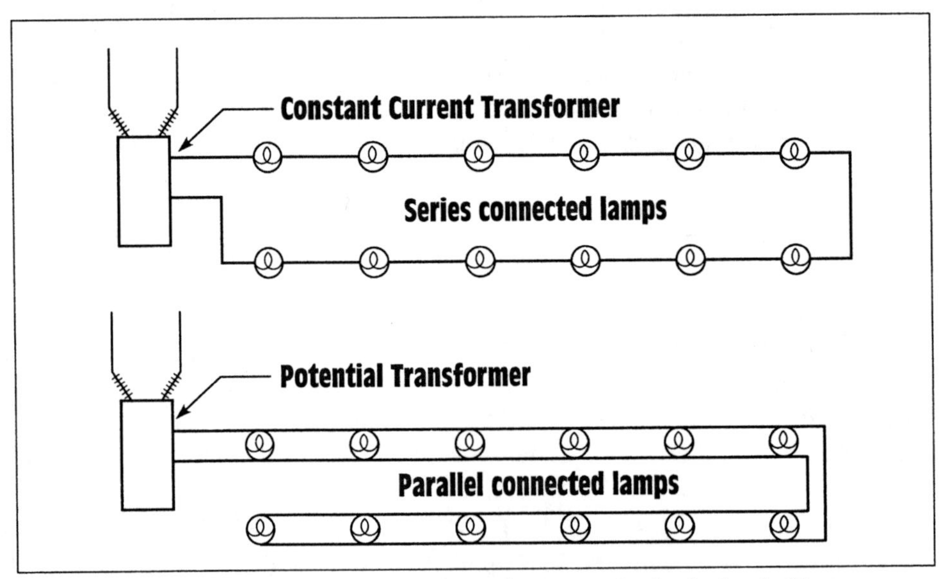

Figure 4-64 Series-connected lamps require only one conductor instead of two.

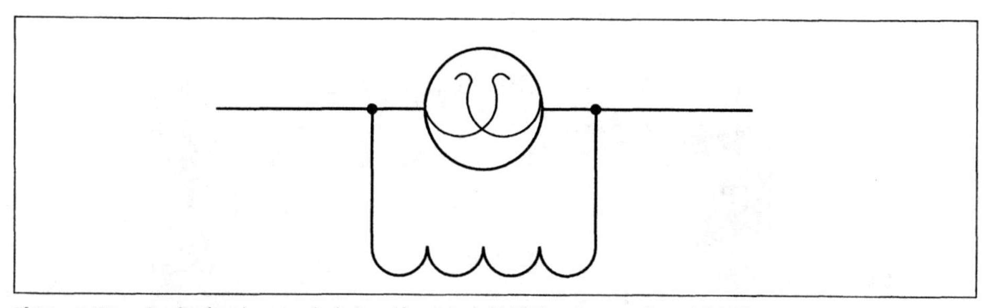

Figure 4-65 An inductor maintains the circuit if the lamp should fail.

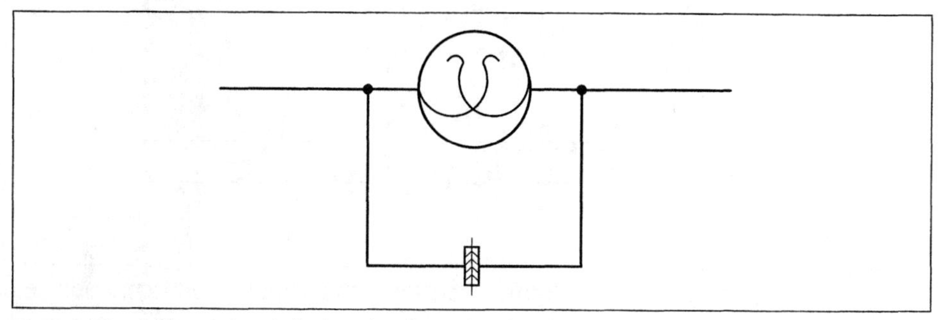

Figure 4-66 A film cut-out device shorts and maintains the circuit if the lamp should fail.

connected in parallel with the lamp, *(Figure 4-65)*. If the lamp should fail, the circuit is continued through the reactor coil. Others use a film cut-out device, *(Figure 4-66)*. The film cut-out device consists of two pieces of metal separated by an insulating film designed to puncture at a predetermined voltage. As long as the lamp is in operation, the voltage drop across the lamp is insufficient to rupture the film. If the lamp should burn out and become an open circuit, the entire circuit voltage is dropped across the film cut-out.

Constant-current transformers contain primary and secondary windings that are movable with respect to each other. Either winding can be made movable. Both windings are mounted on the same core material, *(Figure 4-67)*. Some transformers attach a counterweighted lever to the moving coil to help balance the weight of the movable coil.

The constant-current regulator operates by producing a magnetic field in the secondary windings that is in opposition to the magnetic field produced in the primary winding. If the load current of the secondary increases, the magnetic repulsion increases and the two coils move further apart, *(Figure 4-68)*. This produces a greater amount of flux leakage between the two

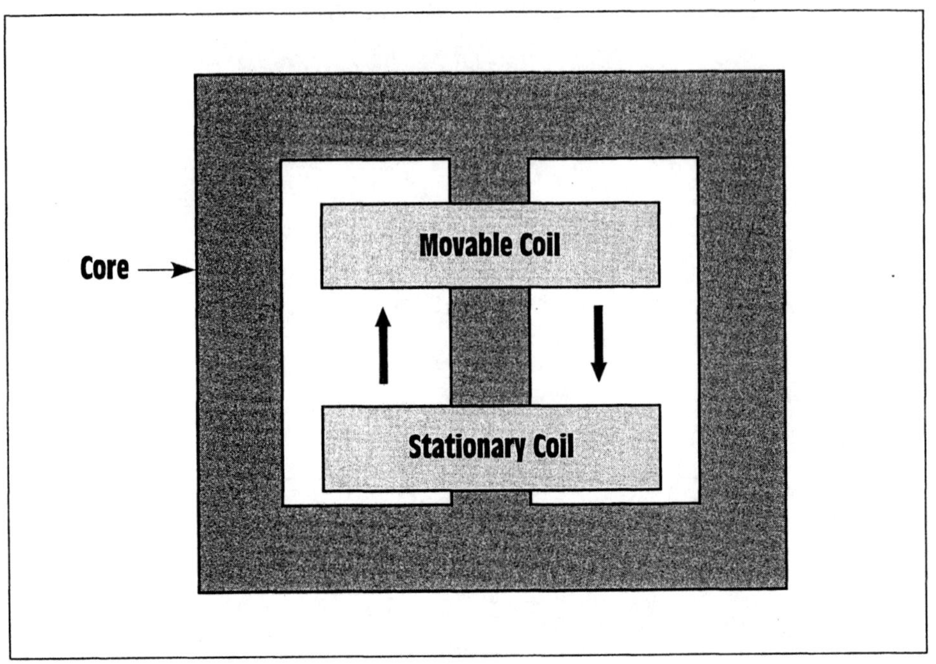

Figure 4-67 Constant-current transformers contain both a stationary and a movable coil.

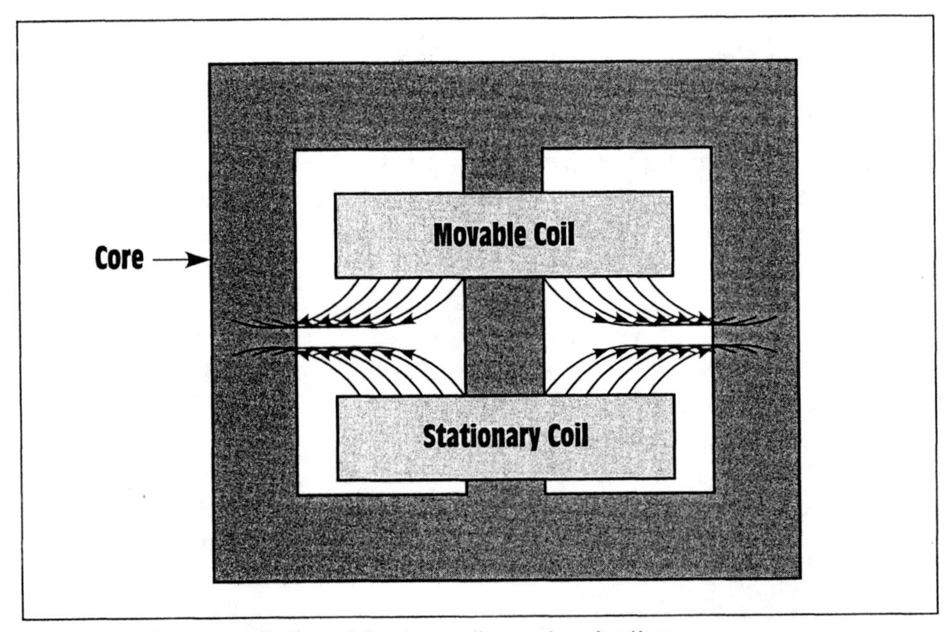

Figure 4-68 The magnetic flux of the two coils repel each other.

windings and causes a reduction in the secondary voltage. If the secondary current should decrease, the magnetic field of the secondary decreases also. This permits the two windings to move closer together which reduces the flux leakage and secondary voltage increases. Most moving coil-type current regulators contain a dashpot mechanism to reduce sudden changes in the spacing between the two coils. This helps reduce any "hunting" action between the two coils.

Summary

1. All values of voltage, current, and impedance in a transformer are proportional to the turns ratio.
2. Transformers can change values of voltage, current, and impedance, but cannot change the frequency.
3. The primary winding of a transformer is connected to the power line.
4. The secondary winding is connected to the load.
5. A transformer that has a lower secondary voltage than primary voltage is a step-down transformer.
6. A transformer that has a higher secondary voltage than primary voltage is a step-up transformer.
7. An isolation transformer has its primary and secondary windings electrically and mechanically separated from each other.
8. When a coil induces a voltage into itself, it is known as self-induction.
9. When a coil induces a voltage into another coil, it is known as mutual induction.
10. Transformers can have very high inrush current when first connected to the power line because of the magnetic domains in the core material.
11. Inductors provide an air gap in their core material that causes the magnetic domains to reset to a neutral position.
12. Inductors are sometimes connected in series with the primary winding of large transformers to reduce initial inrush current.
13. Either winding of a transformer can be used as the primary or secondary as long as its voltage or current ratings are not exceeded.
14. Isolation transformers help filter voltage and current spikes between the primary and secondary side.

15. Polarity dots are often added to schematic diagrams to indicate transformer polarity.
16. Transformers can be connected as additive or subtractive polarity.
17. Transformer impedance is expressed as a percent.
18. Voltage regulation is a percentage based on the change in voltage between the voltage at full load and the voltage at no load.
19. The high-voltage leads of a transformer are marked with an H, and the low-voltage leads are marked with an X.

Review Questions

1. What is a transformer?
2. What are common efficiencies for transformers?
3. What is an isolation transformer?
4. All values of a transformer are proportional to its

   ____________ ____________ .

5. A transformer has a primary voltage of 480 volts and a secondary voltage of 20 volts. What is the turns ratio of the transformer?
6. If the secondary of the transformer in question 5 supplies a current of 9.6 amperes to a load, what is the primary current (disregard excitation current)?
7. Explain the difference between a step-up and a step-down transformer.
8. A transformer has a primary voltage of 240 volts and a secondary voltage of 48 volts. What is the turns ratio of this transformer?
9. A transformer has an output of 750 VA. The primary voltage is 120 V. What is the primary current?
10. A transformer has a turns ratio of 1:6. The primary current is 18 amperes. What is the secondary current?
11. What do the dots shown beside the terminal leads of a transformer represent on a schematic?
12. A transformer has a primary voltage rating of 240 volts and a secondary voltage rating of 80 volts. If the windings were connected subtractive, what voltage would appear across the entire connection?
13. If the windings of the transformer in question 12 were to be connected additive, what voltage would appear across the entire winding?
14. The primary leads of a transformer are labeled 1 and 2. The secondary leads are labeled 3 and 4. If polarity dots are placed beside leads 1 and 4, which secondary lead would be connected to terminal 2 to make the connection additive?

Problems

Refer to *Figure 4-69* to answer the following questions. Find all the missing values.

1.

E_P 120	E_S 24
I_P ______	I_S ______
N_P 300	N_S ______
Ratio ______	$Z = 3\ \Omega$

2.

E_P 240	E_S 320
I_P 0.8533	I_S 0.64
N_P 210	N_S 280
Ratio 1:1.33	$Z = 500\ \Omega$

3.

E_P ______	E_S 160
I_P ______	I_S ______
N_P ______	N_S 80
Ratio 1:2.5	$Z = 12\ \Omega$

4.

E_P 48	E_S 240
I_P ______	I_S ______
N_P 220	N_S ______
Ratio ______	$Z = 360\ \Omega$

5.

E_P ______	E_S ______
I_P 16.5	I_S 3.25
N_P ______	N_S 450
Ratio ______	$Z = 56\ \Omega$

6.

E_P 480	E_S ______
I_P ______	I_S ______
N_P 275	N_S 525
Ratio ______	$Z = 1.2\ k\Omega$

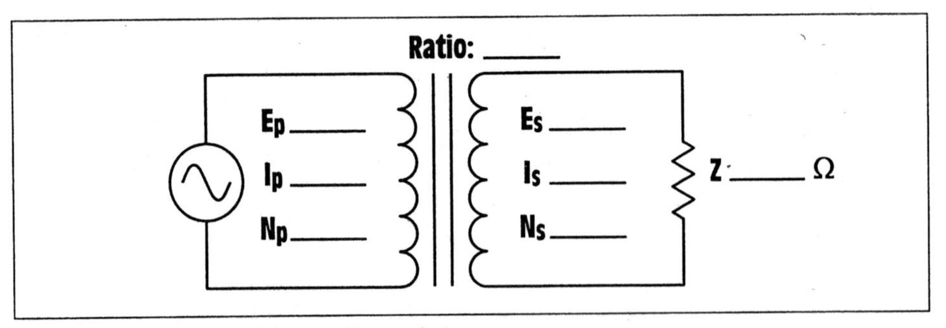

Figure 4-69 Practice problems 1 through 6.

Refer to *Figure 4-70* to answer the following questions. Find all the missing values .

7.

E_P 208	E_{S1} 320	E_{S2} 120	E_{S3} 24
I_P ______	I_{S1} ______	I_{S2} ______	I_{S3} ______
N_P 800	N_{S1} ______	N_{S2} ______	N_{S3} ______
	Ratio 1:	Ratio 2:	Ratio 3:
	R_1 12 kΩ	R_2 6 Ω	R_3 8 Ω

8.

E_P 277	E_{S1} 480	E_{S2} 208	E_{S3} 120
I_P ______	I_{S1} ______	I_{S2} ______	I_{S3} ______
N_P 350	N_{S1} ______	N_{S2} ______	N_{S3} ______
	Ratio 1:	Ratio 2:	Ratio 3:
	R_1 200 Ω	R_2 60 Ω	R_3 24 Ω

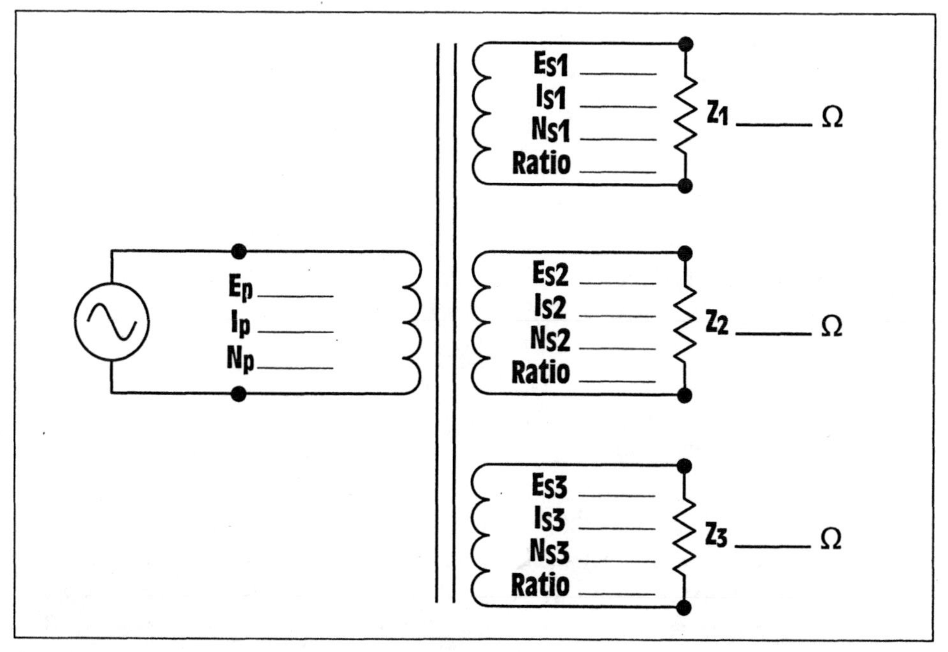

Figure 4-70 Practice problems 7 through 8.

5

Autotransformers

Objectives

After studying this unit, you should be able to

- Discuss the operation of an autotransformer
- List differences between isolation transformers and autotransformers
- Compute values of voltage, current, and turns ratios for autotransformers
- Connect an autotransformer for operation

The word *auto* means self. An autotransformer is literally a *self-transformer.* It uses the same winding as both primary and secondary transformers. Recall that the definition of a primary winding is a winding that is connected to the source of power, and the definition of a secondary winding is a winding that is connected to a load. Autotransformers have very high efficiencies, most in the range of 95% to 98%.

In *Figure 5-1,* the entire winding is connected to the power source, and part of the winding is connected to the load. All the turns of wire form the primary while part of the turns form the secondary. Since the secondary part of the winding contains less turns than the primary section, the secondary will produce less voltage. This autotransformer is a step-down transformer.

In *Figure 5-2,* the primary section is connected across part of a winding and the secondary is connected across the entire winding. The secondary

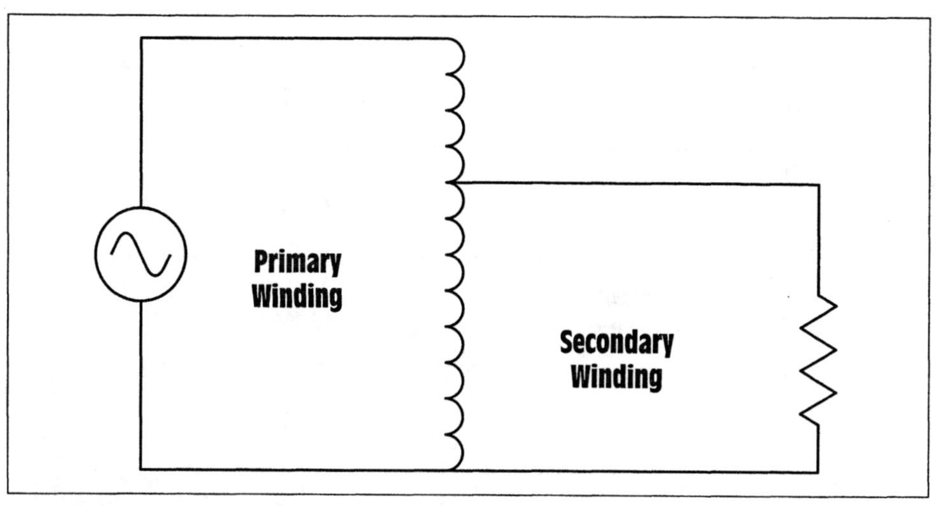

Figure 5-1 Autotransformer used as a step-down transformer.

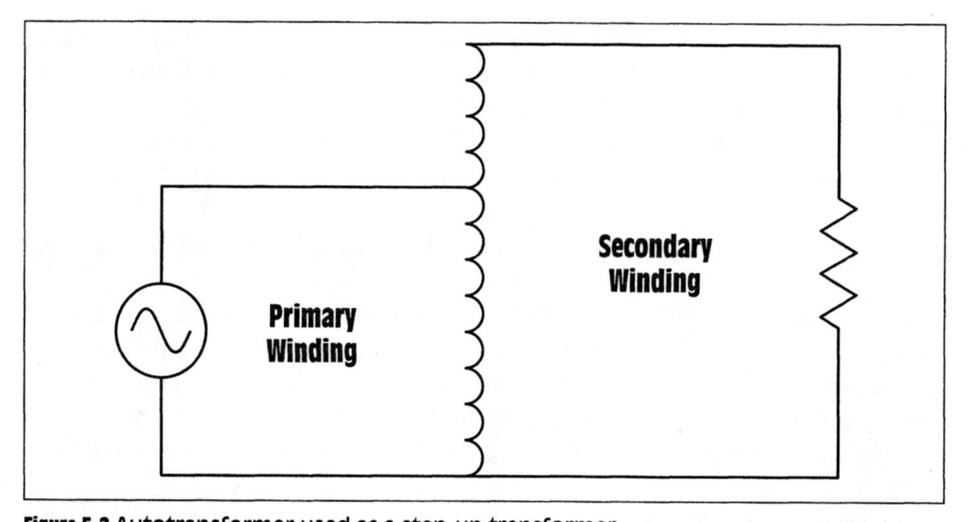

Figure 5-2 Autotransformer used as a step-up transformer.

section contains more windings than the primary. This type of transformer is a step-up transformer. (Notice that autotransformers, like isolation transformers, can be used as step-up or step-down transformers.)

Determining Voltage Values

Autotransformers are not limited to a single secondary winding. Many autotransformers have multiple taps to provide different voltages as shown in *Figure 5-3.* In this example, there are 40 turns of wire between taps A and B, 80 turns of wire between taps B and C, 100 turns of wire between taps C

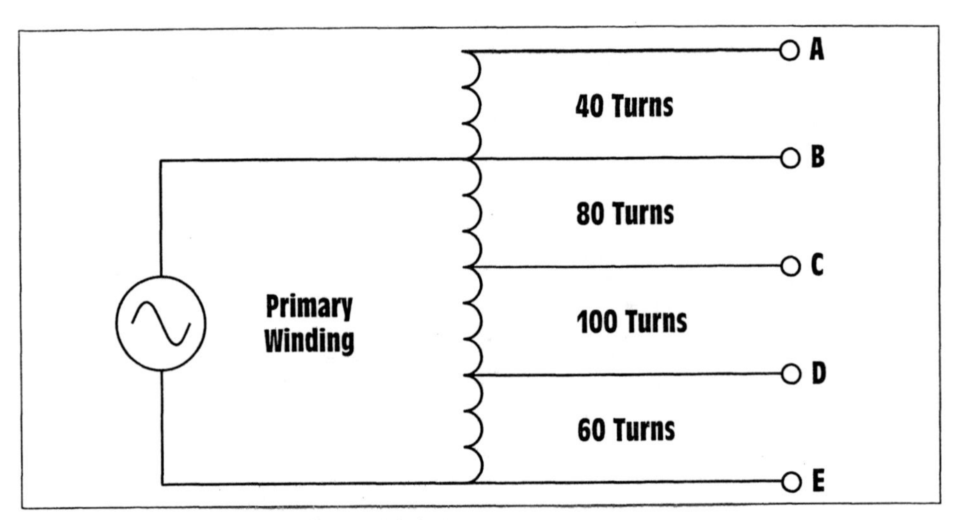

Figure 5-3 Autotransformer with multiple taps.

and D, and 60 turns of wire between taps D and E. The primary section of the windings is connected between taps B and E. It will be assumed that the primary is connected to a 120-volt source.

There is generally more than one method that can be employed to determine values of a transformer. Since the number of turns between each tap is known, the volts-per-turn method will be used in this example. **The volts-per-turn for any transformer is determined by the primary winding.** In *Figure 5-3*, the primary winding is connected across taps B and E. The primary turns are, therefore, the sum of the turns between taps B and E (80 + 100 + 60 = 240 turns). Since 120 volts is connected across 240 turns, this transformer will have a volts-per-turn ratio of 0.5 (240 turns/120 volts = 0.5 volts-per-turn). To determine the amount of voltage between each set of taps is a simple matter of multiplying the number of turns by the volts-per-turn.

A–B: (40 turns x 0.5 = 20 volts)
A–C: (120 turns x 0.5 = 60 volts)
A–D: (220 turns x 0.5 = 110 volts)
A–E: (280 turns x 0.5 = 140 volts)
B–C: (80 turns x 0.5 = 40 volts)
B–D: (180 turns x 0.5 = 90 volts)
B–E: (240 turns x 0.5 = 120 volts)
C–D: (100 turns x 0.5 = 100 volts)
C–E: (160 turns x 0.5 = 80 volts)
D–E: (60 turns x 0.5 = 30 volts)

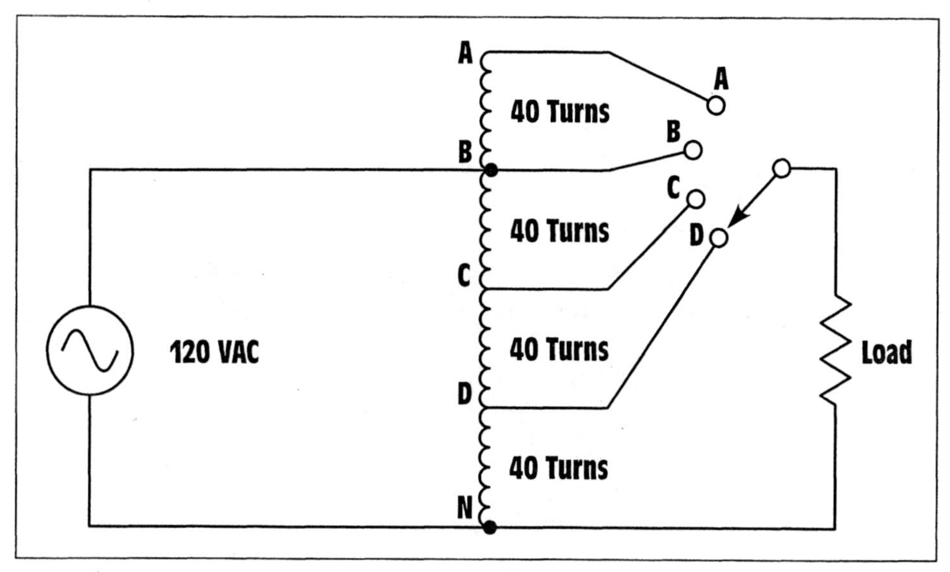

Figure 5-4 Determining voltage and current values.

Using Transformer Formulas

The values of voltage and current for autotransformers can also be determined by using standard transformer formulas. The primary winding of the transformer shown in *Figure 5-4* is between points B and N, and has a voltage of 120 volts applied to it. If the turns of wire are counted between points B and N, it can be seen there are 120 turns of wire. Now assume that the selector switch is set to point D. The load is now connected between points D and N. The secondary of this transformer contains 40 turns of wire. If the amount of voltage applied to the load is to be computed, the following formula can be used.

$$\frac{E_P}{E_S} = \frac{N_P}{N_S}$$

$$\frac{120}{E_S} = \frac{120}{40}$$

$$120\ E_S = 4800$$

$$E_S = 40 \text{ volts}$$

Assume that the load connected to the secondary has an impedance of 10 Ω. The amount of current flow in the secondary circuit can be computed using the formula

$$I = \frac{E}{Z}$$

$$I = \frac{40}{10}$$

$$I = 4 \text{ amps}$$

The primary current can be computed by using the same formula that was used to compute primary current for an isolation-type transformer.

$$\frac{E_P}{E_S} = \frac{I_S}{I_P}$$

$$\frac{120}{40} = \frac{4}{I_P}$$

$$I_P = 1.333 \text{ amps}$$

The amount of power input and output for the autotransformer must also be the same.

Primary	Secondary
120 x 1333 = 160 volt amps	40 x 4 = 160 volt amps

Now assume that the rotary switch is connected to point A. The load is now connected to 160 turns of wire. The voltage applied to the load can be computed by

$$\frac{E_P}{E_S} = \frac{N_P}{N_S}$$

$$\frac{120}{E_S} = \frac{120}{60}$$

$$120\ E_S = 19{,}200$$

$$E_S = 160 \text{ volts}$$

The amount of secondary current can be computed using the formula

$$I = \frac{E}{Z}$$

$$I = \frac{160}{10}$$

$$I = 16 \text{ amps}$$

The primary current can be computed using the formula

$$\frac{E_P}{E_S} = \frac{I_S}{I_P}$$

$$\frac{120}{160} = \frac{16}{I_P}$$

$$120\ I_P = 2560$$

$$I_P = 21.333 \text{ amps}$$

The answers can be checked by determining if the power in and power out are the same.

Primary	Secondary
120 x 21.333 = 2560 volt amps	160 x 16 = 2560 volt amps

Current Relationships

An autotransformer with a 2:1 turns ratio is shown in *Figure 5-5*. It is assumed that a voltage of 480 volts is connected across the entire winding. Since the transformer has a turns ratio of 2:1, a voltage of 240 volts will be supplied to the load. Ammeters connected in series with each winding indicate the current flow in the circuit. It is assumed that the load produces a

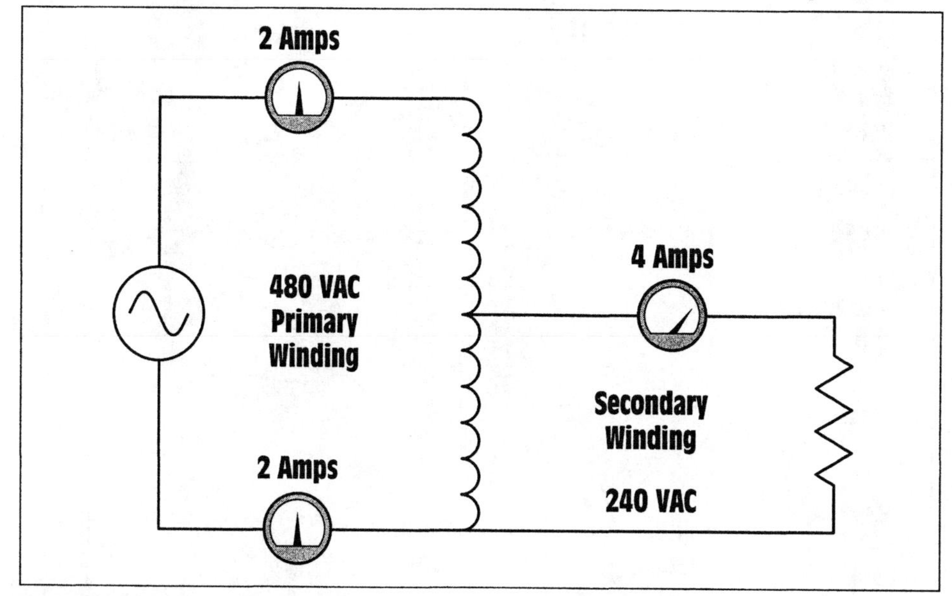

Figure 5-5 Current divides between primary and secondary.

current flow of 4 amperes on the secondary. Note that a current flow of 2 amperes is supplied to the primary.

$$I_{Primary} = \frac{I_{Secondary}}{Ratio}$$

$$I_P = \frac{4}{2}$$

$$I_P = 2 \text{ amperes}$$

If the rotary switch shown in *Figure 5-4* were to be removed and replaced with a sliding tap that made contact directly to the transformer winding, the turns ratio could be adjusted continuously. This type of transformer is commonly referred to as a Variac or Powerstat depending on the manufacturer. A variable autotransformer is shown in *Figure 5-6.* The windings are wrapped around a tape-wound torroid core inside a plastic case. The top of the windings have been milled flat, similar to a commutator. A carbon brush makes contact with the windings. When the brush is moved across the windings the turns ratio changes, which changes the output voltage. This type of autotransformer provides a very efficient means of controlling AC voltage.

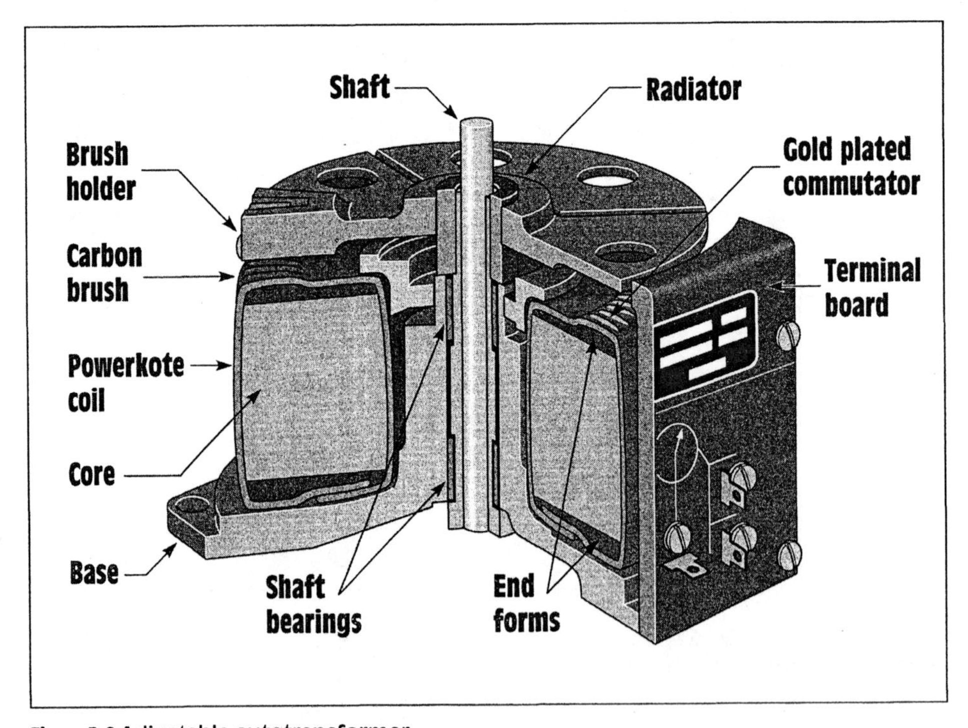

Figure 5-6 Adjustable autotransformer.

Autotransformers are often used by power companies to provide a small increase or decrease to line voltage. They help provide voltage regulation to large power lines. A 600-MVA (mega-volt-amp) three-phase autotransformer is shown in *Figure 5-7*. Notice the cooling fins located on the side of the transformer. These fins are actually hollow tubes through which dielectric oil is circulated to provide cooling.

The autotransformer does have one disadvantage. Since the load is connected to one side of the power line, there is no line isolation between the incoming power and the load. This can cause problems with certain types of equipment and must be a consideration when designing a power system.

Figure 5-7 600 MVA autotransformer (Courtesy of Houston Lighting and Power.)

Summary

1. The autotransformer has only one winding that is used as both the primary and secondary.
2. Autotransformers have efficiencies that range from about 95% to 98%.
3. Values of voltage, current, and turns can be computed in the same manner as an isolation transformer.
4. Autotransformers can be step-up or step-down transformers.
5. Autotransformers can be made to provide a variable-output voltage by connecting a sliding tap to the windings.
6. Autotransformers have the disadvantage of no line isolation between primary and secondary.
7. One of the simplest ways of computing values of voltage for an autotransformer when the turns are known is to use the volts-per-turn method.

Review Questions

1. An AC power source is connected across 325 turns of an autotransformer, and the load is connected across 260 turns. What is the turns ratio of this transformer?
2. Is the transformer in question 1 a step-up or step-down transformer?
3. An autotransformer has a turns ratio of 3.2:1. A voltage of 208 volts is connected across the primary. What is the voltage of the secondary?
4. A load impedance of 52 Ω is connected to the secondary winding of the transformer in question 3. How much current will flow in the secondary?
5. How much current will flow in the primary of the transformer in question 4?
6. The autotransformer shown in *Figure 5-3* has the following number of turns between windings: A–B (120 turns), B–C (180 turns), C–D (250 turns), and D–E (300 turns). A voltage of 240 volts is connected across B and E. Find the voltages between each of the following points:

   A–B ____ A–C ____ A–D ____ A–E ____ B–C ____ B–D ____

   B–E ____ C–D ____ C–E ____ D–E ____

Problems

Refer to the transformer shown in *Figure 5-8* to answer the following questions.

1. Assume that a voltage of 208 volts is applied across terminals B and E. How much voltage is across each of the following terminals?

   A–B ____ A–C ____ A–D ____ A–E ____ B–C ____ B–D ____

   B–E ____ C–D ____ C–E ____ D–E ____

2. Assume a voltage of 120 volts is connected across terminals B and E, and that a load impedance of 20 Ω is connected across terminals A and C. What is the secondary and primary current?

   I_P ____ I_S ____

3. What is the turns ratio of the winding between points B and E as compared to the winding between points D and E?
4. Assume that a voltage of 480 volts is connected across terminals A and E. What is the voltage across terminals C and D?
5. The transformer shown in *Figure 5-8* is supplying 325 volt-amperes to a load. The primary voltage is 240 volts. What is the primary current?

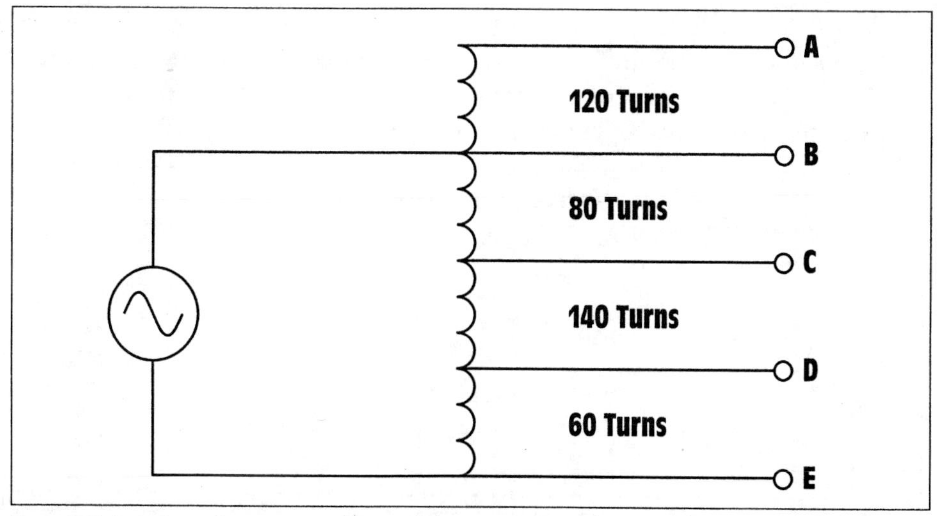

Figure 5-8 Autotransformer practice problems.

6

Current Transformers

Objectives

After studying this unit, you should be able to

- Discuss the operation of a current transformer
- Describe how current transformers differ from voltage transformers
- Discuss safety precautions that should be observed when using current transformers
- Connect a current transformer in a circuit

Current transformers differ from voltage transformers in that the primary winding is generally part of the power line. The primary winding of a current transformer must be connected in series with the load, *(Figure 6-1)*.

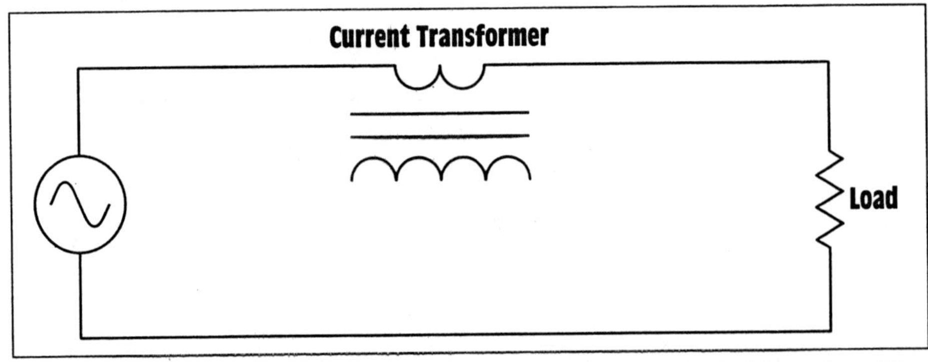

Figure 6-1 The primary winding of a current transformer is connected in series with a load.

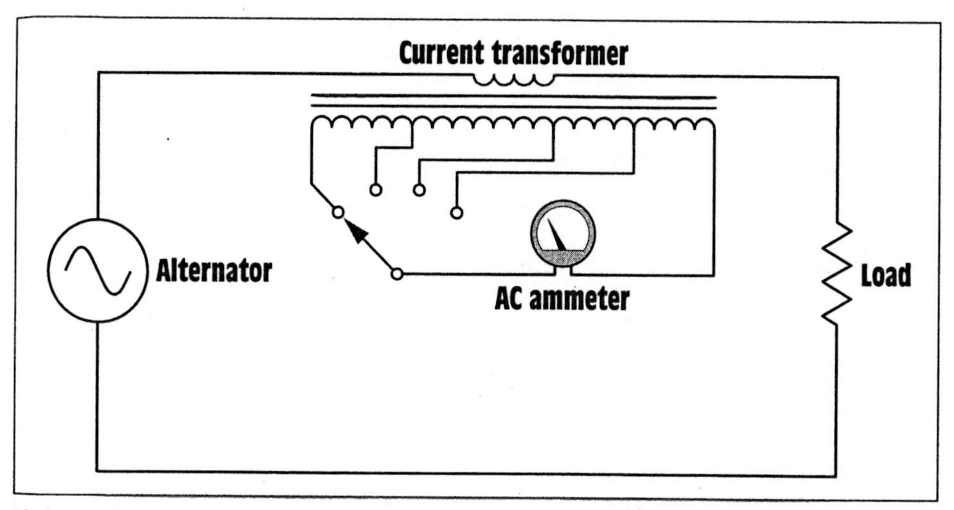

Figure 6-2 A current transformer is used to change the range of an AC ammeter

Current transformers are used to change the full-scale range of AC ammeters. Most in-line ammeters (ammeters that must be connected directly into the line) that have multiple-range values use a current transformer to provide the different ranges, *(Figure 6-2)*. The full-scale value of the ammeter is changed by adjusting the turns ratio. Assume that the ammeter in *Figure 6-2* is to provide range values of 5 A, 2.5 A, 1 A, and 0.5 A. Also assume that the meter movement requires a current flow of 100 ma. (0.100) to deflect the meter full scale and that the primary of the current transformer contains 5 turns of wire. Transformer formulas can be used to determine the number of secondary turns needed to produce the desired ranges.

Turns needed for a full-scale range of 5 amperes.

$$\frac{N_P}{N_S} = \frac{I_S}{I_P}$$

$$\frac{5}{N_S} = \frac{0.1}{5}$$

$$0.1\ N_S = 25$$

$$N_S = 250 \text{ turns}$$

Turns needed for a full-scale range of 2.5 amperes.

$$\frac{5}{N_S} = \frac{0.1}{2.5}$$

$$0.1\ N_S = 12.5$$

$$N_S = 125 \text{ turns}$$

Turns needed for a full-scale range of 1 ampere.

$$\frac{5}{N_S} = \frac{0.1}{1}$$

$$0.1\ N_S = 5$$

$$N_S = 50 \text{ turns}$$

Turns needed for a full-scale range of 0.5 ampere.

$$\frac{5}{N_S} = \frac{0.1}{0.5}$$

$$0.1\ N_S = 2.5$$

$$N_S = 25 \text{ turns}$$

An inline ammeter that can be set for different scale values is shown in *Figure 6-3*.

Figure 6-3 Inline ammeter.

When a large amount of AC current must be measured, a different type of current transformer is connected in the power line. These transformers have ratios that start at 200:5 and can have ratios of several thousand to five. This type of current transformers, generally referred to in industry as "CTs", have a standard secondary-current rating of 5 amps AC. They are designed to be operated with a 5-amp AC ammeter connected directly to their secondary winding, which produces a short circuit. CTs are designed to operate with the secondary winding shorted. **The secondary winding of a CT should never be opened when there is power applied to the primary. This will cause the transformer to produce a step-up in voltage which could be high enough to kill anyone who comes in contact with it.**

A current transformer of this type is basically a torroid transformer, a transformer constructed with a hollow core, similar to a donut in that it has a hole in the middle, *(Figure 6-4)*. When current transformers are used, the main power line is inserted through the opening in the transformer, *(Figure 6-5)*. The power line acts as the primary of the transformer and is considered to be one turn.

The turns ratio of the transformer can be changed by looping the power wire through the opening in the transformer to produce a primary winding of more than one turn. For example, assume a current transformer has a ratio of 600:5. If the primary power wire is inserted through the opening, it

Figure 6-4 Current transformer (CT) with a ratio of 1500:5. (Courtesy of Square D Company)

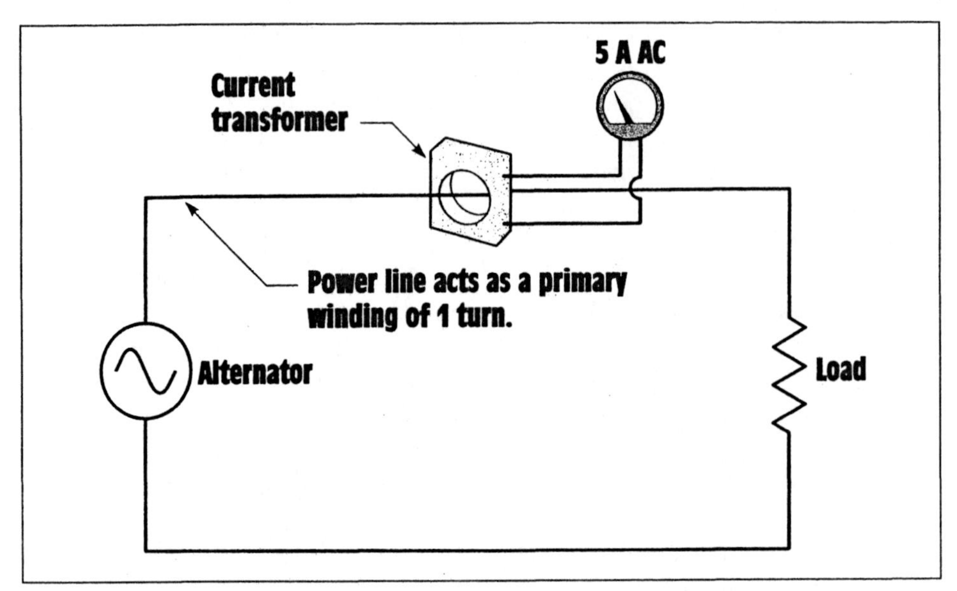

Figure 6-5 Current transformer used to change the scale factor of an AC ammeter.

will require a current of 600 amps to deflect the meter full scale. If the primary power conductor is looped around and inserted through the window a second time, the primary now contains two turns of wire instead of one, *(Figure 6-6)*. It now requires 300 amps of current flow in the primary to

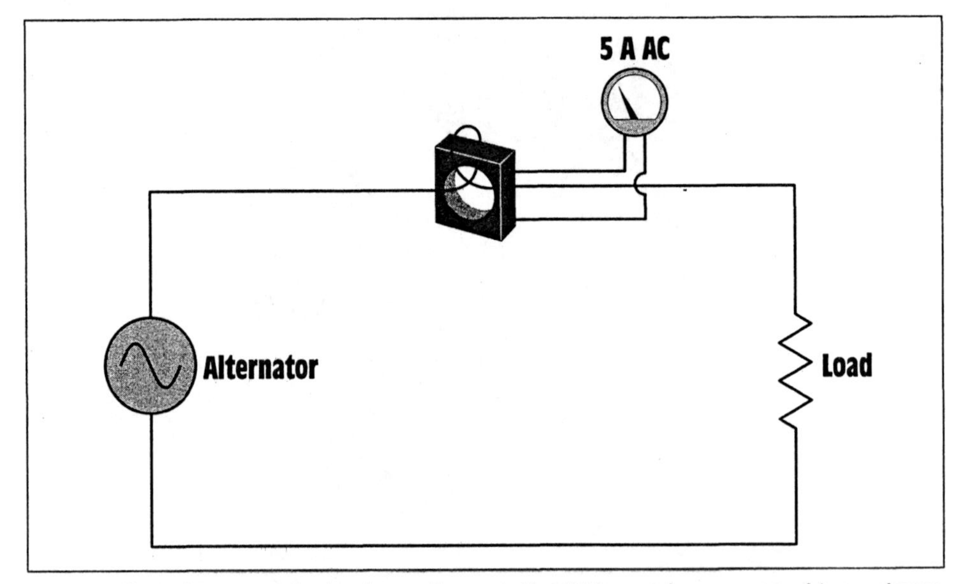

Figure 6-6 The primary conductor loops through the CT to produce a second turn, changing the turns ratio.

deflect the meter full scale. If the primary conductor is looped through the opening a third time, it would require only 200 amps of current flow to deflect the meter full scale.

Clamp-On Ammeters

Many electricians use the clamp-on-type of AC ammeter, *(Figure 6-7 A, B, and C)*. To use this meter, the jaw of the meter is clamped around one of the conductors supplying power to the load, *(Figure 6-8)*. The meter is clamped around only one of the lines. If the meter is clamped around more than one line, the magnetic fields of the wires cancel each other and the meter indicates zero.

This type of meter uses a current transformer to operate. The jaw of the meter is part of the core material of the transformer. When the meter is connected around the current-carrying wire, the changing magnetic field produced by the AC current induces a voltage into the current transformer. The strength of the magnetic field and its frequency determine the amount of voltage induced in the current transformer. Since 60 Hz is a standard frequency throughout the United States and Canada, the amount of induced voltage is proportional to the strength of the magnetic field.

The clamp-on-type ammeter can have different range settings by changing

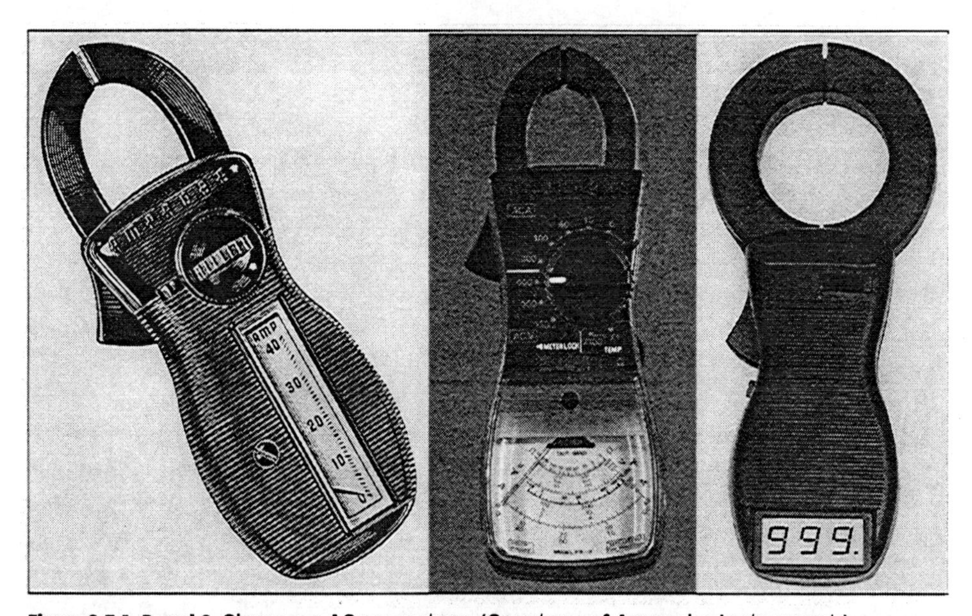

Figure 6-7 A, B, and C Clamp-on AC ammeters. (Courtesy of Amprobe Instrument.)

Figure 6-8 The clamp-on ammeter connects around only one conductor.

the turns ratio of the secondary of the transformer just as the inline ammeter does. The primary of the transformer is the conductor that the movable jaw is connected around. If the ammeter is connected around one wire, the primary has one turn of wire as compared to the turns of the secondary. The turns ratio can be changed in the same manner as changing the ratio of the CT. If two turns of wire are wrapped around the jaw of the ammeter, *Figure 6-9*, the primary winding now contains two turns instead of one, and the turns ratio of the transformer is changed. The ammeter will now indicate double the amount of the current in the circuit. The reading on the scale of the meter would have to be divided by two to get the correct reading. The ability to change the turns ratio of a clamp-on ammeter can be very useful for measuring low currents.

Changing the turns ratio is not limited to wrapping two turns of wire around the jaw of the ammeter. Any number of turns can be wrapped around the jaw and the reading divided by that number. A problem with many clamp-on-type ammeters is that their lowest scale value is too high to accurately measure low-current values. An ammeter with a full-scale range

Figure 6-9 Looping the conductor around the jaw of the ammeter changes the ratio.

of 0–6 amperes would not be able to measure accurately a current flow of 0.2 amperes. The answer to this problem is to wrap multiple turns of wire around the jaw of the ammeter to change the secondary-scale value. If 10 turns of wire are wrapped around the jaw of a 0–6 ampere meter, the meter will indicate a full-scale value of 0.6 ampere. The meter could now accurately measure a current of 0.2 ampere. In the field, however, it is not often that there is enough slack wire to make 10 wraps around the jaw of an ammeter. A simple device can be constructed to overcome this problem. Assume that 10 turns of wire is wound around a piece of nonconductive material, such as a piece of plastic pipe, *(Figure 6-10)*. Plastic tape is then used to prevent the wire from slipping off the plastic core. If alligator clips are attached to the ends of the wire, the device can be inserted in series with the load in a similar manner to using an inline ammeter. The jaw of the ammeter can then be inserted through the opening in the plastic pipe, *(Figure 6-11)*. The current value is read by moving the decimal one place to the left. A full-scale value of 0–6 amperes becomes a full-scale value of 0–0.6 ampere, or a 0–30 ampere scale becomes a 0–3 ampere scale.

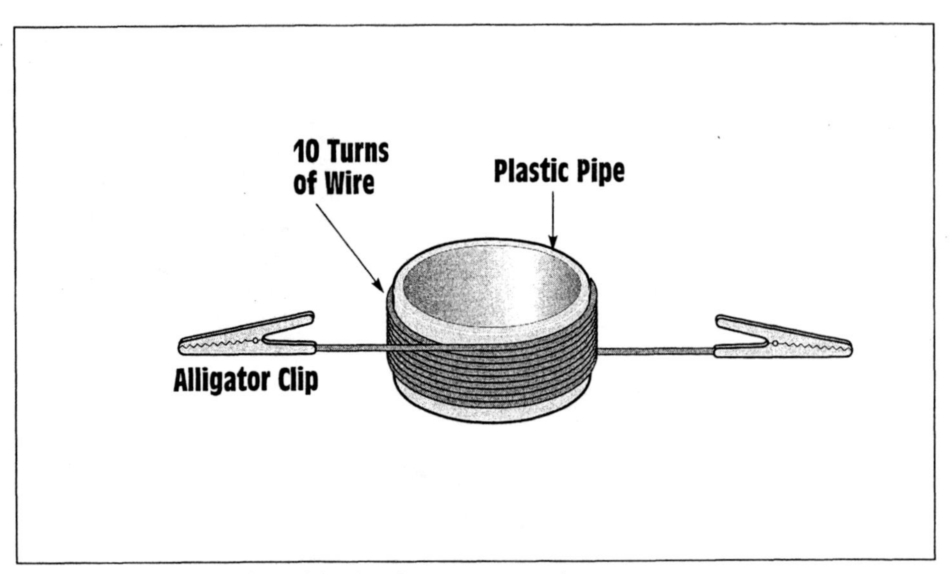

Figure 6-10 A simple scale divider for clamp-on ammeters.

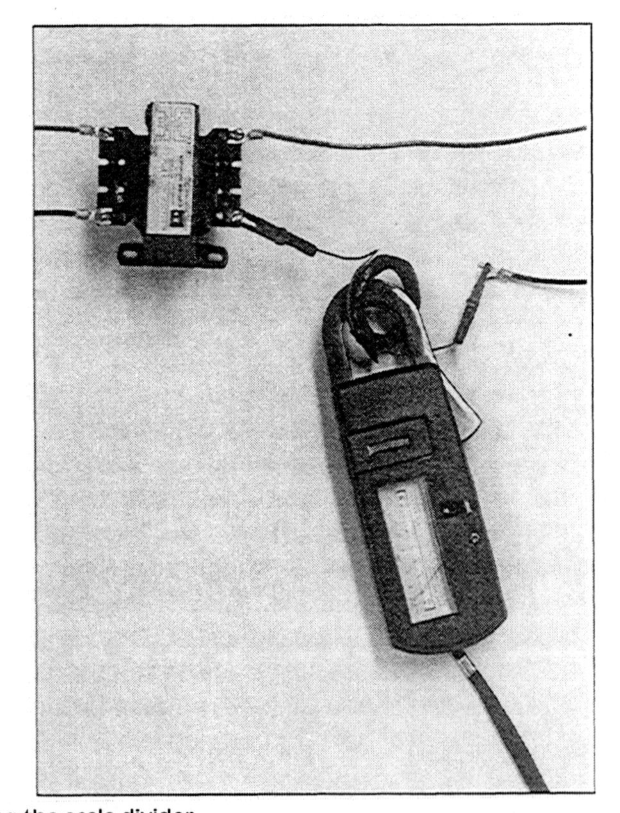

Figure 6-11 Using the scale divider.

Summary

1. Current transformers have their primary winding connected in series with a load.
2. Current transformers are often used to provide multiple-scale values for inline AC ammeters.
3. Current transformers are often referred to as CTs.
4. CTs are used to measure large amounts of AC current.
5. CTs have a standard secondary-current value of 5 amperes.
6. CTs are designed to be operated with their secondary winding shorted.
7. The short circuit connected across the secondary of the CT should never be removed when power is connected to the circuit because the secondary voltage can become very high.
8. Many clamp-on AC ammeters operate on the principle of a current transformer.
9. The movable jaw of the clamp-on ammeter is the core of the transformer.
10. The secondary-current value of a current transformer can be changed by changing the turns of wire of the primary.

Review Questions

1. Explain the difference in connection between the primary winding of a voltage transformer and the primary winding of a current transformer.
2. What is the standard-current rating for the secondary winding of a CT?
3. Why should the secondary winding of a CT never be disconnected from its load when there is current flow in the primary?
4. A current transformer has a ratio of 600:5. If three loops of wire are wound through the transformer core, how much primary current is required to produce 5 amperes of current in the secondary winding?
5. Assume that a primary current of 75 amperes flows through the windings of the transformer in question 4. How much current will flow in the secondary winding?
6. What type of core is generally used in the construction of a CT?
7. A current transformer has 4 turns of wire in its primary winding. How many turns of wire are needed in the secondary winding to produce a current of 2 amperes when a current of 60 amperes flows in the primary winding?
8. At 1500:5 CT develops 3 volts across the primary winding. If the secondary should be disconnected from its load, how much voltage would be developed across the secondary terminals?
9. A CT has a current flow of 80 amperes in its primary winding and a current of 2 amperes in its secondary winding. What is the ratio of the CT?
10. What is the most common use for a CT?

7

Three-Phase Circuits

Objectives

After studying this unit, you should be able to

- Discuss the differences between three-phase and single-phase voltages
- Discuss the characteristics of delta and wye connections
- Compute voltage and current values for delta and wye circuits
- Connect delta and wye circuits and make measurements with measuring instruments
- Compute the amount of capacitance needed to correct the power factor of a three-phase motor

Most of the electrical power generated in the world today is three-phase. Three-phase power was first conceived by Nikola Tesla. In the early days of electrical power generation, Tesla not only led the battle concerning whether the nation should be powered with low-voltage direct current or high-voltage alternating current, but he also proved that three-phase power was the most efficient way that electricity could be produced, transmitted, and consumed.

7-1 Three-Phase Circuits

There are several reasons why three-phase power is superior to single-phase power.

1. The horsepower rating of three-phase motors and the kilovolt-amp rating of three-phase transformers are about 150% greater than for single-phase motors or transformers with a similar frame size.
2. The power delivered by a single-phase system pulsates *(Figure 7-1)*. The power falls to zero three times during each cycle. The power delivered by a three-phase circuit pulsates, but it never falls to zero *(Figure 7-2)*. In a three-phase system, the power delivered to the load is the same at any instant. This produces superior operating characteristics for three-phase motors.
3. In a balanced three-phase system, the conductors need be only about 75% the size of conductors for a single-phase two-wire system of the same kVA (kilovolt-amp) rating. This savings helps offset the cost of supplying the third conductor required by three-phase systems.

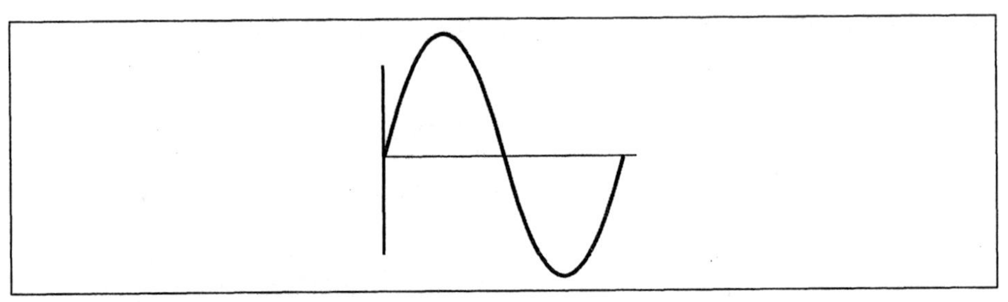

Figure 7-1 Single-phase power falls to zero three times each cycle.

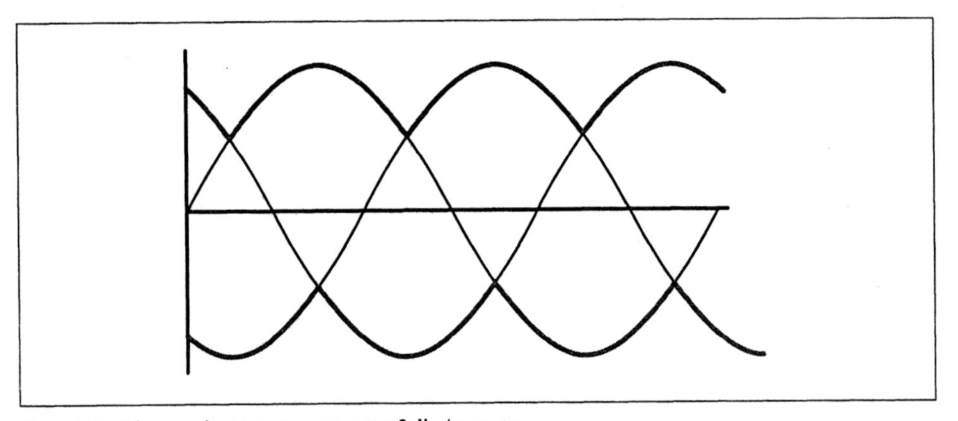

Figure 7-2 Three-phase power never falls to zero.

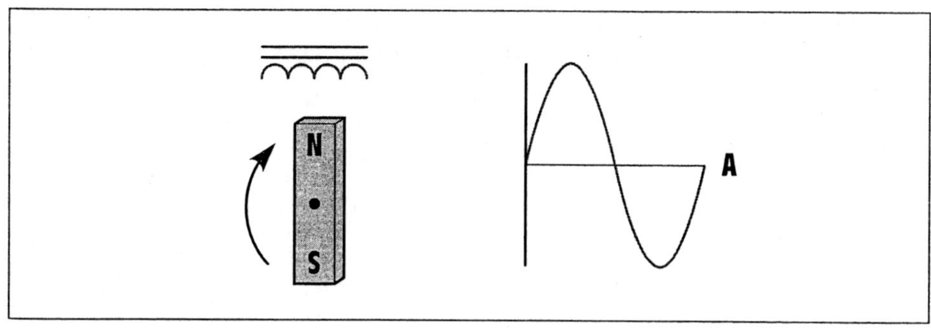

Figure 7-3 Producing a single-phase voltage.

A single-phase alternating voltage can be produced by rotating a magnetic field through the conductors of a stationary coil as shown in *Figure 7-3*.

Since alternate polarities of the magnetic field cut through the conductors of the stationary coil, the induced voltage will change polarity at the same speed as the rotation of the magnetic field. The alternator shown in *Figure 7-3* is single-phase because it produces only one AC voltage.

If three separate coils are spaced 120° apart as shown in *Figure 7-4*, three voltages 120° out of phase with each other will be produced when the mag-

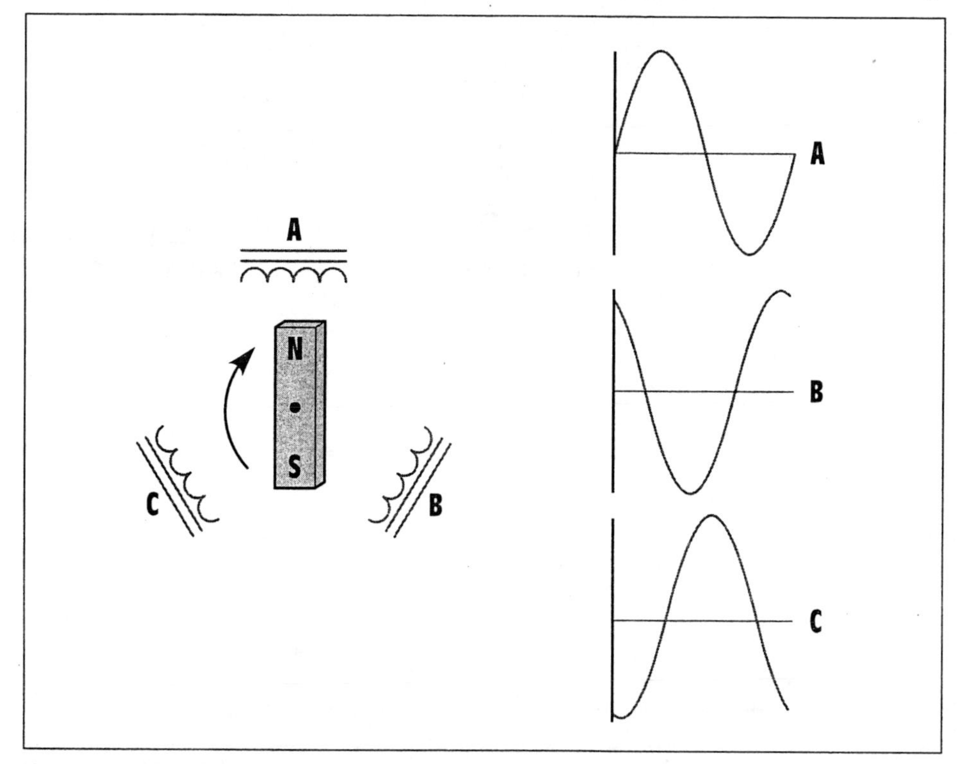

Figure 7-4 The voltages of a three-phase system are 120° out of phase with each other.

netic field cuts through the coils. This is the manner in which a three-phase voltage is produced. There are two basic three-phase connections: the wye, or star, and the delta.

7-2 Wye Connections

The **wye**, or **star**, **connection** is made by connecting one end of each of the three-phase windings together as shown in *Figure 7-5*. The voltage measured across a single winding, or phase, is known as the **phase voltage** as shown in *Figure 7-6*. The voltage measured between the lines is known as the line-to-line voltage, or simply as the **line voltage**.

In *Figure 7-7*, ammeters have been placed in the phase winding of a wye-connected load and in the line that supplies power to the load. Voltmeters have been connected across the input to the load and across the phase. A line voltage of 208 V has been applied to the load. Notice that the voltmeter connected across the lines indicates a value of 208 V, but the voltmeter connected across the phase indicates a value of 120 V.

In a wye-connected system, the line voltage is higher than the phase voltage by a factor of the square root of 3 (1.732). Two formulas

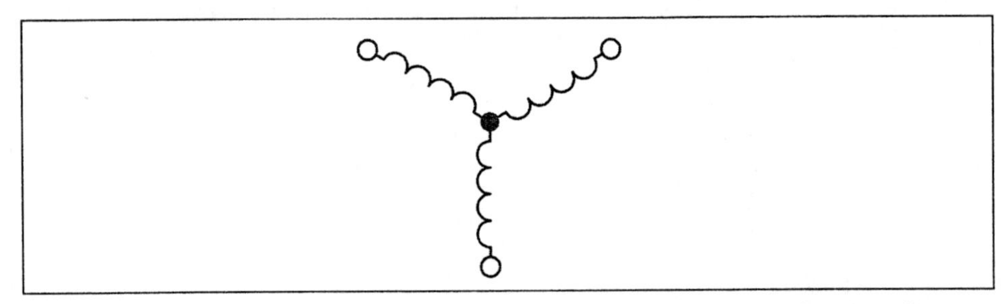

Figure 7-5 A wye connection is formed by joining one end of each winding together.

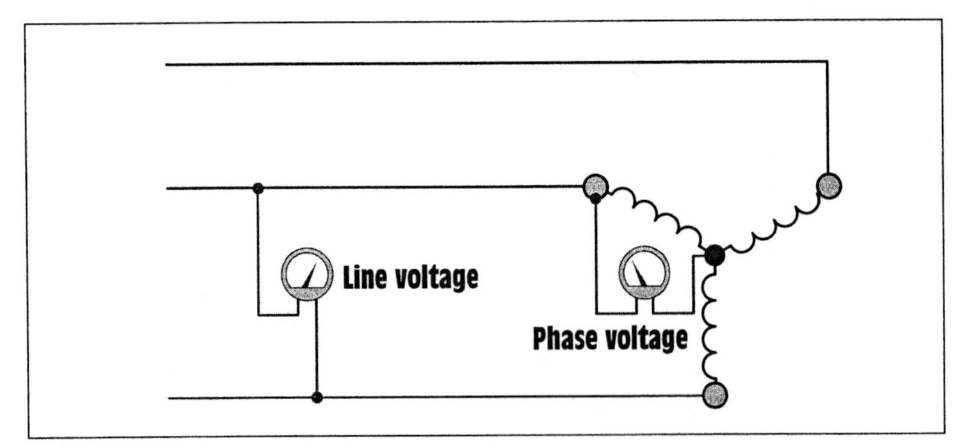

Figure 7-6 Line and phase voltages are different in a wye connection.

used to compute the voltage in a wye-connected system are

$$E_{Line} = E_{Phase} \times 1.732$$

and

$$E_{Phase} = \frac{E_{Line}}{1.732}$$

Notice in *Figure 7-7* that 10 A of current flow in both the phase and the line. **In a wye-connected system, phase current and line current are the same.**

$$I_{Line} = I_{Phase}$$

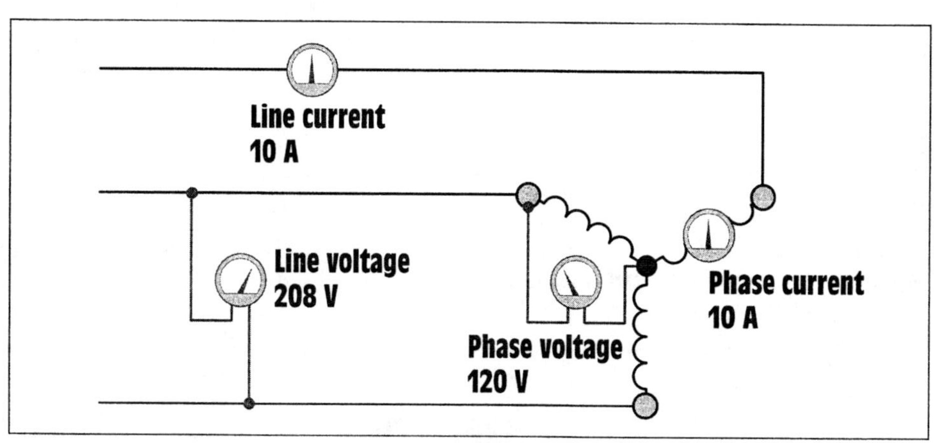

Figure 7-7 Line current and phase current are the same in a wye connection.

Voltage Relationships in a Wye Connection

Many students of electricity have difficulty at first understanding why the line voltage of the wye connection used in this illustration is 208 V instead of 240 V. Since line voltage is measured across two phases that have a voltage of 120 V each, it would appear that the sum of the two voltages should be 240 V. One cause of this misconception is that many students are familiar with the 240/120-V connection supplied to most homes. If voltage is measured across the two incoming lines, a voltage of 240 V will result. If voltage is measured from either of the two lines to the neutral, a voltage of 120 V will be seen. The reason for this is that this is a single-phase connection derived from the center tap of a transformer *(Figure 7-8)*. If the center tap is used as a common point, the two line voltages on either side of it will be 180° apart and opposite in polarity *(Figure 7-9)*. The vector sum of these two voltages would be 240 V.

Three-phase voltages are 120° apart, not 180°. If the three voltages are drawn 120° apart, it will be seen that the vector sum of these voltages is 208 V

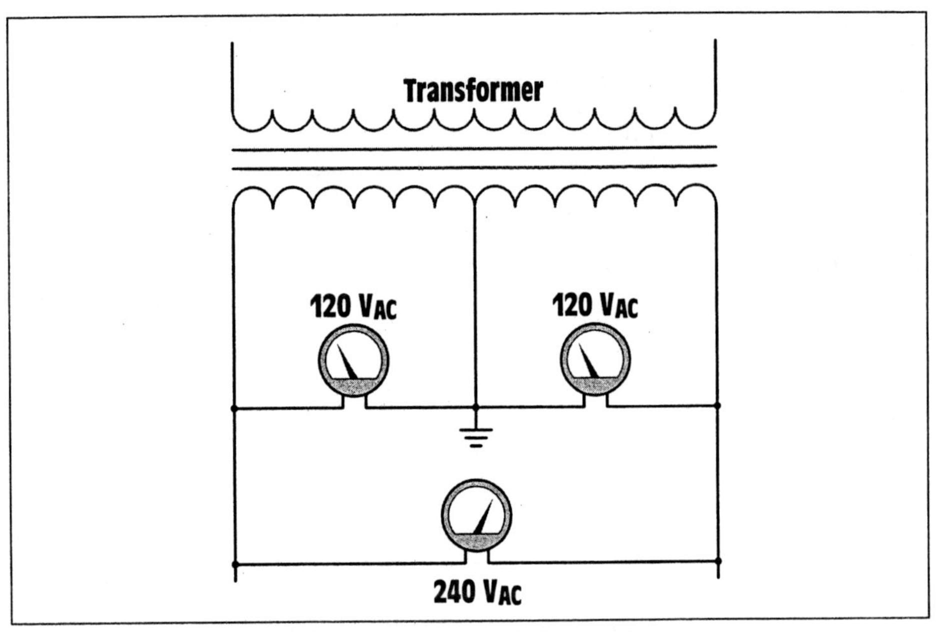

Figure 7-8 Single-phase transformer with grounded center tap.

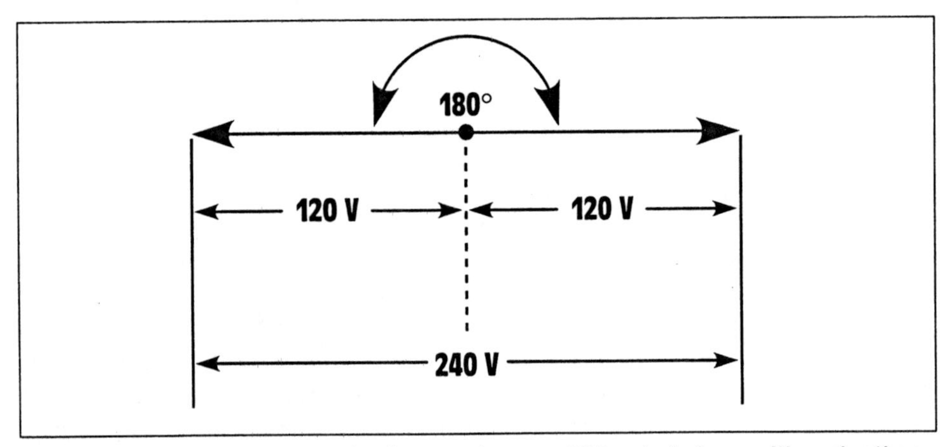

Figure 7-9 The voltages of a single-phase system are 180° out of phase with each other.

(Figure 7-10). Another illustration of vector addition is shown in *Figure 7-11.* In this illustration two phase-voltage vectors are added, and the resultant is drawn from the starting point of one vector to the end point of the other. The parallelogram method of vector addition for the voltages in a wye-connected three-phase system is shown in *Figure 7-12.*

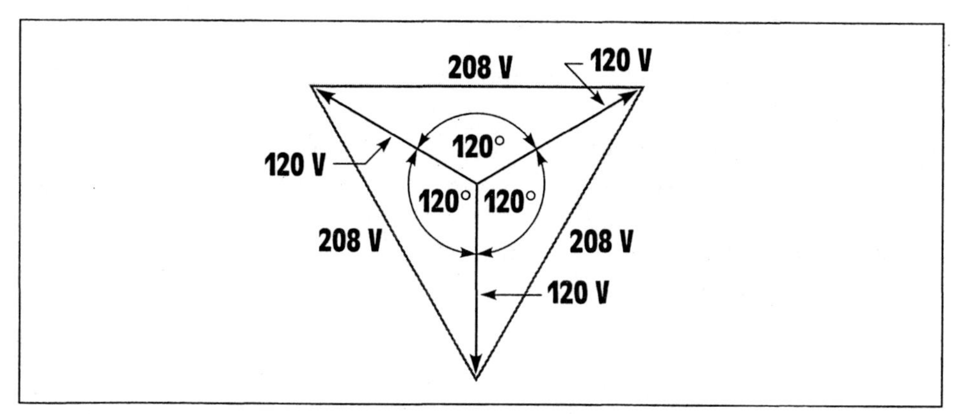

Figure 7-10 Vector sum of the voltages in a three-phase wye connection.

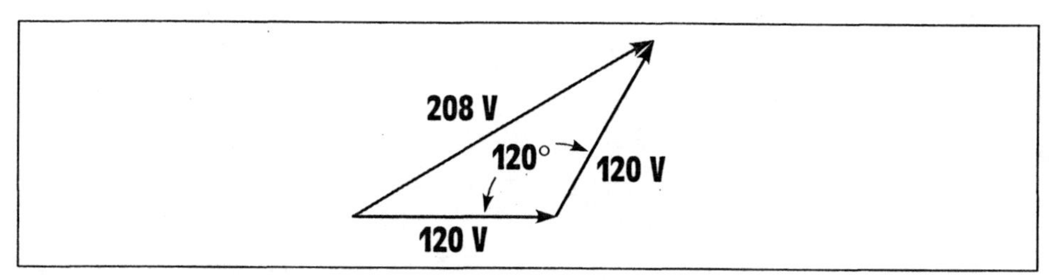

Figure 7-11 Adding voltage vectors of two-phase voltage values.

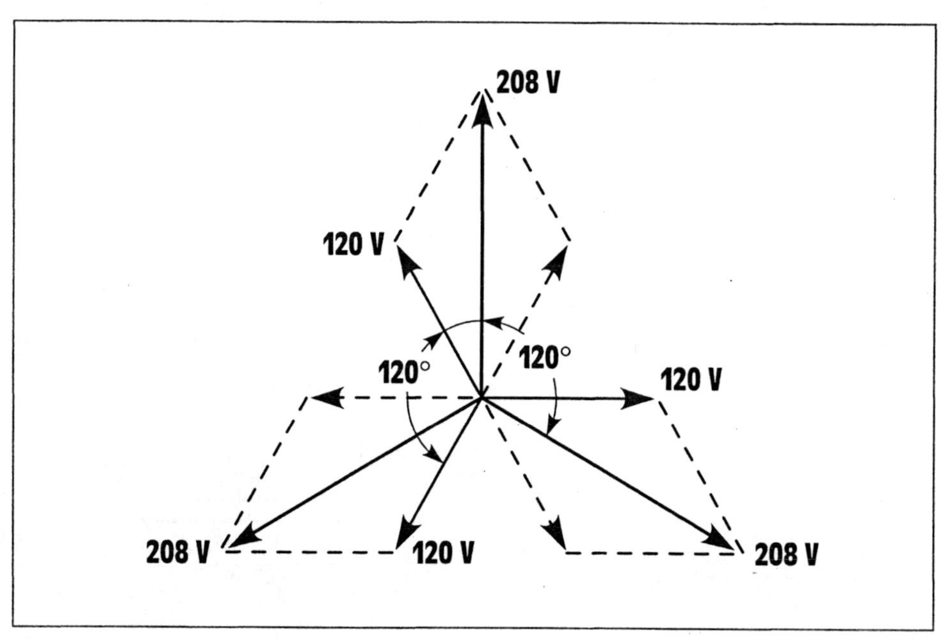

Figure 7-12 The parallelogram method of adding three-phase vectors.

7-3 Delta Connections

In *Figure 7-13*, three separate inductive loads have been connected to form a **delta connection**. This connection receives its name from the fact that a schematic diagram of this connection resembles the Greek letter delta (Δ). In *Figure 7-14*, voltmeters have been connected across the lines and across the phase. Ammeters have been connected in the line and in the phase. **In a delta connection, line voltage and phase voltage are the same**. Notice that both voltmeters indicate a value of 480 V.

$$E_{Line} = E_{Phase}$$

The line current and phase current, however, are different. **The line current of a delta connection is higher than the phase current by a factor of the square root of 3 (1.732)**. In the example shown, it is assumed that each of the phase windings has a current flow of 10 A. The current in each of the lines, however, is 17.32 A. The reason for this differ-

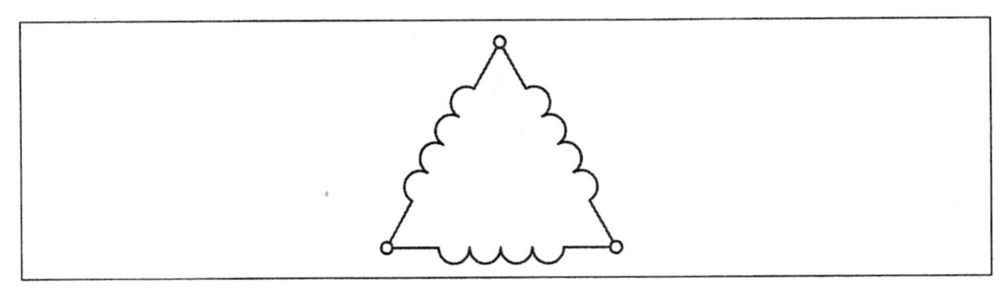

Figure 7-13 Three-phase delta connection.

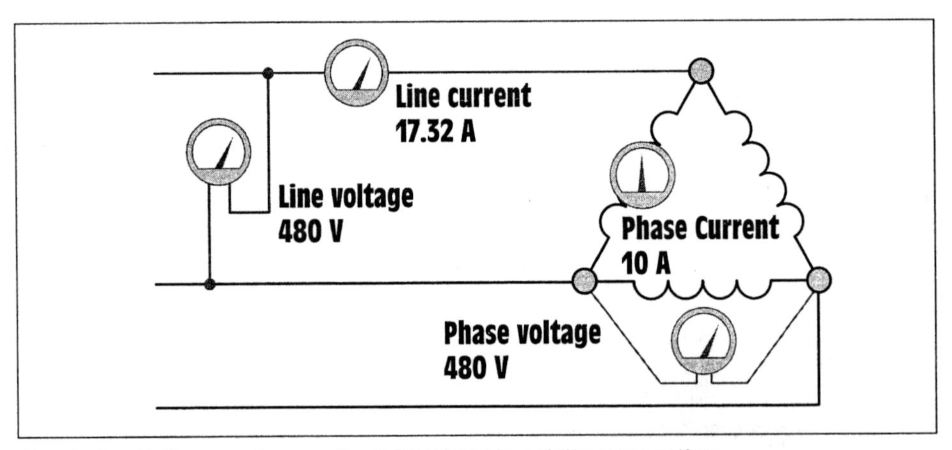

Figure 7-14 Voltage and current relationships in a delta connection.

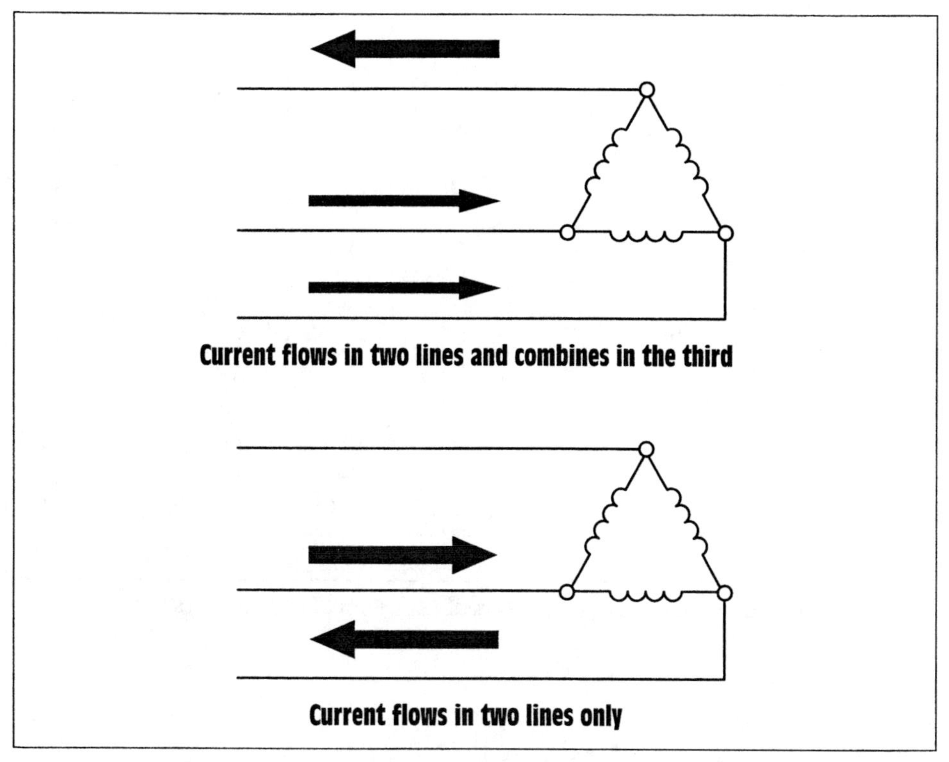

Figure 7-15 Division of currents in a delta connection.

ence in current is that current flows through different windings at different times in a three-phase circuit. During some periods of times, current will flow between two lines only. At other times, current will flow from two lines to the third *(Figure 7-15)*. The delta connection is similar to a parallel connection because there is always more than one path for current flow. Since these currents are 120° out of phase with each other, vector addition must be used when finding the sum of the currents *(Figure 7-16)*. Formulas for determining the current in a delta connection are

$$I_{Line} = I_{Phase} \times 1.732$$

and

$$I_{Phase} = \frac{I_{Line}}{1.732}$$

7-4 Three-Phase Power

Students sometimes become confused when computing values of power in three-phase circuits. One reason for this confusion is there are actually two formulas that can be used. If *line* values of voltage and current are known, the apparent power of the circuit can be computed using the formula

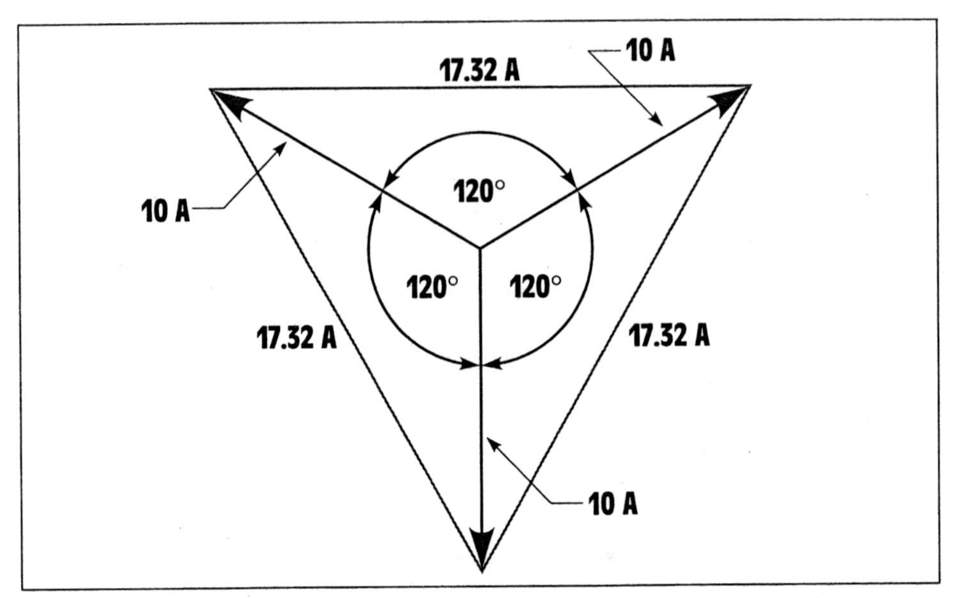

Figure 7-16 Vector addition is used to compute the sum of the currents in a delta connection.

$$VA = \sqrt{3} \times E_{Line} \times I_{Line}$$

If the *phase* values of voltage and current are known, the apparent power can be computed using the formula

$$VA = 3 \times E_{Phase} \times I_{Phase}$$

Notice that in the first formula, the line values of voltage and current are multiplied by the square root of 3. In the second formula, the phase values of voltage and current are multiplied by 3. The first formula is used more because it is generally more convenient to obtain line values of voltage and current since they can be measured with a voltmeter and clamp-on ammeter.

7-5 Watts and VARs

Watts and VARs can be computed in a similar manner. **Three-phase watts** can be computed by multiplying the apparent power by the power factor

$$P = \sqrt{3} \times E_{Line} \times I_{Line} \times PF$$

or

$$P = 3 \times E_{Phase} \times I_{Phase} \times PF$$

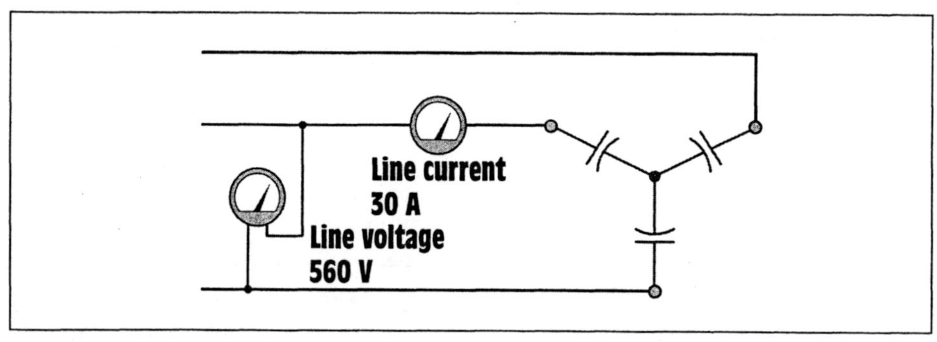

Figure 7-17 Pure capacitive three-phase load.

(**Note:** When computing the power of a pure resistive load, the voltage and current are in phase with each other and the power factor is 1.)

Three-phase VARs can be computed in a similar manner, except that voltage and current values of a pure reactive load are used. For example, a pure capacitive load is shown in *Figure 7-17.* In this example, it is assumed that the line voltage is 480 V and the line current is 30 A. Capacitive VARs can be computed using the formula

$$VARs_C = \sqrt{3} \times E_{Line\ (Capacitive)} \times I_{Line\ (Capacitive)}$$

$$VARs_C = 1.732 \times 560 \times 30$$

$$VARs_C = 29{,}097.6$$

7-6 Three-Phase Circuit Calculations

In the following examples, values of line and phase voltage, line and phase current, and power will be computed for different types of three-phase connections.

Example 1

A wye-connected three-phase alternator supplies power to a delta-connected resistive load *(Figure 7-18).* The alternator has a line voltage of 480 V. Each resistor of the delta load has 8 Ω of resistance. Find the following values.

$E_{L(Load)}$ — line voltage of the load

$E_{P(Load)}$ — phase voltage of the load

$I_{P(Load)}$ — phase current of the load

$I_{L(Load)}$ — line current to the load

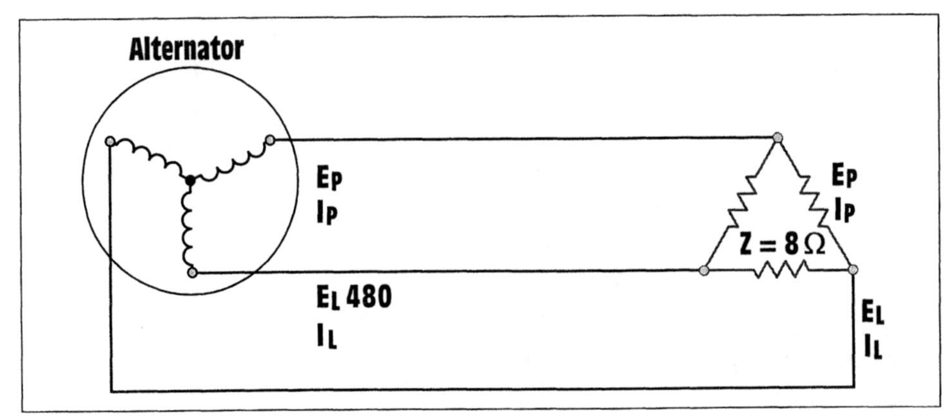

Figure 7-18 Computing three-phase values. Example circuit 1.

$I_{L(Alt)}$ — line current delivered by the alternator

$I_{P(Alt)}$ — phase current of the alternator

$E_{P(Alt)}$ — phase voltage of the alternator

P — true power

Solution

The load is connected directly to the alternator. Therefore, the line voltage supplied by the alternator is the line voltage of the load.

$$E_{L(Load)} = 480\text{ V}$$

The three resistors of the load are connected in a delta connection. In a delta connection, the phase voltage is the same as the line voltage.

$$E_{P(Load)} = E_{L(Load)}$$

$$E_{P(Load)} = 480\text{ V}$$

Each of the three resistors in the load is one phase of the load. Now that the phase voltage is known (480 V), the amount of phase current can be computed using Ohm's law.

$$I_{P(Load)} = \frac{E_{P(Load)}}{Z}$$

$$I_{P(Load)} = \frac{480}{8}$$

$$I_{P(Load)} = 60\text{ A}$$

The three load resistors are connected as a delta with 60 A of current flow in each phase. The line current supplying a delta connection must be 1.732 times greater than the phase current.

$$I_{L(Load)} = I_{P(Load)} \times 1.732$$

$$I_{L(Load)} = 60 \times 1.732$$

$$I_{L(Load)} = 103.92 \text{ A}$$

The alternator must supply the line current to the load or loads to which it is connected. In this example, only one load is connected to the alternator. Therefore, the line current of the load will be the same as the line current of the alternator.

$$I_{L(Alt)} = 103.92 \text{ A}$$

The phase windings of the alternator are connected in a wye connection. In a wye connection, the phase current and line current are equal. The phase current of the alternator will, therefore, be the same as the alternator line current.

$$I_{P(Alt)} = 103.92 \text{ A}$$

The phase voltage of a wye connection is less than the line voltage by a factor of the square root of 3. The phase voltage of the alternator will be

$$E_{P(Alt)} = \frac{E_{L(Alt)}}{1.732}$$

$$E_{P(Alt)} = \frac{480}{1.732}$$

$$E_{P(Alt)} = 277.13 \text{ V}$$

In this circuit, the load is pure resistive. The voltage and current are in phase with each other, which produces a unity power factor of 1. The true power in this circuit will be computed using the formula

$$P = 1.732 \times E_{L(Alt)} \times I_{L(Alt)} \times PF$$

$$P = 1.732 \times 480 \times 103.92 \times 1$$

$$P = 86{,}394.93 \text{ W}$$

Example 2

A delta-connected alternator is connected to a wye-connected resistive load *(Figure 7-19)*. The alternator produces a line voltage of 240 V and the resistors have a value of 6 Ω each. Find the following values.

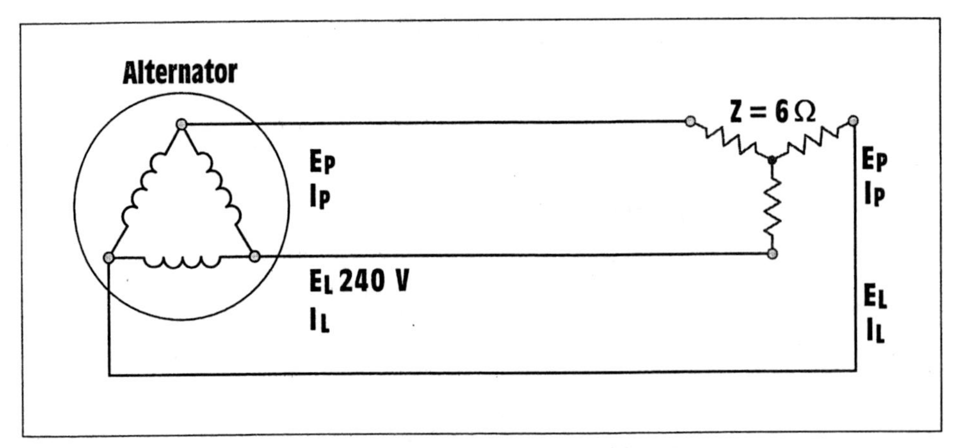

Figure 7-19 Computing three-phase values. Example problem 2.

$E_{L(Load)}$ — line voltage of the load

$E_{P(Load)}$ — phase voltage of the load

$I_{P(Load)}$ — phase current of the load

$I_{L(Load)}$ — line current to the load

$I_{L(Alt)}$ — line current delivered by the alternator

$I_{P(Alt)}$ — phase current of the alternator

$E_{P(Alt)}$ — phase voltage of the alternator

P — true power

Solution

As was the case in the previous example, the load is connected directly to the output of the alternator. The line voltage of the load must, therefore, be the same as the line voltage of the alternator.

$$E_{L(Load)} = 240 \text{ V}$$

The phase voltage of a wye connection is less than the line voltage by a factor of 1.732.

$$E_{P(Load)} = \frac{240}{1.732}$$

$$E_{P(Load)} = 138.57 \text{ V}$$

Each of the three 6-Ω resistors is one phase of the wye-connected load. Since the phase voltage is 138.57 V, this voltage is applied to each of the three resistors. The amount of phase current can now be determined using Ohm's law.

$$I_{P(Load)} = \frac{E_{P(Load)}}{Z}$$

$$I_{P(Load)} = \frac{138.57}{6}$$

$$I_{P(Load)} = 23.1\ A$$

The amount of line current needed to supply a wye-connected load is the same as the phase current of the load.

$$I_{L(Load)} = 23.1\ A$$

Only one load is connected to the alternator. The line current supplied to the load is the same as the line current of the alternator.

$$I_{L(Alt)} = 23.1\ A$$

The phase windings of the alternator are connected in delta. In a delta connection the phase current is less than the line current by a factor of 1.732.

$$I_{P(Alt)} = \frac{I_{L(Alt)}}{1.732}$$

$$I_{P(Alt)} = \frac{23.1}{1.732}$$

$$I_{P(Alt)} = 13.34\ A$$

The phase voltage of a delta is the same as the line voltage.

$$E_{P(Alt)} = 240\ V$$

Since the load in this example is pure resistive, the power factor has a value of unity, or 1. Power will be computed by using the line values of voltage and current.

$$P = 1.732 \times E_L \times I_L \times PF$$

$$P = 1.732 \times 240 \times 23.1 \times 1$$

$$P = 9{,}602.21\ W$$

Example 3

The phase windings of an alternator are connected in wye. The alternator produces a line voltage of 440 V and supplies power to two resistive loads. One load contains resistors with a value of 4 Ω each, connected in wye. The

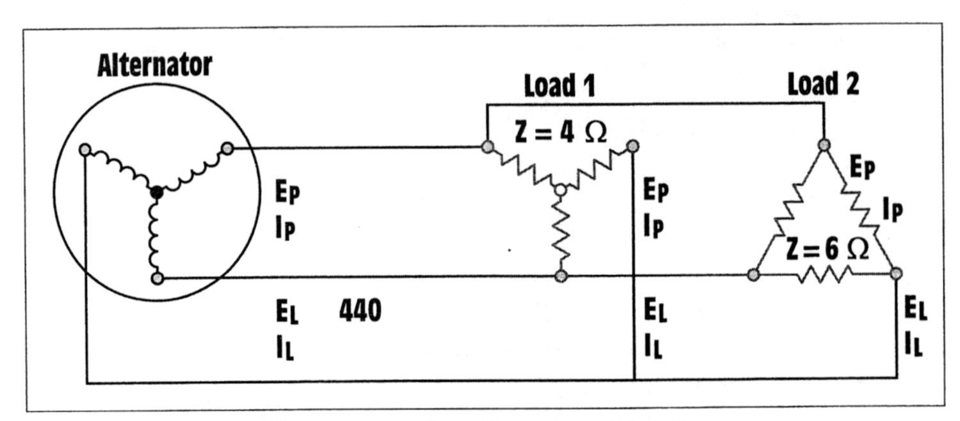

Figure 7-20 Computing three-phase values. Example problem 3.

second load contains resistors with a value of 6 Ω each, connected in delta *(Figure 7-20)*. Find the following circuit values.

$E_{L(Load\ 2)}$ — line voltage of load 2

$E_{P(Load\ 2)}$ — phase voltage of load 2

$I_{P(Load\ 2)}$ — phase current of load 2

$I_{L(Load\ 2)}$ — line current to load 2

$E_{P(Load\ 1)}$ — phase voltage of load 1

$I_{P(Load\ 1)}$ — phase current of load 1

$I_{L(Load\ 1)}$ — line current to load 1

$I_{L(Alt)}$ — line current delivered by the alternator

$I_{P(Alt)}$ — phase current of the alternator

$E_{P(Alt)}$ — phase voltage of the alternator

P — true power

Solution

Both loads are connected directly to the output of the alternator. The line voltage for both loads 1 and 2 will be the same as the line voltage of the alternator.

$$E_{L(Load\ 2)} = 440\ V$$

$$E_{L(Load\ 1)} = 440\ V$$

Load 2 is connected as a delta. The phase voltage will be the same as the line voltage.

$$E_{P(Load\ 2)} = 440\ V$$

Each of the resistors that constitutes a phase of load 2 has a value of 6 Ω. The amount of phase current can be found using Ohm's law.

$$I_{P(Load\ 2)} = \frac{E_{P(Load\ 2)}}{Z}$$

$$I_{P(Load\ 2)} = \frac{440}{6}$$

$$I_{P(Load\ 2)} = 73.33\ A$$

The line current supplying a delta-connected load is 1.732 times greater than the phase current. The amount of line current needed for load 2 can be computed by increasing the phase current value by 1.732.

$$I_{L(Load\ 2)} = I_{P(Load\ 2)} \times 1.732$$

$$I_{L(Load\ 2)} = 73.33 \times 1.732$$

$$I_{L(Load\ 2)} = 127.01\ A$$

The resistors of load 1 are connected to form a wye. The phase voltage of a wye connection is less than the line voltage by a factor of 1.732.

$$E_{P(Load\ 1)} = \frac{E_{L(Load\ 1)}}{1.732}$$

$$E_{P(Load\ 1)} = \frac{440}{1.732}$$

$$E_{P(Load\ 1)} = 254.04\ V$$

Now that the voltage applied to each of the 4-Ω resistors is known, the phase current can be computed using Ohm's law.

$$I_{P(Load\ 1)} = \frac{E_{P(Load\ 1)}}{Z}$$

$$I_{P(Load\ 1)} = \frac{254.04}{4}$$

$$I_{P(Load\ 1)} = 63.51\ A$$

The line current supplying a wye-connected load is the same as the phase current. Therefore, the amount of line current needed to supply load 1 is

$$I_{L(Load\ 1)} = 63.51\ A$$

The alternator must supply the line current needed to operate both loads. In this example, both loads are resistive. The total line current supplied by the alternator will be the sum of the line currents of the two loads.

$$I_{L(Alt)} = I_{L(Load\ 1)} + I_{L(Load\ 2)}$$

$$I_{L(Alt)} = 63.51 + 127.01$$

$$I_{L(Alt)} = 190.52\ A$$

Since the phase windings of the alternator in this example are connected in a wye, the phase current will be the same as the line current.

$$I_{P(Alt)} = 190.52\ A$$

The phase voltage of the alternator will be less than the line voltage by a factor of 1.732.

$$E_{P(Alt)} = \frac{440}{1.732}$$

$$E_{P(Alt)} = 254.04\ V$$

Both of the loads in this example are resistive and have a unity power factor of 1. The total power in this circuit can be found by using the line voltage and total line current supplied by the alternator.

$$P = 1.732 \times E_L \times I_L \times PF$$

$$P = 1.732 \times 440 \times 190.52 \times 1$$

$$P = 145{,}191.48\ W$$

Example 4

A wye-connected three-phase alternator with a line voltage of 560 V supplies power to three different loads *(Figure 7-21)*. The first load is formed by three resistors with a value of 6 Ω each, connected in a wye. The second load comprises three inductors with an inductive reactance of 10 Ω each, connected in delta, and the third load comprises three capacitors with a capacitive reactance of 8 Ω each, connected in wye. Find the following circuit values.

$E_{L(Load\ 3)}$ — line voltage of load 3 (capacitive)

$E_{P(Load\ 3)}$ — phase voltage of load 3 (capacitive)

$I_{P(Load\ 3)}$ — phase current of load 3 (capacitive)

$I_{L(Load\ 3)}$ — line current to load 3 (capacitive)

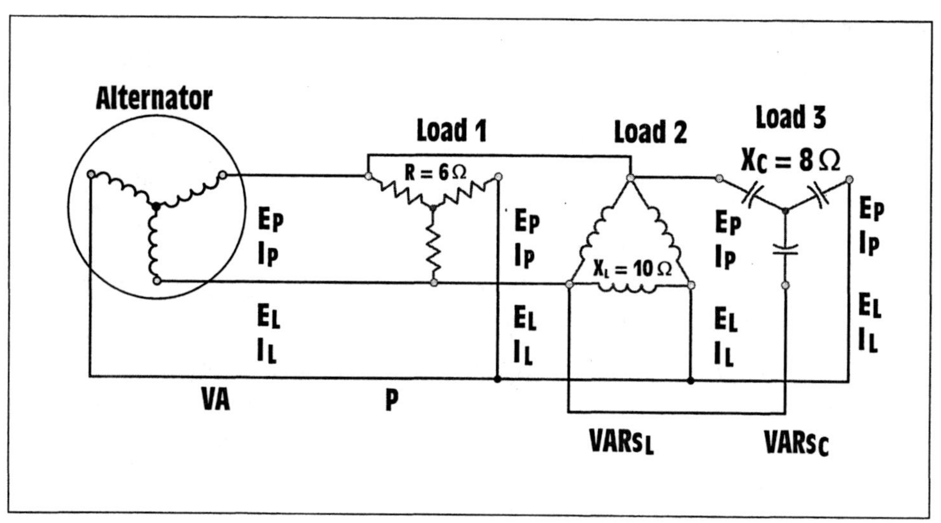

Figure 7-21 Computing three-phase values. Example problem 4.

$E_{L(Load\ 2)}$ — line voltage of load 2 (inductive)

$E_{P(Load\ 2)}$ — phase voltage of load 2 (inductive)

$I_{P(Load\ 2)}$ — phase current of load 2 (inductive)

$I_{L(Load\ 2)}$ — line current to load 2 (inductive)

$E_{L(Load\ 1)}$ — line voltage of load 1 (resistive)

$E_{P(Load\ 1)}$ — phase voltage of load 1 (resistive)

$I_{P(Load\ 1)}$ — phase current of load 1 (resistive)

$I_{L(Load\ 1)}$ — line current to load 1 (resistive)

$I_{L(Alt)}$ — line current delivered by the alternator

$E_{P(Alt)}$ — phase voltage of the alternator

P — true power

$VARs_L$ — reactive power of the inductive load

$VARs_C$ — reactive power of the capacitive load

VA — apparent power

PF — power factor

Solution

All three loads are connected to the output of the alternator. The line voltage connected to each load is the same as the line voltage of the alternator.

$$E_{L(Load\ 3)} = 560\ V$$

$$E_{L(Load\ 2)} = 560\ V$$

$$E_{L(Load\ 1)} = 560\ V$$

7-7 Load 3 Calculations

Load 3 is formed from three capacitors with a capacitive reactance of 8 Ω each, connected in a wye. Since this load is wye-connected, the phase voltage will be less than the line voltage by a factor of 1.732.

$$E_{P(Load\ 3)} = \frac{E_{L(Load\ 3)}}{1.732}$$

$$E_{P(Load\ 3)} = \frac{560}{1.732}$$

$$E_{P(Load\ 3)} = 323.33\ V$$

Now that the voltage applied to each capacitor is known, the phase current can be computed using Ohm's law.

$$I_{P(Load\ 3)} = \frac{E_{P(Load\ 3)}}{X_C}$$

$$I_{P(Load\ 3)} = \frac{323.33}{8}$$

$$I_{P(Load\ 3)} = 40.42\ A$$

The line current required to supply a wye-connected load is the same as the phase current.

$$I_{L(Load\ 3)} = 40.42\ A$$

The reactive power of load 3 can be found using a formula similar to the formula for computing apparent power. Since load 3 is pure capacitive, the current and voltage are 90° out of phase with each other, and the power factor is zero.

$$VARs_C = 1.732 \times E_{L(Load\ 3)} \times I_{L(Load\ 3)}$$

$$VARs_C = 1.732 \times 560 \times 40.42$$

$$VARs_C = 39{,}204.17$$

7-8 Load 2 Calculations

Load 2 comprises three inductors connected in a delta with an inductive reactance of 10 Ω each. Since the load is connected in delta, the phase voltage will be same as the line voltage.

$$E_{L(Load\ 2)} = 560\ V$$

The phase current can be computed by using Ohm's law.

$$I_{P(Load\ 2)} = \frac{E_{P(Load\ 2)}}{X_L}$$

$$I_{P(Load\ 2)} = \frac{560}{10}$$

$$I_{P(Load\ 2)} = 56\ A$$

The amount of line current needed to supply a delta-connected load is 1.732 times greater than the phase current of the load.

$$I_{L(Load\ 2)} = I_{P(Load\ 2)} \times 1.732$$

$$I_{L(Load\ 2)} = 56 \times 1.732$$

$$I_{L(Load\ 2)} = 96.99\ A$$

Since load 2 is made up of inductors, the reactive power can be computed using the line values of voltage and current supplied to the load.

$$VARs_L = 1.732 \times E_{L(Load\ 2)} \times I_{L(Load\ 2)}$$

$$VARs_L = 1.732 \times 560 \times 96.99$$

$$VARs_L = 94{,}072.54$$

7-9 Load 1 Calculations

Load 1 consists of three resistors with a resistance of 6 Ω each, connected in wye. In a wye connection the phase voltage is less than the line voltage by a factor of 1.732. The phase voltage for load 1 will be the same as the phase voltage for load 3.

$$E_{P(Load\ 1)} = 323.33\ V$$

The amount of phase current can now be computed using the phase voltage and the resistance of each phase.

$$I_{P(Load\ 1)} = \frac{E_{P(Load\ 1)}}{R}$$

$$I_{P(Load\ 1)} = \frac{323.33}{6}$$

$$I_{P(Load\ 1)} = 53.89\ A$$

Since the resistors of load 1 are connected in a wye, the line current will be the same as the phase current.

$$I_{L(Load\ 1)} = 53.89\ A$$

Since load 1 is pure resistive, true power can be computed using the line- and phase-current values.

$$P = 1.732 \times E_{L(Load\ 1)} \times I_{L(Load\ 1)}$$

$$P = 1.732 \times 560 \times 53.89$$

$$P = 52{,}267\ W$$

7-10 Alternator Calculations

The alternator must supply the line current for each of the loads. In this problem, however, the line currents are out of phase with each other. To find the total line current delivered by the alternator, vector addition must be used. The current flow in load 1 is resistive and in phase with the line voltage. The current flow in load 2 is inductive and lags the line voltage by 90°. The current flow in load 3 is capacitive and leads the line voltage by 90°. A formula similar to the formula used to find total current flow in an RLC parallel circuit can be employed to find the total current delivered by the alternator.

$$I_{L(Alt)} = \sqrt{I_{L(Load\ 1)}^2 + (I_{L(Load\ 2)} - I_{L(Load\ 3)})^2}$$

$$I_{L(Alt)} = \sqrt{53.89^2 + (96.99 - 40.42)^2}$$

$$I_{L(Alt)} = 78.13\ A$$

The apparent power can now be found using the line voltage and current values of the alternator.

$$VA = 1.732 \times E_{L(Alt)} \times I_{L(Alt)}$$

$$VA = 1.732 \times 560 \times 78.13$$

$$VA = 75{,}779.85$$

The circuit power factor is the ratio of apparent power and true power.

$$PF = \frac{W}{VA}$$

$$PF = \frac{52{,}267}{75{,}779.85}$$

$$PF = 69\%$$

7-11 Power Factor Correction

Correcting the power factor of a three-phase circuit is similar to the procedure used to correct the power factor of a single-phase circuit.

Example 5

A three-phase motor is connected to a 480-V, 60-Hz line. A clamp-on ammeter indicates a running current of 68 A at full load, and a three-phase wattmeter indicates a true power of 40,277 W. Compute the motor power factor first. Then find the amount of capacitance needed to correct the power factor to 95%. Assume that the capacitors used for power-factor correction are to be connected in wye, and the capacitor bank is then to be connected in parallel with the motor.

Solution

First find the amount of apparent power in the circuit.

$$VA = 1.732 \times E_L \times I_L$$

$$VA = 1.732 \times 480 \times 68$$

$$VA = 56{,}532.48$$

The motor power factor can be computed by dividing the true power by the apparent power.

$$PF = \frac{P}{VA}$$

$$PF = \frac{40{,}277}{56{,}532.48}$$

$$PF = 71.2\%$$

The inductive VARs in the circuit can be computed using the formula

$$VARs_L = \sqrt{VA^2 - P^2}$$

$$VARs_L = \sqrt{56{,}532.48^2 - 40{,}277^2}$$

$$VARs_L = 39{,}669.69$$

If the power factor is to be corrected to 95%, the apparent power at 95% power factor must be found. This can be done using the formula

$$VA = \frac{P}{PF}$$

$$VA = \frac{40{,}277}{0.95}$$

$$VA = 42{,}396.84$$

The amount of inductive VARs needed to produce an apparent power of 42,396.84 VA can be found using the formula

$$VARs_L = \sqrt{VA^2 - P^2}$$

$$VARs_L = \sqrt{42{,}396.84^2 - 40{,}277^2}$$

$$VARs_L = 13{,}238.4$$

To correct the power factor to 95%, the inductive VARs must be reduced from 39,669.69 to 13,238.4. This can be done by connecting a bank of capacitors in the circuit that will produce a total of 26,431.29 capacitive VARs (39,669.69 – 13,238.4 = 26,431.29). This amount of capacitive VARs will reduce the inductive VARs to the desired amount.

Now that the amount of capacitive VARs needed to correct the power factor is known, the amount of line current supplying the capacitor bank can be computed using the formula

$$I_L = \frac{VARs_C}{E_L \times 1.732}$$

$$I_L = \frac{26{,}431.29}{480 \times 1.732}$$

$$I_L = 31.79 \text{ A}$$

The capacitive load bank is to be connected in a wye. Therefore, the phase current will be the same as the line current. The phase voltage, however, will be less than the line voltage by a factor of 1.732, or 277.14 V. Ohm's law can be used to find the amount of capacitive reactance needed to produce a phase current of 31.79 A with an applied voltage of 277.14 V.

$$X_C = \frac{E_P}{I_P}$$

$$X_C = \frac{277.14}{31.79}$$

$$X_C = 8.72\ \Omega$$

The amount of capacitance needed to produce a capacitive reactance of 8.72 Ω can now be computed.

$$C = \frac{1}{2\pi F X_C}$$

$$C = \frac{40}{377 \times 8.72}$$

$$C = 304.2\ \mu F$$

When a bank of wye-connected capacitors with a value of 304.2 μF each are connected in parallel with the motor, the power factor will be corrected to 95%.

Summary

1. The voltages of a three-phase system are 120° out of phase with each other.
2. The two types of three-phase connections are wye and delta.
3. Wye connections are characterized by the fact that one terminal of each of the devices is connected together.
4. In a wye connection, the phase voltage is less than the line voltage by a factor of 1.732. The phase current and line current are the same.
5. In a delta connection, the phase voltage is the same as the line voltage. The phase current is less than the line current by a factor of 1.732.

Review Questions

1. How many degrees out of phase with each other are the voltages of a three-phase system?
2. What are the two main types of three-phase connections?
3. A wye-connected load has a voltage of 480 V applied to it. What is the voltage dropped across each phase?

4. A wye-connected load has a phase current of 25 A. How much current is flowing through the lines supplying the load?
5. A delta connection has a voltage of 560 V connected to it. How much voltage is dropped across each phase?
6. A delta connection has 30 A of current flowing through each phase winding. How much current is flowing through each of the lines supplying power to the load?
7. A three-phase load has a phase voltage of 240 V and a phase current of 18 A. What is the apparent power of this load?
8. If the load in question 7 is connected in a wye, what would be the line voltage and line current supplying the load?
9. An alternator with a line voltage of 2400 V supplies a delta-connected load. The line current supplied to the load is 40 A. Assuming the load is a balanced three-phase load, what is the impedance of each phase?
10. What is the apparent power of the circuit in question 9?

Problems

1. Refer to the circuit shown in *Figure 7-18* to answer the following questions. But assume that the alternator has a line voltage of 240 V and the load has an impedance of 12 Ω per phase. Find all the missing values.

$E_{P(A)}$ ________	$E_{P(L)}$ ________
$I_{P(A)}$ ________	$I_{P(L)}$ ________
$E_{L(A)}$ 240	$E_{L(L)}$ ________
$I_{L(A)}$ ________	$I_{L(L)}$ ________
P ________	$Z_{(Phase)}$ 12 Ω

2. Refer to the circuit shown in *Figure 7-19* to answer the following questions. But assume that the alternator has a line voltage of 4160 V, and the load has an impedance of 60 Ω per phase. Find all the missing values.

$E_{P(A)}$ ______	$E_{P(L)}$ ______
$I_{P(A)}$ ______	$I_{P(L)}$ ______
$E_{L(A)}$ 4160	$E_{L(L)}$ ______
$I_{L(A)}$ ______	$I_{L(L)}$ ______
P ______	$Z_{(Phase)}$ 60 Ω

3. Refer to the circuit shown in *Figure 7-20* to answer the following questions. But assume that the alternator has a line voltage of 560 V. Load 1 has an impedance of 5 Ω per phase, and load 2 has an impedance of 8 Ω per phase. Find all the missing values.

$E_{P(A)}$ ______	$E_{P(L1)}$ ______	$E_{P(L2)}$ ______
$I_{P(A)}$ ______	$I_{P(L1)}$ ______	$I_{P(L2)}$ ______
$E_{L(A)}$ 560	$E_{L(L1)}$ ______	$E_{L(L2)}$ ______
$I_{L(A)}$ ______	$I_{L(L1)}$ ______	$I_{L(L2)}$ ______
P ______	$Z_{(PHASE)}$ 5 Ω	$Z_{(PHASE)}$ 8 Ω

4. Refer to the circuit shown in *Figure 7-21* to answer the following questions. But assume that the alternator has a line voltage of 480 V. Load 1 has a resistance of 12 Ω per phase. Load 2 has an inductive reactance of 16 Ω per phase, and load 3 has a capacitive reactance of 10 Ω per phase. Find all the missing values.

$E_{P(A)}$ ______	$E_{P(L1)}$ ______	$E_{P(L2)}$ ______	$E_{P(L3)}$ ______
$I_{P(A)}$ ______	$I_{P(L1)}$ ______	$I_{P(L2)}$ ______	$I_{P(L3)}$ ______
$E_{L(A)}$ 480	$E_{L(L1)}$ ______	$E_{L(L2)}$ ______	$E_{L(L3)}$ ______
$I_{L(A)}$ ______	$I_{L(L1)}$ ______	$I_{L(L2)}$ ______	$I_{L(L3)}$ ______
VA ______	$R_{(Phase)}$ 12 Ω	$X_{L(Phase)}$ 16 Ω	$X_{C(Phase)}$ 10 Ω
	P ______	$VARs_L$ ______	$VARs_C$ ______

8

Three-Phase Transformers

Objectives

After studying this unit, you should be able to

- Discuss the operation of three-phase transformers
- Connect three single-phase transformers to form a three-phase bank
- Calculate voltage and current values for a three-phase transformer connection
- Connect two single-phase transformers to form a three-phase open delta connection
- Discuss the characteristics of an open delta connection

Three-phase transformers are used throughout industry to change values of three-phase voltage and current. Since three-phase power is the major way in which power is produced, transmitted, and used, an understanding of how three-phase transformer connections are made is essential. This unit will discuss different types of three-phase transformer connections, and present examples of how values of voltage and current for these connections are computed.

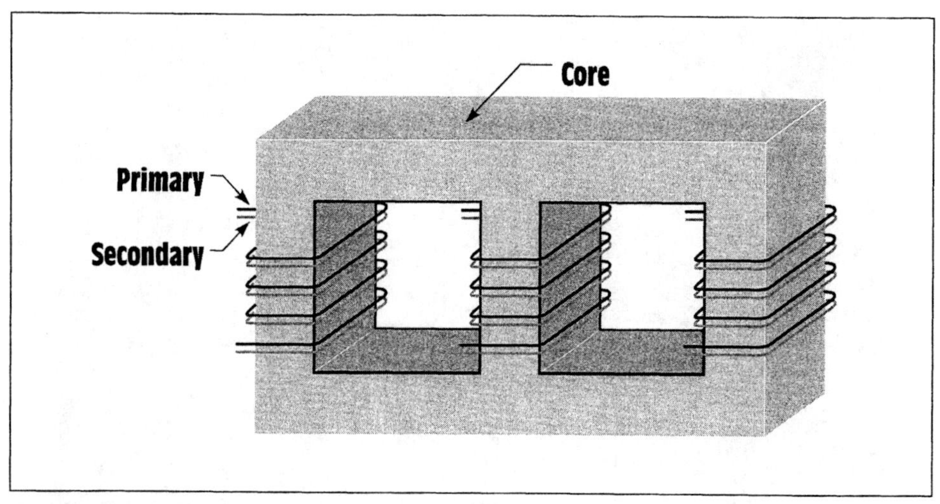

Figure 8-1 Basic construction of a three-phase transformer.

8-1 Three-Phase Transformers

A three-phase transformer is constructed by winding three single-phase transformers on a single core *(Figure 8-1)*. A photograph of a three-phase transformer is shown in *Figure 8-2*. The transformer is shown before it is mounted in an enclosure, which will be filled with a **dielectric oil**. The dielectric oil performs several functions. Since it is a dielectric, it provides electrical insulation between the windings and the case. It is also used to help provide cooling and to prevent the formation of moisture, which can deteriorate the winding insulation.

Three-Phase Transformer Connections

Three-phase transformers are connected in delta or wye configurations. A **wye-delta** transformer, for example, has its primary winding connected in a wye and its secondary winding connected in a delta *(Figure 8-3)*. A **delta-wye** transformer would have its primary winding connected in delta and its secondary connected in wye *(Figure 8-4)*.

Connecting Single-Phase Transformers into a Three-Phase Bank

If three-phase transformation is needed and a three-phase transformer of the proper size and turns ratio is not available, three single-phase transformers can be connected to form a **three-phase bank**. When three single-phase transformers are used to make a three-phase transformer bank,

Figure 8-2 Three-phase pyranol-filled transformer.

their primary and secondary windings are connected in a wye or delta connection. The three transformer windings in *Figure 8-5* have been labeled A, B, and C. One end of each primary lead is labeled H_1, and the other end is labeled H_2. One end of each secondary lead is labeled X_1, and the other end is labeled X_2.

Figure 8-6 shows three single-phase transformers labeled A, B, and C. The primary leads of each transformer have been labeled H_1 and H_2, and the secondary leads are labeled X_1 and X_2. The schematic diagram of *Figure 8-5* will be used to connect the three single-phase transformers into a three-phase wye-delta connection as shown in *Figure 8-7*.

The primary winding will be tied into a wye connection first. The

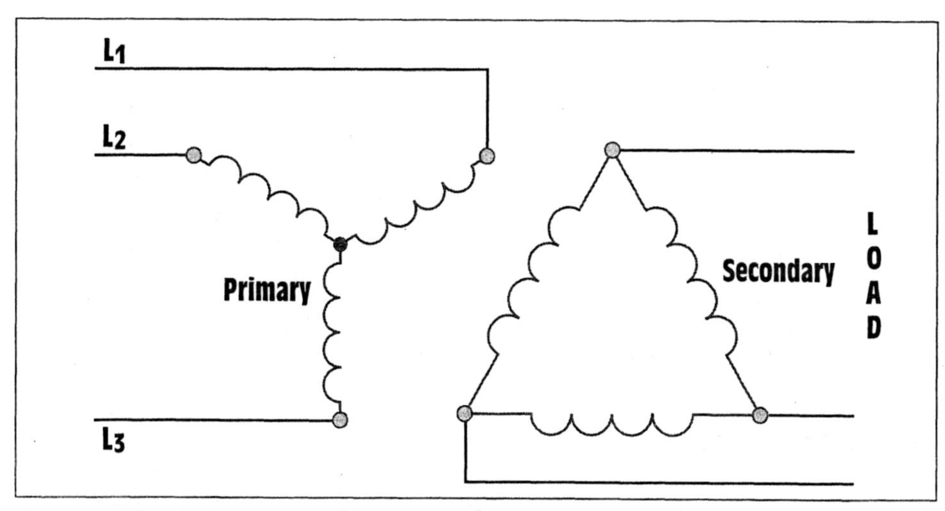

Figure 8-3 Wye-delta connected three-phase transformer.

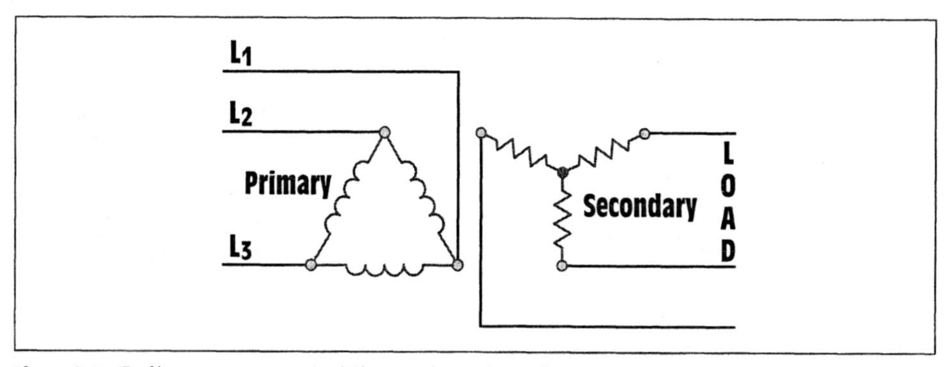

Figure 8-4 Delta-wye connected three-phase transformer.

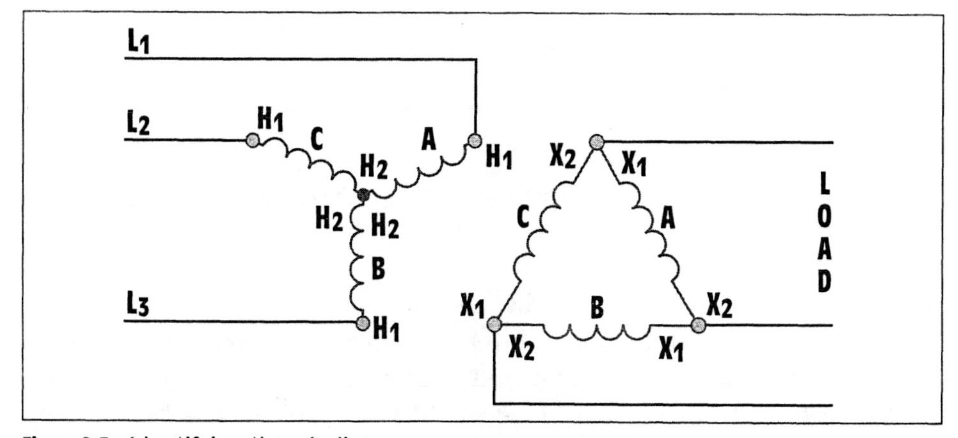

Figure 8-5 Identifying the windings.

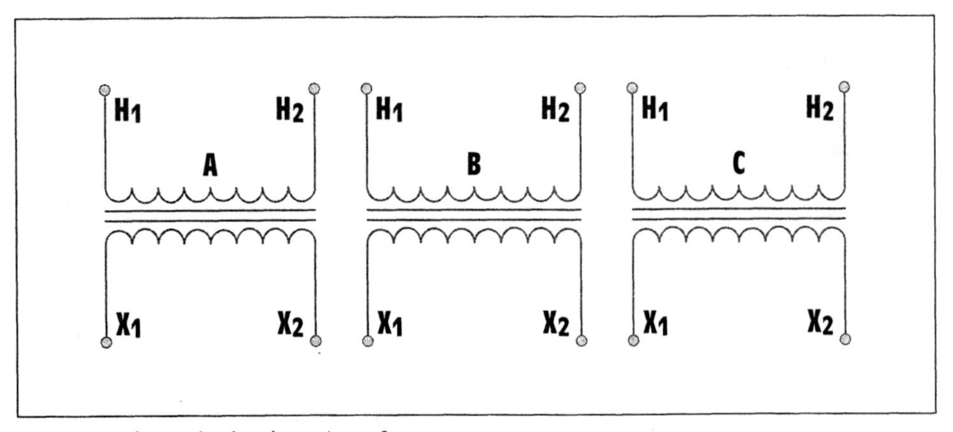

Figure 8-6 Three single-phase transformers.

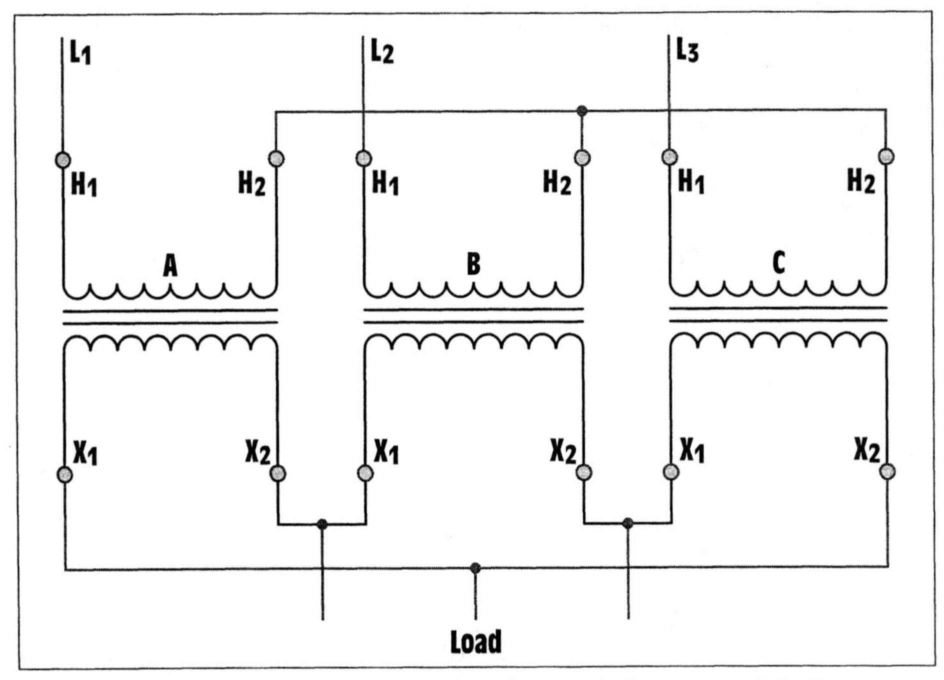

Figure 8-7 Connecting three single-phase transformers to form a wye-delta three-phase bank.

schematic in *Figure 8-5* shows that the H_2 leads of primary windings are connected together, and the H_1 lead of each winding is open for connection to the incoming power line. Notice in *Figure 8-7* that the H_2 leads of each primary winding have been connected together, and the H_1 lead of each winding has been connected to the incoming power line.

Figure 8-5 shows that the X_1 lead of transformer A is connected to the X_2 lead of transformer C. Notice that this same connection has been made in *Figure 8-7*. The X_1 lead of transformer B is connected to the X_2 lead of transformer A, and the X_1 lead of transformer C is connected to the X_2 lead of transformer B. The load is connected to the points of the delta connection.

Although *Figure 8-5* illustrates the proper schematic symbology for a three-phase transformer connection, some electrical schematics and wiring diagrams do not illustrate three-phase transformer connections in this manner. One type of diagram, called the **one-line diagram**, would illustrate a delta-wye connection as shown in *Figure 8-8*. These diagrams are generally used to show the main power distribution system of a large industrial plant. The one-line diagram in *Figure 8-9* shows the main power to the plant and the transformation of voltages to different subfeeders. Notice that each transformer shows whether the primary and secondary are connected as a wye or delta, and the secondary voltage of the subfeeder.

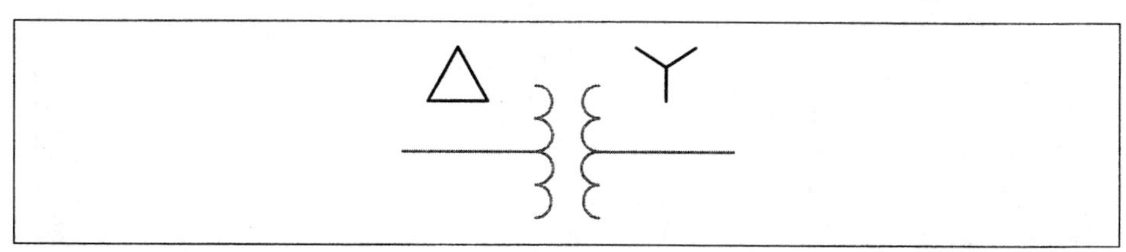

Figure 8-8 One-line diagram symbol used to represent a delta-wye three-phase transformer connection.

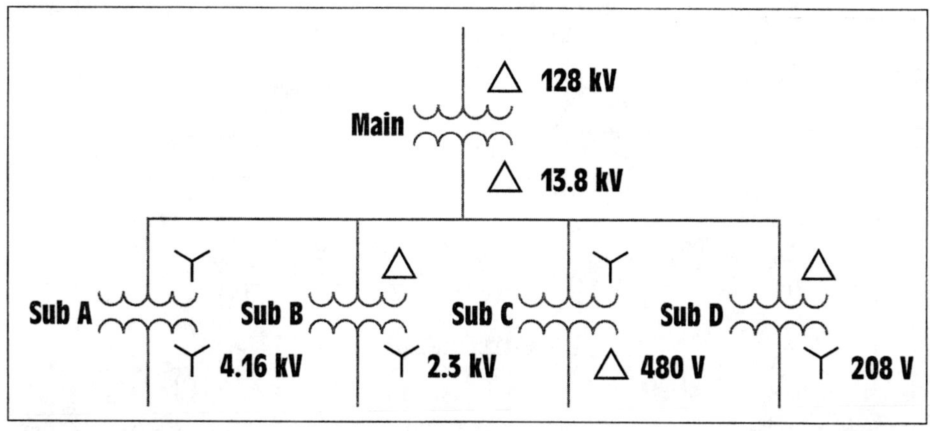

Figure 8-9 One-line diagrams are generally used to show the main power distribution of a plant.

8-2 Closing a Delta

Delta connections should be checked for proper polarity before making the final connection and applying power. If the phase winding of one transformer is reversed, an extremely high current will flow when power is applied. Proper phasing can be checked with a voltmeter as shown in *Figure 8-10*. If power is applied to the transformer bank before the delta connection is closed, the voltmeter should indicate 0 volts. If one phase winding has been reversed, however, the voltmeter will indicate double the amount of voltage. For example, assume that the output voltage of a delta secondary is 240 volts. If the voltage is checked before the delta is closed, the voltmeter should indicate a voltage of 0 volts if all windings have been phased properly. If one winding has been reversed, however, the voltmeter will indicate a voltage of 480 volts (240 + 240). This test will confirm whether a phase winding has been reversed, but it will not indicate whether the reversed winding is located in the primary or secondary. If either primary or secondary windings have been reversed, the voltmeter will indicate double the output voltage.

It should be noted, however, that a voltmeter is a high-impedance device. It is not unusual for a voltmeter to indicate some amount of voltage before the delta is closed, especially if the primary has been connected as a wye and the secondary as a delta. When this is the case, however, the voltmeter will generally indicate close to the normal output voltage if the connection is correct and double the output voltage if the connection is incorrect.

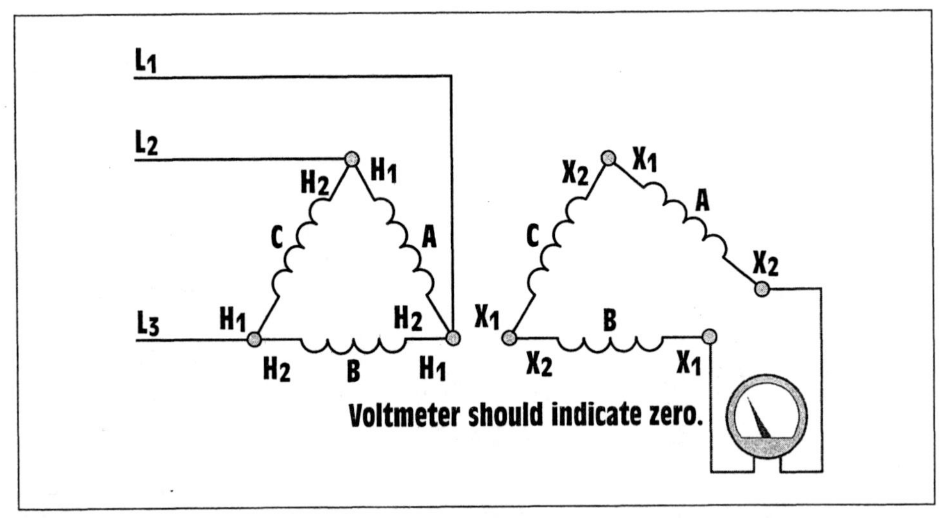

Figure 8-10 Testing for proper transformer polarity before closing the delta.

8-3 Three-Phase Transformer Calculations

When computing the values of voltage and current for three-phase transformers, the formulas used for making transformer calculations and three-phase calculations must be followed. Another very important rule that must be understood is that **only phase values of voltage and current can be used when computing transformer values**. When three-phase transformers are connected as a wye or delta, the primary and secondary windings themselves become the phases of a three-phase connection. This is true whether a three-phase transformer is used or whether three single-phase transformers are employed to form a three-phase bank. Refer to transformer A in *Figure 8-6*. All transformation of voltage and current takes place between the primary and secondary windings. Since these windings form the phase values of the three-phase connection, only phase and not line values can be used when calculating transformed voltages and currents.

Example 1

A three-phase transformer connection is shown in *Figure 8-11*. Three single-phase transformers have been connected to form a wye-delta bank. The primary is connected to a three-phase line of 13,800 volts, and the secondary voltage is 480. A three-phase resistive load with an impedance of 2.77 Ω per phase is connected to the secondary of the transformer. The following values will be computed for this circuit.

$E_{P(Primary)}$ — phase voltage of the primary
$E_{P(Secondary)}$ — phase voltage of the secondary
Ratio — turns ratio of the transformer
$E_{P(Load)}$ — phase voltage of the load bank
$I_{P(Load)}$ — phase current of the load bank
$I_{L(Secondary)}$ — secondary line current
$I_{P(Secondary)}$ — phase current of the secondary
$I_{P(Primary)}$ — phase current of the primary
$I_{L(Primary)}$ — line current of the primary

The primary windings of the three single-phase transformers have been connected to form a wye connection. In a wye connection, the phase voltage is less than the line voltage by a factor of 1.732. Therefore, the phase value of the primary voltage can be computed using the formula

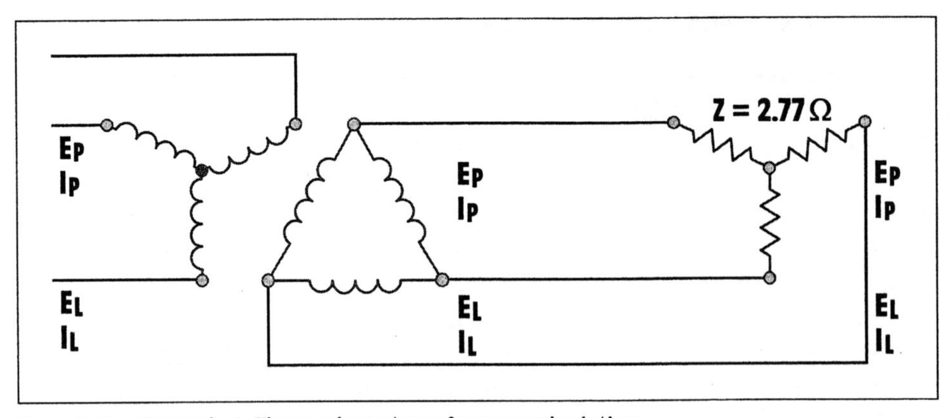

Figure 8-11 Example 1: Three-phase transformer calculation.

$$E_{P(Primary)} = \frac{E_L}{1.732}$$

$$E_{P(Primary)} = \frac{13,800}{1.732}$$

$$E_{P(Primary)} = 7967.67 \text{ volts}$$

The secondary windings are connected as a delta. In a delta connection, the phase voltage and line voltage are the same.

$$E_{P(Secondary)} = E_{L(Secondary)}$$

$$E_{P(Secondary)} = 480 \text{ volts}$$

The turns ratio can be computed by comparing the phase voltage of the primary with the phase voltage of the secondary.

$$\text{Ratio} = \frac{\text{primary-phase voltage}}{\text{secondaryp-phase voltage}}$$

$$\text{Ratio} = \frac{7967.67}{480}$$

$$\text{Ratio} = 16.6:1$$

The load bank is connected in a wye connection. The voltage across the phase of the load bank will be less than the line voltage by a factor of 1.732.

$$E_{P(Load)} = \frac{E_{L(Load)}}{1.732}$$

$$E_{P(Load)} = \frac{480}{1.732}$$

$$E_{P(Load)} = 277 \text{ volts}$$

Now that the voltage across each of the load resistors is known, the current flow through the phase of the load can be computed using Ohm's law.

$$I_{P(Load)} = \frac{E_{P(Load)}}{R}$$

$$I_{P(Load)} = \frac{277}{2.77}$$

$$I_{P(Load)} = 100 \text{ amperes}$$

Since the load is connected as a wye connection, the line current will be the same as the phase current. Therefore, the line current supplied by the secondary of the transformer is equal to the phase current of the load.

$$I_{L(Secondary)} = 100 \text{ A}$$

The secondary of the transformer bank is connected as a delta. The phase current of the delta is less than the line current by a factor of 1.732.

$$I_{P(Secondary)} = \frac{I_{L(Secondary)}}{1.732}$$

$$I_{P(Secondary)} = \frac{100}{1.732}$$

$$I_{P(Secondary)} = 57.74 \text{ amps}$$

The amount of current flow through the primary can be computed using the turns ratio. Since the primary has a higher voltage than the secondary, it will have a lower current. (Volts x amps input must equal volts x amps output.)

$$I_{P\ (Primary)} = \frac{I_{P\ (Secondary)}}{Ratio}$$

$$I_{P(Primary)} = \frac{57.74}{16.6}$$

$$I_{P(Primary)} = 3.48 \text{ amperes}$$

Recall that all transformed values of voltage and current take place across the phases, the primary has a phase current of 3.48 amps. In a wye connection, the phase current is the same as the line current.

$$I_{L(Primary)} = 3.48 \text{ amps}$$

The transformer connection with all computed values is shown in *Figure 8-12*.

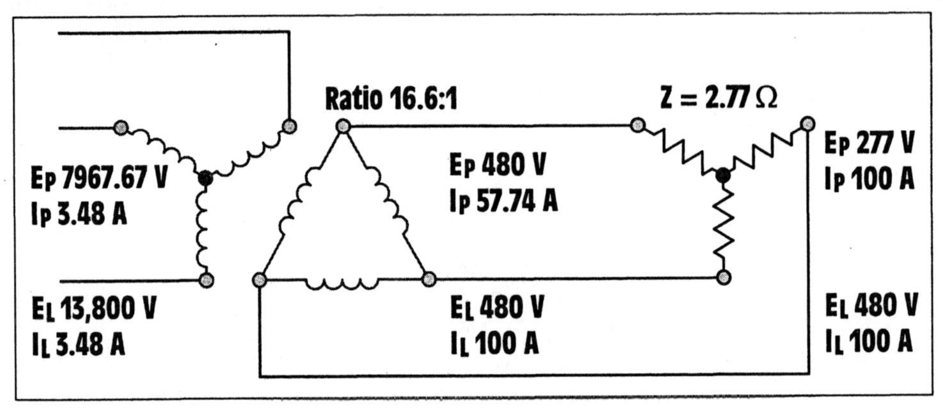

Figure 8-12 Example 1 with all missing values.

Example 2

In the next example, a three-phase transformer is connected in a delta-delta configuration *(Figure 8-13)*. The load is connected as a wye, and each phase has an impedance of 7 Ω. The primary is connected to a line voltage of 4160 volts and the secondary line voltage is 440 volts. The following values will be found:

$E_{P(Primary)}$ — phase voltage of the primary
$E_{P(Secondary)}$ — phase voltage of the secondary
Ratio — turns ratio of the transformer
$E_{P(Load)}$ — phase voltage of the load bank
$I_{P(Load)}$ — phase current of the load bank
$I_{L(Secondary)}$ — secondary line current
$I_{P(Secondary)}$ — phase current of the secondary
$I_{P(Primary)}$ — phase current of the primary
$I_{L(Primary)}$ — line current of the primary

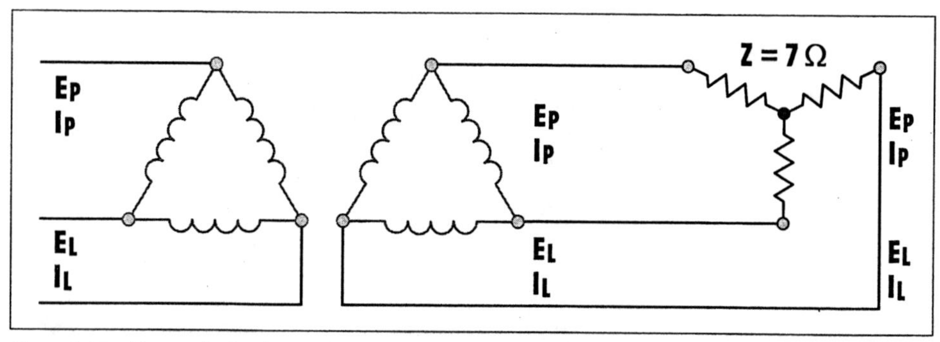

Figure 8-13 Example 2: Three-phase transformer calculation.

The primary is connected as a delta. The phase voltage will be the same as the applied line voltage.

$$E_{P(Primary)} = E_{L(Primary)}$$

$$E_{P(Primary)} = 4160 \text{ volts}$$

The secondary of the transformer is connected as a delta also. Therefore, the phase voltage of the secondary will be the same as the line voltage of the secondary.

$$E_{P(Secondary)} = 440 \text{ volts}$$

All transformer values must be computed using phase values of voltage and current. The turns ratio can be found by dividing the phase voltage of the primary by the phase voltage of the secondary.

$$\text{Ratio} = \frac{\text{Primary-phase voltage}}{\text{Single-phase voltage}}$$

$$\text{Ratio} = \frac{4160}{440}$$

$$\text{Ratio} = 9.45{:}1$$

The load is connected directly to the output of the secondary. The line voltage applied to the load must, therefore, be the same as the line voltage of the secondary.

$$E_{L(LOAD)} = 440 \text{ volts}$$

The load is connected in a wye. The voltage applied across each phase will be less than the line voltage by a factor of 1.732.

$$E_{P(Load)} = \frac{E_{L(Load)}}{1.732}$$

$$E_{P(Load)} = \frac{440}{1.732}$$

$$E_{P(Load)} = 254 \text{ volts}$$

The phase current of the load can be computed using Ohm's law.

$$I_{P(Load)} = \frac{E_{P(Load)}}{Z}$$

$$I_{P(Load)} = \frac{254}{7}$$

$$I_{P(Load)} = 36.29 \text{ amps}$$

The amount of line current supplying a wye-connected load will be the same as the phase current of the load.

$$I_{L(Load)} = 36.29 \text{ amps}$$

Since the secondary of the transformer is supplying current to only one load, the line current of the secondary will be the same as the line current of the load.

$$I_{L(Secondary)} = 36.29 \text{ amps}$$

The phase current in a delta connection is less than the line current by a factor of 1.732.

$$I_{P(Secondary)} = \frac{I_{L(Secondary)}}{1.732}$$

$$I_{P(Secondary)} = \frac{36.29}{1.732}$$

$$I_{P(Secondary)} = 20.95 \text{ amps}$$

The phase current of the transformer primary can now be computed using the phase current of the secondary and the turns ratio.

$$I_{P(Primary)} = \frac{I_{P(Secondary)}}{\text{ratio}}$$

$$I_{P(Primary)} = \frac{20.95}{9.45}$$

$$I_{P(Primary)} = 2.27 \text{ amps}$$

In this example, the primary of the transformer is connected as a delta. The line current supplying the transformer will be higher than the phase current by a factor of 1.732.

$$I_{L(Primary)} = I_{P(Primary)} \times 1.732$$

$$I_{L(Primary)} = 2.27 \times 1.732$$

$$I_{L(Primary)} = 3.93 \text{ amps}$$

The circuit with all computed values is shown in *Figure 8-14*.

8-4 Open Delta Connection

The **open-delta** transformer connection can be made with only two transformers instead of three *(Figure 8-15)*. This connection is often used when the amount of three-phase power needed is not excessive, such as in a small business. It should be noted that the output power of an open-delta connection is only 87% of the rated power of the two transformers. For example,

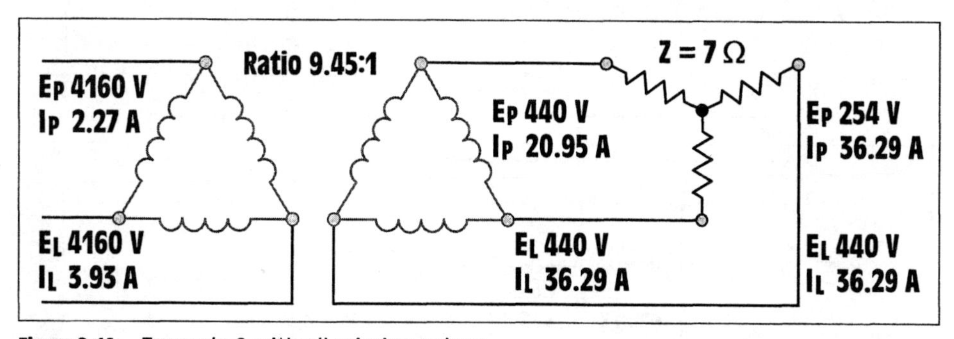

Figure 8-14 Example 2 with all missing values.

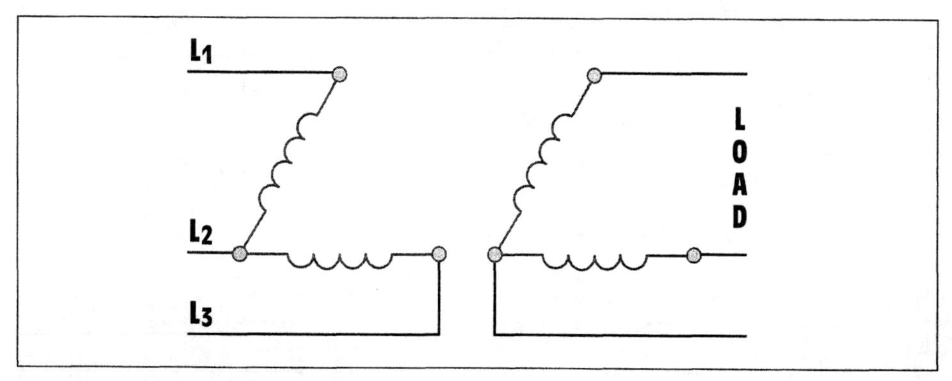

Figure 8-15 Open-delta connection.

assume two transformers, each having a capacity of 25 kVA (kilo-volt-amperes), are connected in an open-delta connection. The total output power of this connection is 43.5 kVA (50 kVA x 0.87 = 43.5 kVA).

Another figure given for this calculation is 58%. This percentage assumes a closed-delta bank containing three transformers. If three 25-kVA transformers were connected to form a closed-delta connection, the total output power would be 75 kVA (3 x 25 kVA = 75 kVA). If one of these transformers were removed and the transformer bank operated as an open-delta connection, the output power would be reduced to 58% of its original capacity of 75 kVA. The output capacity of the open-delta bank is 43.5 kVA (75 kVA x 0.58 = 43.5 kVA).

The voltage and current values of an open-delta connection are computed in the same manner as a standard delta-delta connection when three transformers are employed. The voltage and current rules for a delta connection must be used when determining line and phase values of voltage and current.

T-Connected Transformers

Another connection involving the use of two transformers to supply three-phase power is the **T-connection**, *(Figure 8-16)*. In this connection, one transformer is generally referred to as the main transformer and the other is

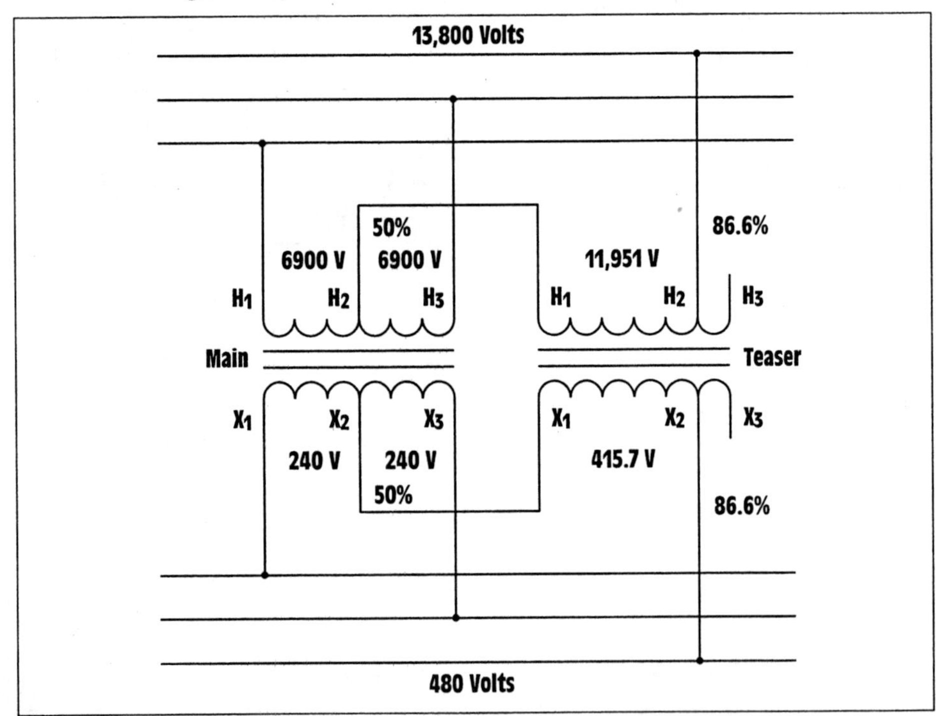

Figure 8-16 T-connected transformers.

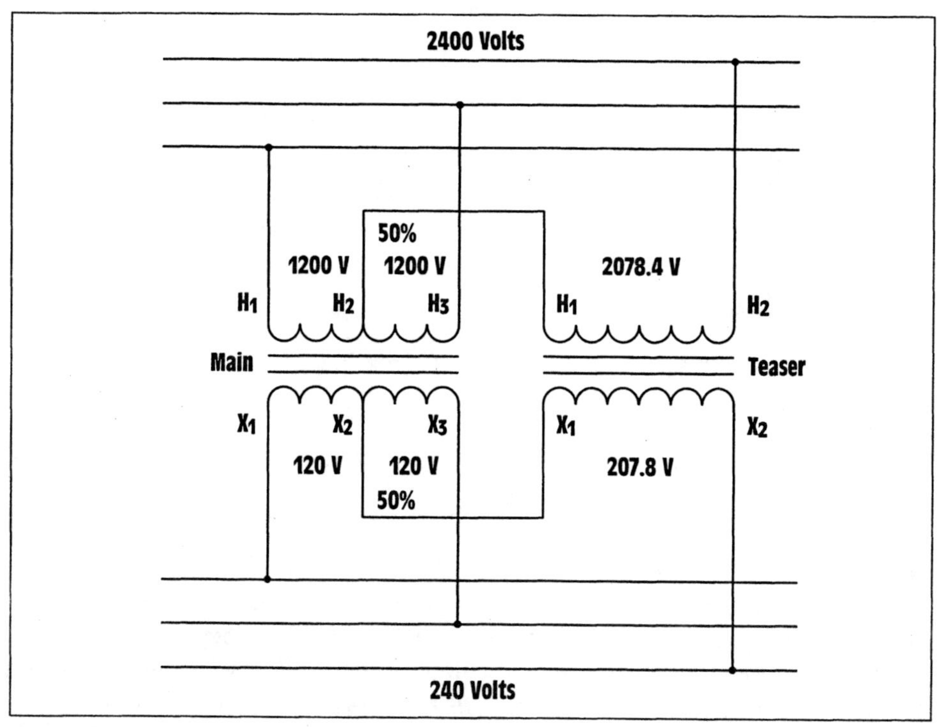

Figure 8-17 T-connected transformers with the same voltage ratings.

called the teaser transformer. The main transformer must contain a center or 50% tap for both the primary and secondary winding, and it is preferred that the teaser transformer contain an 86.6% voltage tap for both the primary and secondary winding. Although the 86.6% tap is preferred, the connection can be made with a teaser transformer that has the same voltage rating as the main transformer. In this instance the teaser transformer is operated at reduced flux, *(Figure 8-17)*. This connection permits two transformers to be connected T instead of open delta in the event that one transformer of a delta-delta bank should fail.

Transformers intended for use as T-connected transformers are often specially wound for the purpose, and both transformers are often contained in the same case. When making the T connection, the main transformer is connected directly across the power line. One primary lead of the teaser transformer is connected to the center tap of the main transformer and the 86.6% tap is connected to the power line. The same basic connection is made for the secondary. A vector diagram illustrating the voltage relationships of the T connection is shown in *Figure 8-18*. The greatest advantage of the T connection over the open-delta connection is that it maintains a better phase balance. The greatest disadvantage of the T connection is that one

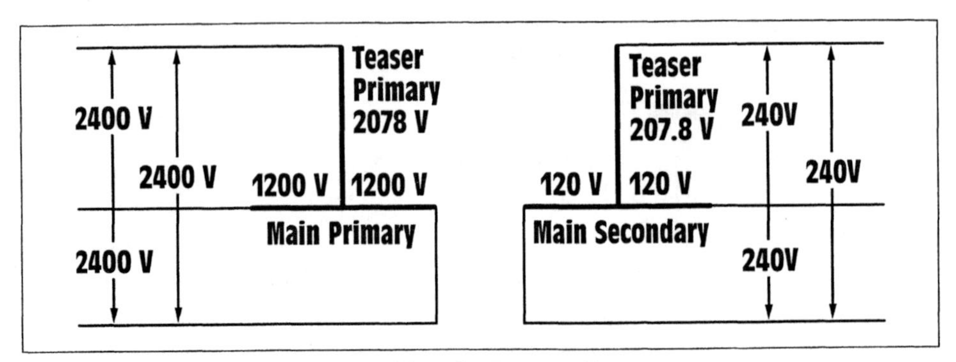

Figure 8-18 Vector voltage relationships of a T connection.

transformer must contain a center tap of both its primary and secondary windings.

Scott Connection

The **Scott connection** is used to convert three-phase power into two-phase power using two single-phase transformers. The Scott connection is very similar to the T connection in that one transformer called the main transformer, must have a center or 50% tap, and the second or teaser transformer, must have an 86.6% tap on the primary side. The difference between the Scott and T connections lies in the connection of the secondary windings, *Figure 8-19*. In the Scott connection, the secondary windings of each transformer provide the phases of a two-phase system. The voltages of the secondary windings are 90° out of phase with each other. The Scott connection is generally used to provide two-phase power for the operation of two-phase motors.

Zig-Zag Connection

The **zig-zag** or **interconnected-wye** transformer is primarily used for grounding purposes; to establish a neutral point for the grounding of fault currents. The zig-zag connection is basically a three-phase autotransformer with windings divided into six equal parts, *(Figure 8-20)*. In the event of a fault current, the zig-zag connection forces the current to flow equally in the three legs of the autotransformer, offering minimum impedance to the flow of fault current. A schematic diagram of the zig-zag connection is shown in *Figure 8-21*.

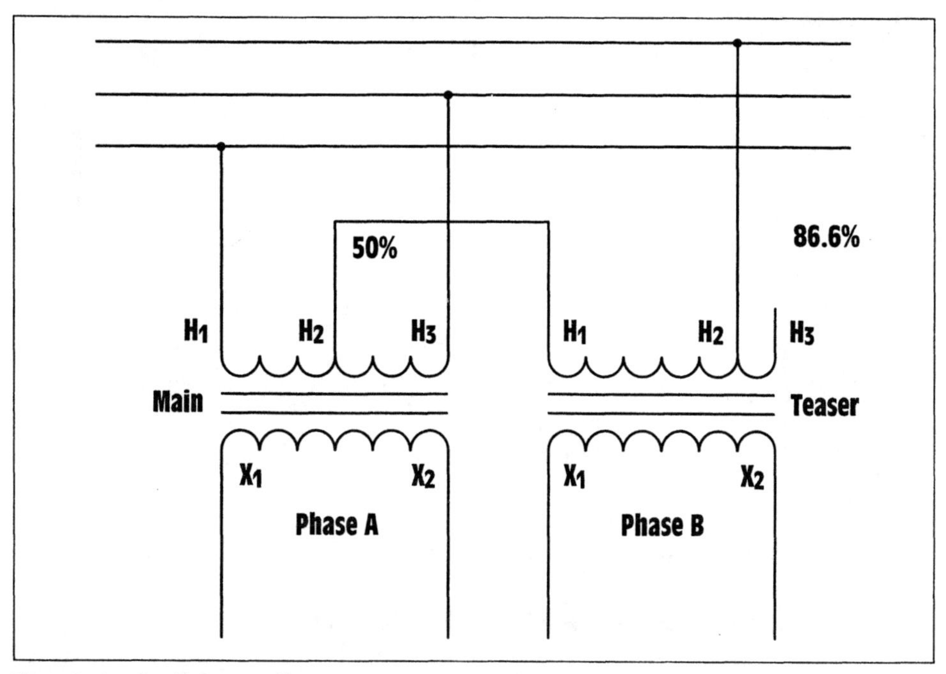

Figure 8-19 Scott Connection.

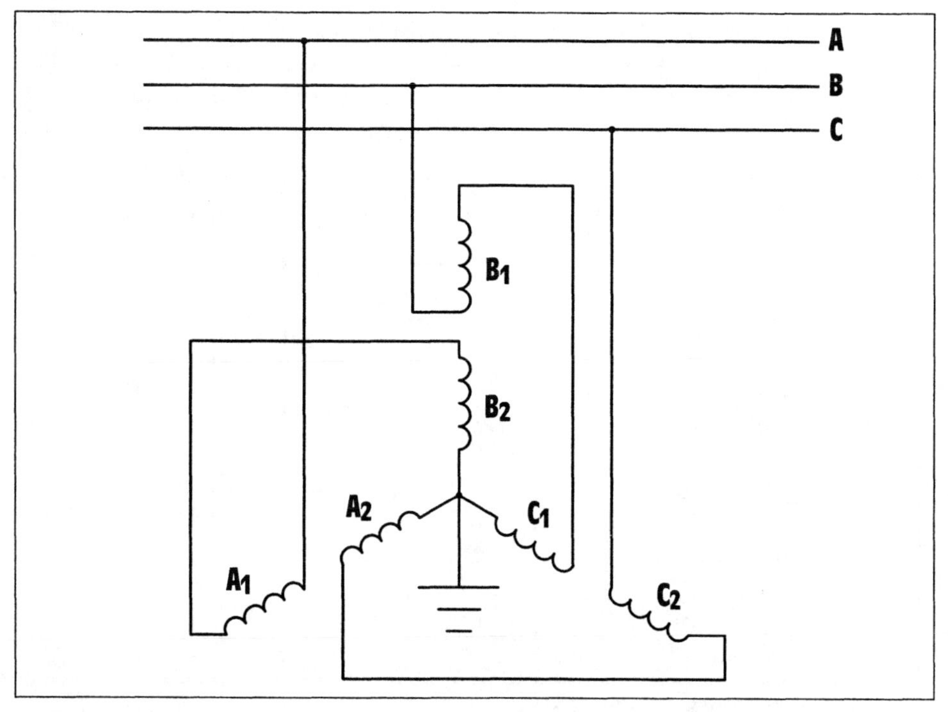

Figure 8-20 Zig-zag connection.

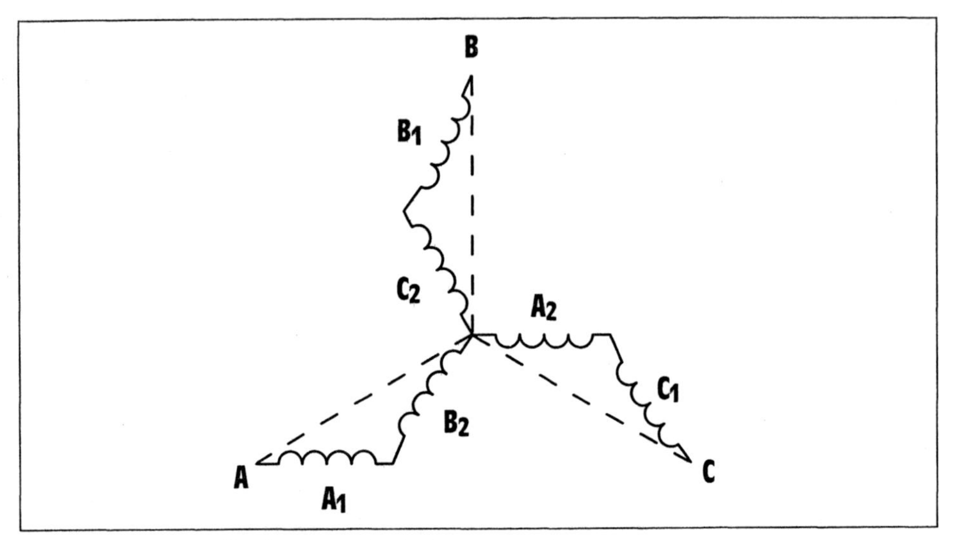

Figure 8-21 Schematic diagram of a zig-zag connection.

Three-Phase to Six-Phase Connections

There are some instances when it is desirable to have a power system with more than three phases. A good example of this is when it is necessary to convert or rectify alternating current into direct current with a minimum amount of ripple (pulsations of voltage). Power supplies that produce a low amount of ripple require less filtering. One of the most common three-phase to six-phase connections is the **diametrical connection**, *(Figure 8-22)*. The diametrical connection is preferred because it requires only one low-

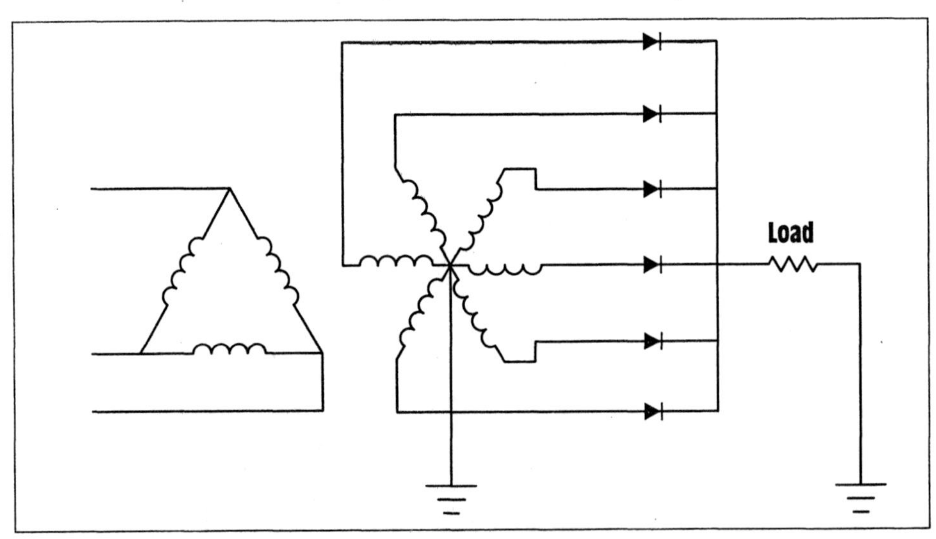

Figure 8-22 Diametrical connection.

voltage winding on each transformer. If these windings are center tapped, a neutral conductor can be provided for the six-phase output, permitting half-wave rectification to be used. The high-voltage windings can be connected in wye or delta, but delta is preferred because it helps to reduce harmonics in the secondary winding. A schematic diagram of a diametrical connection with a delta-connected primary and three-phase half-wave rectifier is shown in *Figure 8-23*.

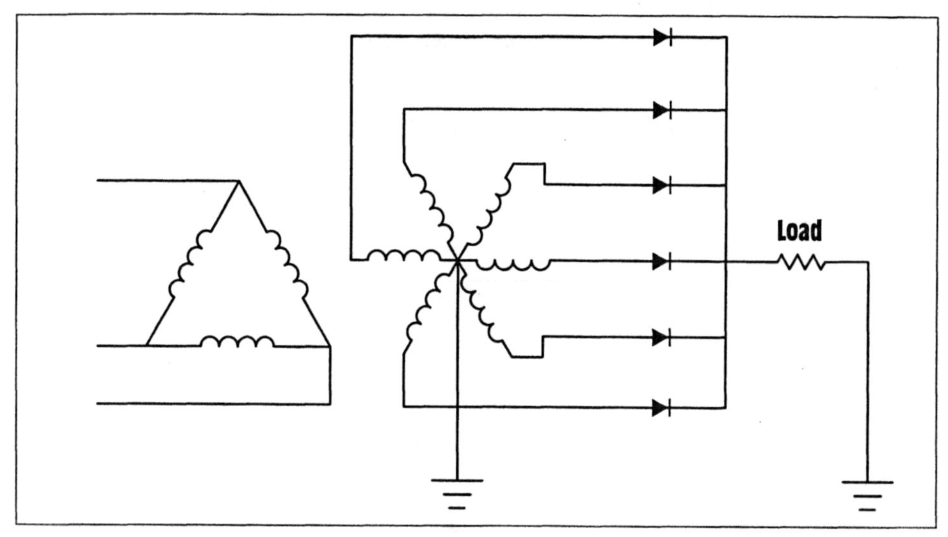

Figure 8-23 Schematic diagram of a diametrically connected transformer with six-phase half-wave rectifier.

Double-Delta Connection

Another three-phase to six-phase connection is the **double delta**. The transformers used in this connection must have two secondary windings that provide equal voltage. The secondary windings are connected in such a manner that two delta connections are formed. The two delta windings are reverse connected so they have an angular difference of 180° with respect to each other, *(Figure 8-24)*. A schematic drawing showing the double-delta connection is shown in *Figure 8-25*. The primary windings may be connected in a wye or delta.

Double-Wye Connection

The **double-wye** connection is very similar to the double-delta connection in that the transformers must have two secondary windings that produce equal voltage, *(Figure 8-26)*. This connection is made by connecting

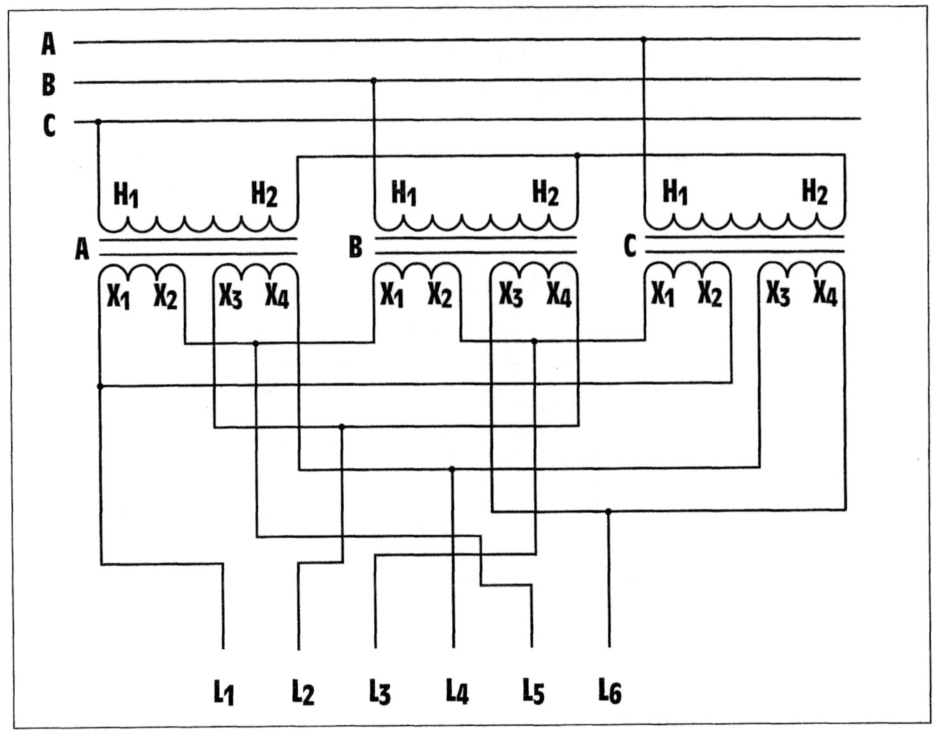

Figure 8-24 Double-delta connection.

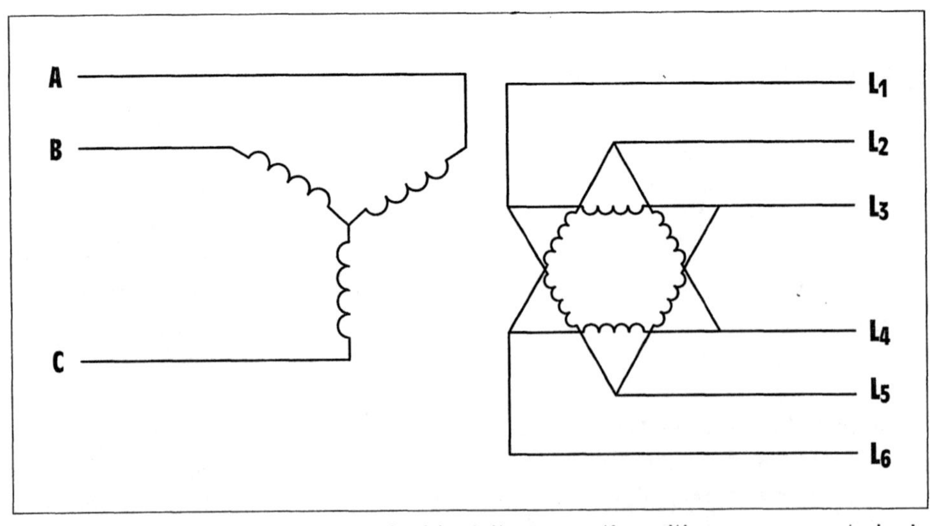

Figure 8-25 Schematic diagram of a double-delta connection with a wye-connected primary.

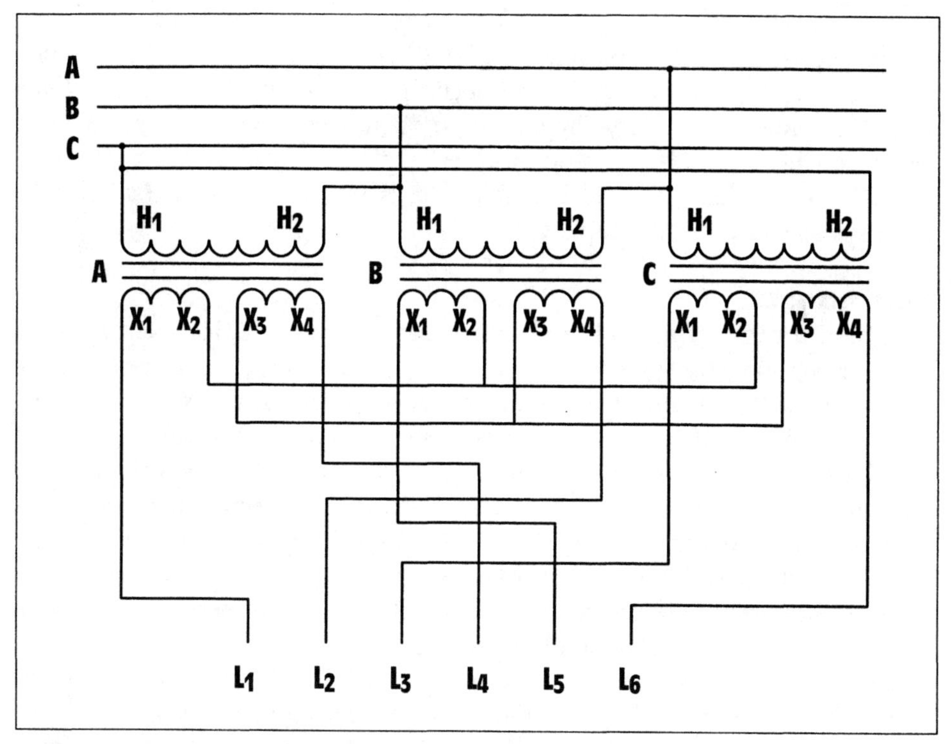

Figure 8-26 Double-wye connection.

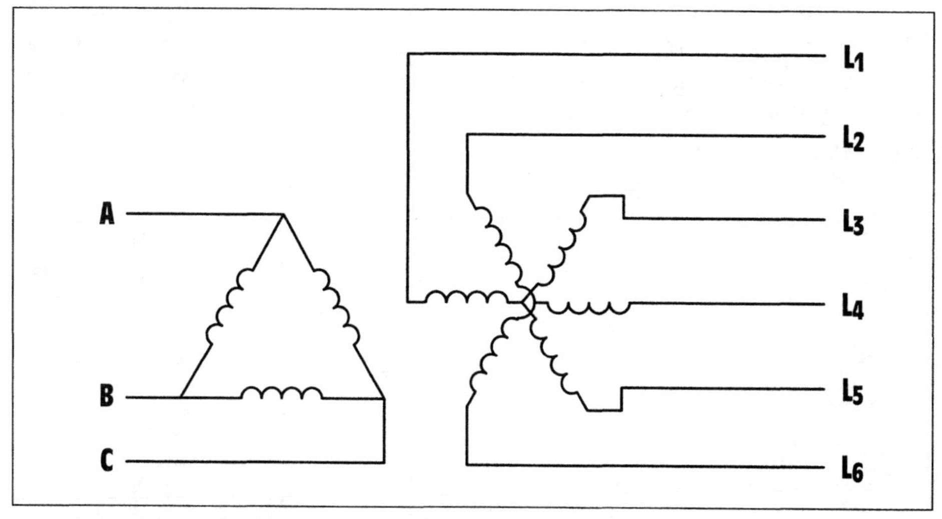

Figure 8-27 Schematic diagram of a double-wye connection.

the secondary windings in two different wyes. The windings are reverse connected so a phase displacement of 180° is formed between the two wye connections. A schematic of this connection is shown in *Figure 8-27.* Notice that this connection is very similar to the diametrical connection except that no center tap connection is possible. The primary windings may be connected wye or delta. The delta connection is preferred because it provides better phase and voltage stability.

Harmonics

A **harmonic** is a multiple of some basic frequency. Since the frequency used throughout the United States and Canada is 60 Hz., a harmonic will be a multiple of 60. All alternating-current power supply systems, such as alternators and transformers, can produce harmonic frequencies. The **fundamental** frequency is the lowest base frequency. In this case the fundamental frequency is 60 Hz, the second harmonic is double the fundamental frequency of 120 Hz, and the third harmonic is three times the fundamental frequency or 180 Hz., and so on.

These harmonics are produced in three-phase transformer systems mainly because of unbalanced conditions on the lines. Harmonics in three-phase systems are generally divided into two categories: triple components and nontriple components. Triple components are harmonics that are multiples of three, such as the third harmonic, the sixth harmonic, the ninth harmonic, and so forth. Harmonics, especially triple harmonics, add to the current in the phase conductors, increasing the heating effect on the conductors.

Although harmonics can add to the heating effect of all conductors in the supply circuit, their effect on phase conductors is generally less than 10%. These harmonic currents can combine in the neutral conductor producing a large increase in neutral current. For this reason, it is generally recommended that the neutral conductor be at least as large as the phase conductors.

Summary

1. Three-phase transformers are constructed by winding three separate transformers on the same core material.
2. Single-phase transformers can be used as a three-phase transformer bank by connecting their primary and secondary windings as either wyes or deltas.
3. When computing three-phase transformer values, the rules for three-phase circuits must be followed as well as the rules for transformers.
4. Phase values of voltage and current must be used when computing the values associated with the transformer.
5. The total power output of a three-phase transformer bank is the sum of the rating of the three transformers.
6. An open-delta connection can be made with the use of only two transformers.
7. When an open-delta connection is used, the total output power is 87% of the sum of the power rating of both transformers.
8. Two transformers connected in T can be used to supply three-phase power.
9. When making a T connection, the main transformer must contain a center tap and it is preferred that the teaser transformer contain an 86.6% voltage tap.
10. The Scott connection is used to change three-phase power into two-phase power.
11. Three-phase transformers may be connected to produce six-phase power.
12. The most common three-phase to six-phase connection is the diametrical connection.
13. The diametrical connection is favored because only one secondary winding is needed on each transformer and if the secondary windings are center tapped, the six-phase output can have a center tap for use as a neutral conductor.
14. The double-delta and double-wye connections require transformers with two secondary windings on each transformer.
15. A center-tap winding is not possible with the double-delta or double-wye connection.

Review Questions

1. How many transformers are needed to make an open-delta connection?
2. Two transformers rated at 100 kVA each are connected in an open-delta connection. What is the total output power that can be supplied by this bank?
3. When computing values of voltage and current for a three-phase transformer, should the lines values of voltage and current be used or the phase values?

Refer to *Figure 8-11* to answer the following questions.

4. Assume a line voltage of 2400 volts is connected to the primary of the three-phase transformer and the line voltage of the secondary is 240 volts. What is the turns ratio of the transformer?
5. Assume the load has an impedance of 3.5 Ω per phase. What is the line current provided by the transformer secondary?
6. How much current is flowing through the secondary winding?
7. How much current is flowing through the primary winding?

Refer to *Figure 8-13* to answer the following questions.

8. Assume a line voltage of 12,470 volts is connected to the primary of the transformer and the line voltage of the secondary is 480 volts. What is the turns ratio of the transformer?
9. Assume the load has an impedance of 6 Ω per phase. What is the secondary line current?
10. How much current is flowing in the secondary winding?
11. How much current is flowing in the primary winding?
12. What is the line current of the primary?

Problems

Refer to the transformer shown in *Figure 8-11* to fill in the following unknown values.

1.

Primary	Secondary	Load
E_P ________	E_P ________	E_P ________
I_P ________	I_P ________	I_P ________
E_L 4160	E_L 440	E_L ________
I_L ________	I_L ________	I_L ________
Ratio	$Z = 3.5\ \Omega$	

2.

Primary	Secondary	Load
E_P ________	E_P ________	E_P ________
I_P ________	I_P ________	I_P ________
E_L 7200	E_L 240	E_L ________
I_L ________	I_L ________	I_L ________
Ratio	$Z = 4\ \Omega$	

Refer to the transformer connection shown in *Figure 8-28* to fill in the following unknown values.

3.

Primary	Secondary	Load
E_P ________	E_P ________	E_P ________
I_P ________	I_P ________	I_P ________
E_L 13,800	E_L 480	E_L ________
I_L ________	I_L ________	I_L ________
Ratio	$Z = 2.5\ \Omega$	

4.

Primary	Secondary	Load
E_P ________	E_P ________	E_P ________
I_P ________	I_P ________	I_P ________
E_L 23,000	E_L 208	E_L ________
I_L ________	I_L ________	I_L ________
Ratio	$Z = 3\ \Omega$	

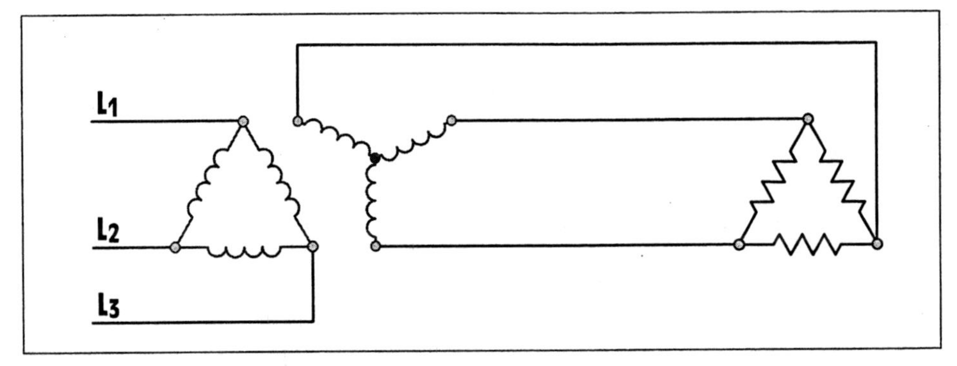

Figure 8-28 Practice problems circuit 2.

9

Single-Phase Loads for Three-Phase Transformers

Objectives

After studying this unit, you should be able to

- Compute values of voltage and current for single-phase loads connected to a three-phase transformer bank
- Discuss different types of three-phase transformer connections generally used to supply single-phase loads
- Connect single-phase loads to a three-phase transformer

When true three-phase loads are connected to a three-phase transformer bank, there are no problems in balancing the currents and voltages of the individual phases. *Figure 9-1* illustrates this condition. In this circuit, a delta-wye three-phase transformer bank is supplying power to a wye-connected three-phase load in which the impedance of each phases is the same. Notice that the amount of current flow in the phases is the same. Although this is the ideal condition, and is certainly desired for all three-phase transformer loads, it is not always possible to obtain a balanced load. Three-phase transformer connections are often used to supply **single-phase loads**, which tend to unbalance the system.

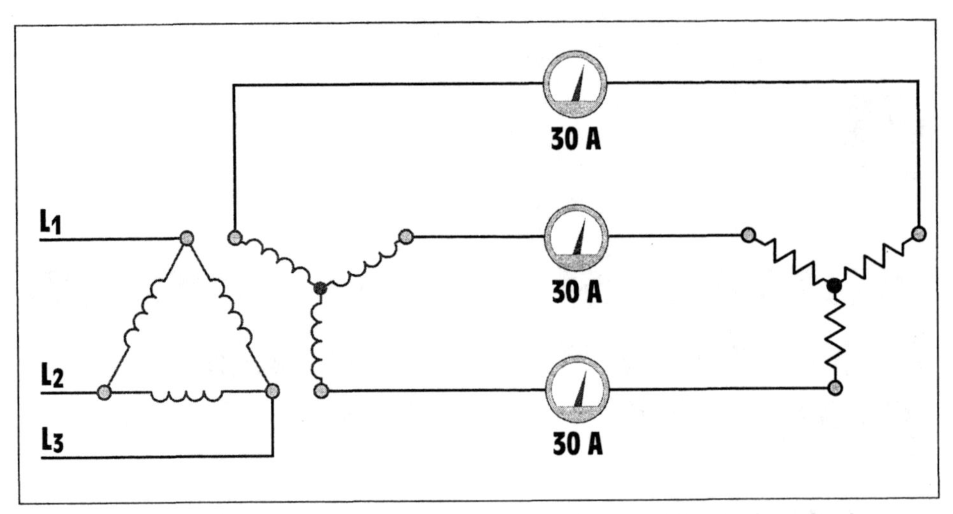

Figure 9-1 Three-phase transformer connected to a balanced three-phase load.

Open-Delta Connection Supplying a Single-Phase Load

The type of three-phase transformer connection used is generally determined by the amount of power needed. When a transformer bank must supply both three-phase and single-phase loads, the utility company will often provide an open-delta connection with one transformer center-tapped as shown in *Figure 9-2*. In this connection, it is assumed that the amount of three-phase power needed is 20 kVA, and the amount of single-phase power needed is 30 kVA. Notice that the transformer that has been center tapped must supply power to both the three-phase and single-phase loads. Since this is an open-delta connection, the transformer bank can be loaded to only 87% of its full capacity when supplying a three-phase load. The rating of the three-phase transformer bank must therefore be 23 kVA (20 kVA/0.87 = 23 kVA). Since the rating of the two transformers can be added to obtain a total output power rating, one transformer is rated at only half the total amount of power needed, or 12 kVA (23 kVA/2 = 11.5 kVA). The transformer that is used to supply power to the three-phase load only is rated at 12 kVA. The transformer that has been center tapped must supply power to both the single-phase and three-phase load. Its capacity will, therefore, be 42 kVA (12 kVA + 30 kVA). Therefore a 45-kVA transformer will be used.

Voltage Values

The connection shown in *Figure 9-2* has a line-to-line voltage of 240 volts. The three voltmeters V_1, V_2, and V_3 have all been connected across the

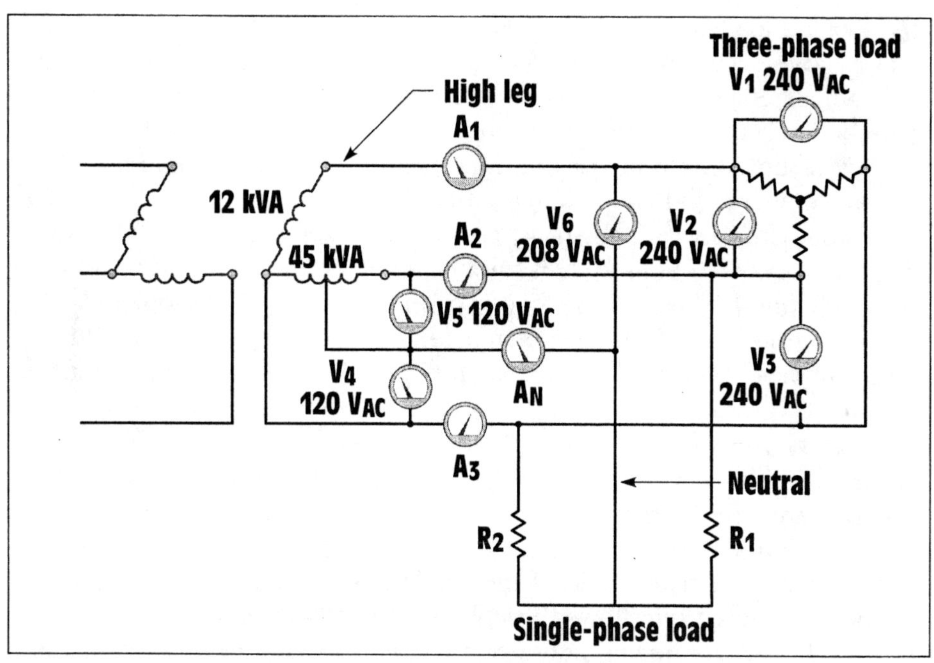

Figure 9-2 Three-phase open-delta transformer supplying both three-phase and single-phase loads.

three-phase lines and should indicate 240 volts each. Voltmeters V_4 and V_5 have been connected between the two lines of the larger transformer and its center tap. These two voltmeters will indicate a voltage of 120 volts each. Notice that it is these two lines and the center tap that are used to supply the single-phase power needed. The center tap of the larger transformer is used as a neutral conductor for the single-phase loads. Voltmeter V_6 has been connected between the center tap of the larger transformer and the line of the smaller transformer. This line is known as a **high leg**, because the voltage between this line and the neutral conductor will be higher than the voltage between the neutral and either of the other two conductors. The high-leg voltage can be computed by increasing the single-phase center-tapped voltage value by 1.732 (squre root of 3). In this case, the high-leg voltage will be 208 V (120 x 1.732 = 208). When this type of connection is employed, the *National Electrical Code*® requires that the high leg be identified by connecting it to an **orange wire** or by **tagging** it at any point that it enters an enclosure with the neutral conductor.

Load Conditions

In the first load condition, it will be assumed that only the three-phase load is in operation and none of the single-phase load is operating. If the three-phase load is operating at maximum capacity, ammeters A_1, A_2, and A_3 will indicate a current flow of 48.1 amps each (20 kVA/240 volts x 1.732 = 48.1 amps). Notice that when only the three-phase load is in operation, the current on each line is balanced.

Now assume that none of the three-phase load is in operation and only the single-phase load is operating. If all the single-phase load is operating at maximum capacity, ammeters A_2 and A_3 will each indicate a value of 125 amps (30 kVA/240 volts = 125 amps). Ammeter A_1 will indicate a current flow of 0 amp because all the load is connected between the other two lines of the transformer connection. Ammeter A_N will also indicate a value of 0 amp. Ammeter A_N is connected in the neutral conductor, and the neutral conductor carries the sum of the unbalanced load between the two phase conductors. Another way of stating this is to say that the neutral conductor carries the difference between the two line currents. Since both of these conductors are now carrying the same amount of current, the difference between them is 0 amp.

Now assume that one side of the single-phase load, resistor R_2, has been opened and no current flows through it. If the other line maintains a current flow of 125 A, the neutral conductor will have a current flow of 125 amps also (125 – 0 = 125).

Now assume that resistor R_2 has a value that will permit a current flow of 50 amps on that phase. The neutral current will now be 75 amps (125 – 50 = 75). Since the neutral conductor carries the sum of the unbalanced load, the size of the neutral conductor never needs to be larger than the largest line conductor.

It will now be assumed that both three-phase and single-phase loads are operating at the same time. If the three-phase load is operating at maximum capacity, and the single-phase load is operating in such a manner that 125 amps flow through resistor R_1 and 50 amps flow through resistor R_2, the ammeters will indicate the following values:

A_1 = 48.1 amps

A_2 = 173.1 amps (48.1 + 125 = 173.1)

A_3 = 98.1 amps (48.1 + 50 = 98.1)

A_N = 75 amps (125 – 50 = 75)

Notice that the smaller of the two transformers is supplying current to only the three-phase load, but the larger transformer must supply current for both the single-phase and three-phase loads.

Although the circuit shown in *Figure 9-2* is the most common method of connecting both three-phase and single-phase loads to an open-delta transformer bank, it is possible to use the high leg to supply power to a single-phase load also. The circuit shown in *Figure 9-3* is a circuit of this type. Resistors R_1 and R_2 are connected to the lines of the transformer that has been center tapped, and resistor R_3 is connected to the line of the other transformer. If the line-to-line voltage is 240 volts, voltmeters V_1 and V_2 will each indicate a value of 120 volts across resistors R_1 and R_2. Voltmeter V_3, however, will indicate that a voltage of 208 volts is applied across resistor R_3.

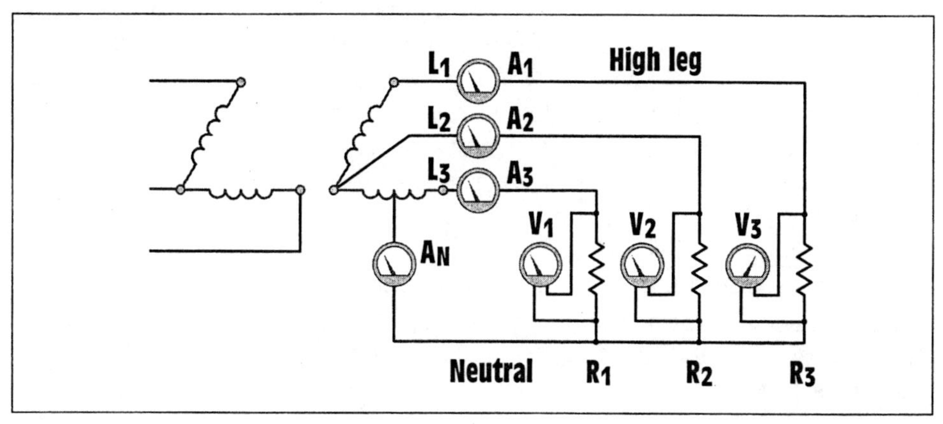

Figure 9-3 High leg supplies a single-phase load.

Calculating Neutral Current

The amount of current flow in the **neutral conductor** will still be the sum of the unbalanced load between lines L_2 and L_3, with the addition of the current flow in the high leg, L_1. To determine the amount of neutral current, use the formula

$$A_N = A_1 + (A_2 - A_3)$$

For example, assume line L_1 has a current flow of 100 amps, line L_2 has a current flow of 75 amps, and line L_3 has a current flow of 50 amps. The amount of current flow in the neutral conductor would be:

$$A_N = A_1 + (A_2 - A_3)$$

$$A_N = 100 + (75 - 50)$$

$$A_N = 100 + 25$$

$$A_N = 125 \text{ amps}$$

In this circuit, it is possible for the neutral conductor to carry more current than any of the three-phase lines. This circuit is more of an example as to why the *National Electrical Code*® requires a high leg to be identified than it is a practical working circuit. It is seldom that the high-leg side of this type of connection will be connected to the neutral conductor.

9-1 Closed Delta with Center Tap

Another three-phase transformer configuration used to supply power to single-phase and three-phase loads is shown in *Figure 9-4*. This circuit is virtually identical to the circuit shown in *Figure 9-2*, with the exception that a third transformer has been added to close the delta. Closing the delta permits more power to be supplied for the operation of three-phase loads. In this circuit, it is assumed that the three-phase load has a power requirement of 75 kVA, and the single-phase load requires an additional 50 kVA. Three 25-kVA transformers could be used to supply the three-phase power needed (25 kVA x 3 = 75 kVA). The addition of the single-phase load, however, requires one of the transformers to be larger. This transformer must supply both the three-phase and single-phase load, which requires it to have a rating of 75 kVA (25 kVA + 50 kVA = 75 kVA). In this circuit, the primary is

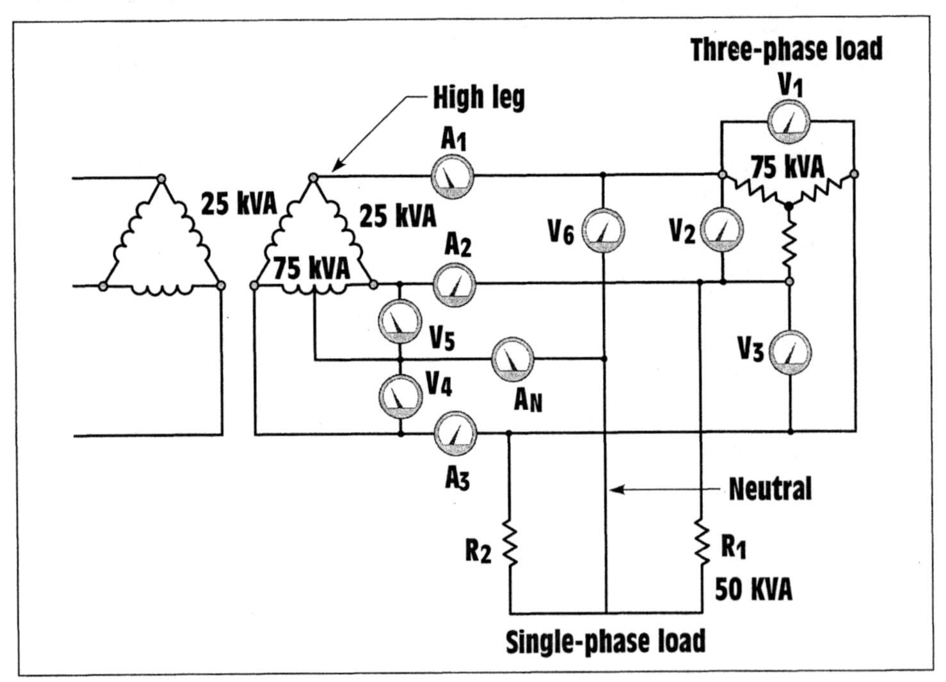

Figure 9-4 Closed-delta connection with high leg.

connected in a delta configuration. Since the secondary side of the transformer bank is a delta connection, either a wye or a delta primary could have been used. This, however, will not be true of all three-phase transformer connections supplying single-phase loads.

9-2 Closed Delta without Center Tap

In the circuit shown in *Figure 9-5,* the transformer bank has been connected in a wye-delta configuration. Notice that there is no transformer

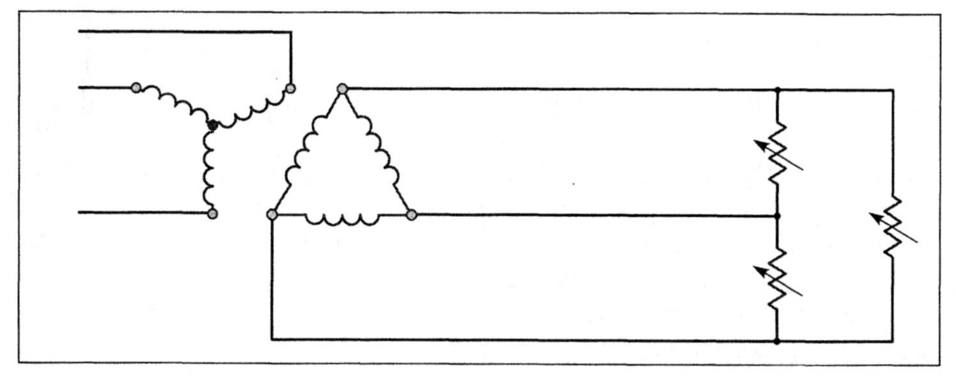

Figure 9-5 Single-phase loads supplied by a wye-delta transformer connection.

secondary with a center-tapped winding. In this circuit, there is no neutral conductor. The three loads have been connected directly across the three-phase lines. Since these three loads are connected directly across the lines, they form a delta-connected load. If these three loads are intended to be used as single-phase loads, they will in all likelihood have changing resistance values. The result of this connection is a three-phase delta-connected load that can be unbalanced in different ways. The amount of current flow in each phase is determined by the impedance of the load and the vectorial relationships of each phase. Each time one of the single-phase loads is altered, the vector relationship changes also. No one phase will become overloaded, however, if the transformer bank has been properly sized for the maximum connected load.

9-3 Delta-Wye Connection with Neutral

The circuit shown in *Figure 9-6* is a three-phase transformer connection with a delta-connected primary and wye-connected secondary. The secondary has been center-tapped to form a neutral conductor. This is one of the most common connections used to provide power for single-phase loads.

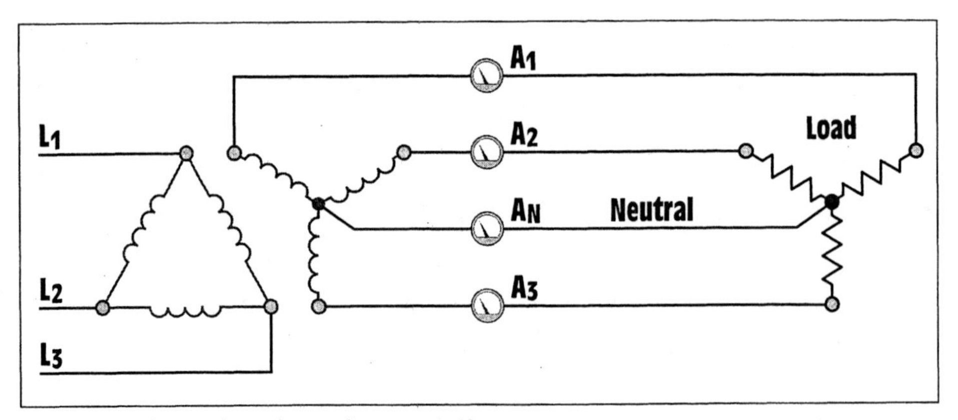

Figure 9-6 Three-phase four-wire connection.

Typical voltages for this type of connection are 208/120 and 480/277. The neutral conductor will carry the vector sum of the unbalanced current. It should be noted, however, in this circuit the sum of the unbalanced current is not the difference between two phases. In the delta connection where one transformer was center tapped to form a neutral conductor, the two lines were 180° out of phase when compared with the center tap. In the wye connection, the lines will be 120° out of phase. When all three lines are carrying the same amount of amperage, the neutral current will be zero.

It should be noted that a wye-connected secondary with center tap can, under the right conditions, experience extreme unbalance problems. **If this transformer connection is powered by a three-phase three-wire system, the primary winding must be connected in a delta configuration.** If the primary is connected as a wye connection, the circuit will become exceedingly unbalanced when load is added to the circuit. Connecting the center tap of the primary to the center tap of the secondary will not solve the unbalance problem if a wye primary is used on a three-wire system.

If the incoming power is a three-phase four-wire system as shown in *Figure 9-7,* however, a wye-connected primary can be used without problem. The neutral conductor connected to the center tap of the primary prevents the unbalance problems. It is a common practice with this type of connection to tie the neutral conductor of both primary and secondary together as shown. When this is done, however, line isolation between the primary and secondary windings is lost.

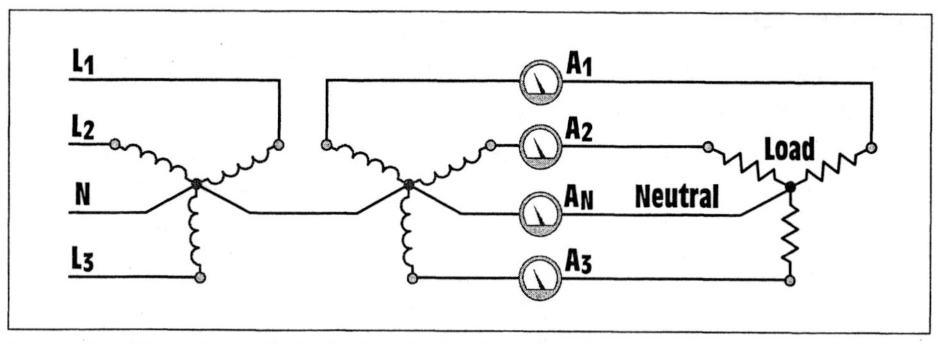

Figure 9-7 Neutral conductor is supplied by the incoming power.

Summary

1. It is common practice to center tap one of the transformers in a delta connection to provide power for single-phase loads. When this is done, the remaining phase connection becomes a high leg.
2. The *National Electrical Code®* requires that a high leg be identified by an orange wire or by tagging.
3. The center connection of a wye is often tapped to provide a neutral conductor for three-phase loads. This produces a three-phase four-wire system. Common voltages produced by this type of connection are 208/120 and 480/277.
4. Transformers should not be connected as a wye-wye unless the incoming power line contains a neutral conductor.

Review Questions

1. How does the *National Electrical Code*® specify that the high leg of a four-wire delta connection be marked?
2. An open-delta three-phase transformer system has one transformer center tapped to provide a neutral for single-phase voltages. If the voltage from line to center tap is 277 volts, what is the high-leg voltage?
3. If a single-phase load is connected across the two line conductors and neutral of the above transformer, and one line has a current of 80 amps and the other line has a current of 68 amps, how much current is flowing in the neutral conductor?
4. A three-phase transformer connection has a delta-connected secondary and one of the transformers has been center tapped to form a neutral conductor. The phase-to-neutral value of the center-tapped secondary winding is 120 volts. If the high leg should become connected to a single-phase load, how much voltage will be applied to that load?
5. A three-phase transformer connection has a delta-connected primary and a wye-connected secondary. The center tap of the wye is used as a neutral conductor. If the line-to-line voltage is 480 volts, what is the voltage between any one phase conductor and the neutral conductor?
6. A three-phase transformer bank has the secondary connected in a wye configuration. The center tap is used as a neutral conductor. If the voltage across any phase conductor and neutral is 120 volts, how much voltage would be applied to a three-phase load connected to the secondary of this transformer bank?
7. A three-phase transformer bank has the primary and secondary windings connected in a wye configuration. The secondary center tap is being used as a neutral to supply single-phase loads. Will connecting the center-tap connection of the secondary to the center-tap connection of the primary permit the secondary voltage to stay in balance when a single-phase load is added to the secondary?
8. Referring to the transformer connection in question 7, if the center tap of the primary is connected to a neutral conductor on the incoming power, will it permit the secondary voltages to be balanced when single-phase loads are added?

10

Transformer Installation

Objectives

After studying this unit, you should be able to

- Determine the correct conductor size for the installation of a transformer in accord with the *National Electrical Code*®
- Determine the proper size of the branch-circuit protective device for transformers of different ratings

This unit will cover the *National Electrical Code*® requirements for the installation of transformers. The two main factors to be discussed will be the selection of the proper size of the branch-circuit protective device, and the selection of the conductor size.

10-1 Transformer Protection

Fuses or circuit breakers can be used as branch-circuit protective devices in circuits supplying power to transformers. In general, fuses are selected for protection of high-voltage circuits because they are less expensive than other types of protection, are extremely reliable, and do not require as much maintenance as circuit breakers. Regardless of the type of protection employed, the ampere ratings will be the same whether using fuses or circuit breakers, and either must be capable of interrupting the maximum available fault current of the circuit. This ability of the fuse or circuit breaker to open under the maximum fault current is expressed as the **interrupting rating** of the device. According to NEC® Section 240-60 (c), fuses must be plainly marked with the interrupting current or ratings if it is other than 10,000 amperes. NEC® Section 240-83 (c) states that circuit breakers must be plainly marked with the interrupting rating if it is other than 5,000 amperes. The National Electrical Code® requires these markings because the minimum interrupting rating of a fuse is 10,000 amperes and the minimum interrupting rating of a circuit breaker is 5,000 amperes.

The minimum interrupting rating permitted for a fuse or circuit breaker in a specific installation is the maximum symmetrical fault current available at the location of the protective device. Power companies will provide this information when requested and will recommend an interrupting rating in excess of this value.

10-2 Determining Interrupting Rating

As stated above, the maximum rating of overcurrent devices for 600 volt, and higher, transformers is set forth in NEC® Table 450-3(a)(1), *Figure 10-1*. To use this table, the percent impedance (%Z) of the transformer must be known. (This value is commonly stamped on the transformer nameplate.) The actual impedance of a transformer is determined by its physical construction, such as the gage of the wire in the winding, the number of turns, type of core material, and magnetic efficiency of the core construction. Percent impedance is an empirical value that can be used to predict transformer performance. It is common practice to use the symbol %Z to represent the percent impedance. Percentages must be converted to decimal form before they can be used in a mathematical formula. When this conversion has been made, the symbol (•Z) will be used to represent the percent

Table 450-3(a)(1).
Transformers Over 600 Volts

	Maximum Rating or Setting for Overcurrent Device				
	Primary		Secondary		
	Over 600 Volts		Over 600 Volts		600 Volts or Below
Transformer Rated Impedance	Circuit Breaker Setting	Fuse Rating	Circuit Breaker Setting	Fuse Rating	Circuit Breaker Setting or Fuse Rating
Not more than 6%	600%	300%	300%	250%	125%
More than 6% and not more than 10%	400%	300%	250%	225%	125%

Figure 10-1 NEC® Table 450-3(a)(1).

impedance in decimal form, called the decimal impedance. The percent value is converted to numeric value by moving the decimal point two places to the left, thus, 5.75% becomes 0.0575. This value has no units as it represents a ratio.

When working with any transformer it is important to keep in mind the full meaning of the terms primary, secondary, high voltage, and low voltage. The primary is the winding connected to a voltage source; the secondary is the winding connected to an electrical load. The source may be connected to either the low-voltage or high-voltage terminals of the transformer. If a person should inadvertently connect a high-voltage source to low-voltage terminals, the transformer would increase the voltage by the ratio of the turns. A 600 V to 200 V transformer would become a 600 V to 1800 V transformer if the connections were reversed. This will not only create a very dangerous situation, but could also result in permanent damage to the transformer because of excessive current flow in the winding. Always be careful when working with transformers and never touch a terminal unless the power source has been disconnected.

The percent impedance is measured by connecting an ammeter across the low-voltage terminals and a variable-voltage source across the high-voltage terminals. This arrangement is shown in *Figure 10-2*. The connection of the

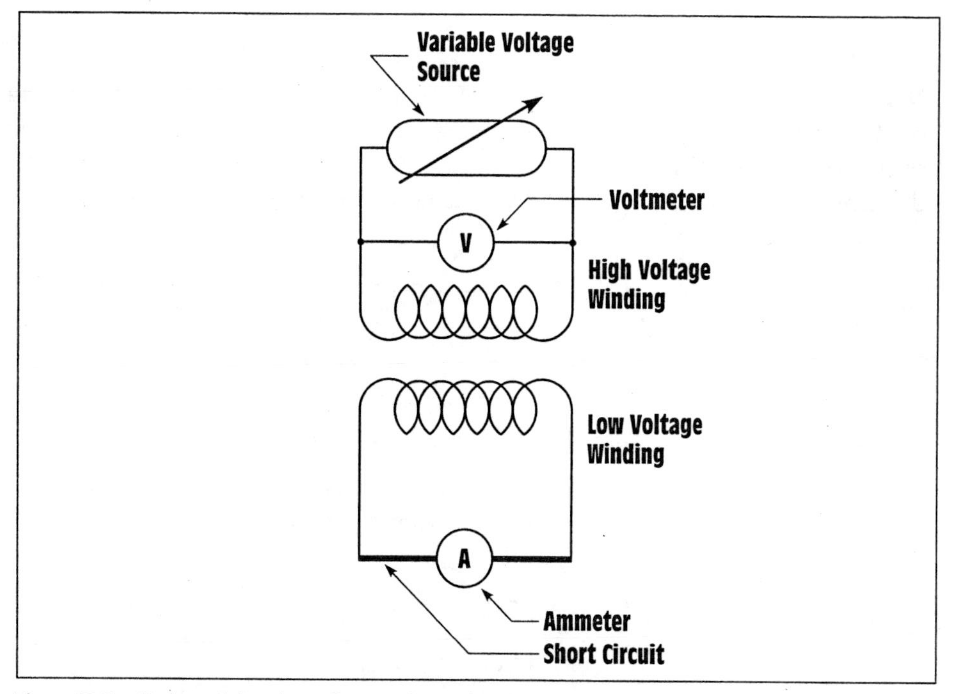

Figure 10-2 Determining transformer impedance.

ammeter is short-circuiting the secondary of the transformer. An ammeter should be chosen that has a scale about twice the range of the value to be measured so the reading will be taken in the middle of the range. If the current to be measured is expected to be about 30 amperes, a meter with a 0–60-amp range would be ideal. Using a meter with a range under 40 amperes or over 100 amperes may not permit an accurate reading.

After the connections have been made the voltage is increased until the ammeter indicates the rated full-load current of the secondary (low-voltage winding). The value of the source voltage is then used to calculate the decimal impedance (•Z). The •Z is found by determining the ratio of the source voltage compared to the rated voltage of the high-voltage winding.

Example 1

Assume the transformer shown in *Figure 10-2* is a 2400/480 volt 15-kVA transformer. To determine impedance of the transformer, first compute the full-load current rating of the secondary winding.

$$I = \frac{VA}{E}$$

$$I = \frac{15{,}000}{480}$$

$$I = 31.25\ A$$

Next, increase the source voltage connected to the high-voltage winding until a current of 31.25 amperes flows in the low-voltage winding. For the purpose of this example assume that voltage value is 138 volts. Finally, determine the ratio of applied voltage as compared to the rated voltage.

$$\bullet Z = \frac{\text{Source Voltage}}{\text{Rated Voltage}}$$

$$\bullet Z = \frac{138}{2400}$$

$$\bullet Z = 0.0575$$

To change the decimal value to %Z, move the decimal point two places to the right and add a % sign. This is the same as multiplying the decimal value by 100.

$$\%Z = 5.75\%$$

Transformer impedance is a major factor in determining the amount of voltage drop a transformer will exhibit between no load and full load and in determining the amount of current flow in a short-circuit condition. When transformer impedance is known, it is possible to calculate the maximum possible short-circuit current. This would be a worse case scenario and the available short-circuit current would decrease as the length of the connecting wires increased the impedance. The formulas shown below can be used to calculate the short-circuit current value when the transformer impedance is known.

$$\text{(Single Phase) } I_{SC} = \frac{VA}{E \times \bullet Z}$$

$$\text{(Three Phase) } I_{SC} = \frac{VA}{E \times \sqrt{3} \times \bullet Z}$$

Since the formula for finding the rated current for a single-phase transformer is

$$I = \frac{VA}{E}$$

and the formula for finding the rated current for a three-phase transformer is

$$I = \frac{VA}{E \times \sqrt{3}}$$

the short-circuit current can be determined by dividing the rated secondary current by the decimal impedance of the transformer.

$$I_{SC} = \frac{I_{Secondary}}{\bullet Z}$$

The short-circuit current for the transformer in the previous example would be

$$I_{SX} = \frac{31.25}{0.0575}$$

$$I_{SC} = 543.5 \text{ amperes}$$

10-3 Determining Transformer Fuse or Circuit Breaker Size

The transformer impedance value is also used to determine the fuse or circuit breaker size for the primary and secondary windings. It will be assumed that the transformer shown in *Figure 10-2* is a step-down transformer and the 2400-volt winding is used as the primary and the 480-volt winding is used as the secondary. NEC® Section 450-3(a)(1) indicates that the size of the protective device for a primary over 600 volts, having an impedance of 6% or less, is 300% of the rated current. The rated current for the primary winding in this example is

$$I = \frac{15{,}000}{2400}$$

$$I = 6.25 \text{ A}$$

The fuse or circuit breaker size will be 6.25 x 3.00 = 18.75 amperes. Exception No. 1 of NEC® Table 450-3(a)(1) permits the next higher rating to be used if the computed value does not correspond to one of the standard fuse sizes listed in NEC® Section 240-6. The next higher standard size is 20 amperes.

NEC® Table 450-3(a)(1) indicates that if the secondary voltage is 600 volts or less, the fuse size will be set at 125% of the rated secondary current. In this example, the fuse size will be

$$31.25 \times 1.25 = 39 \text{ amperes}$$

A 40-ampere fuse will be used as the secondary short-circuit protective device, *Figure 10-3.*

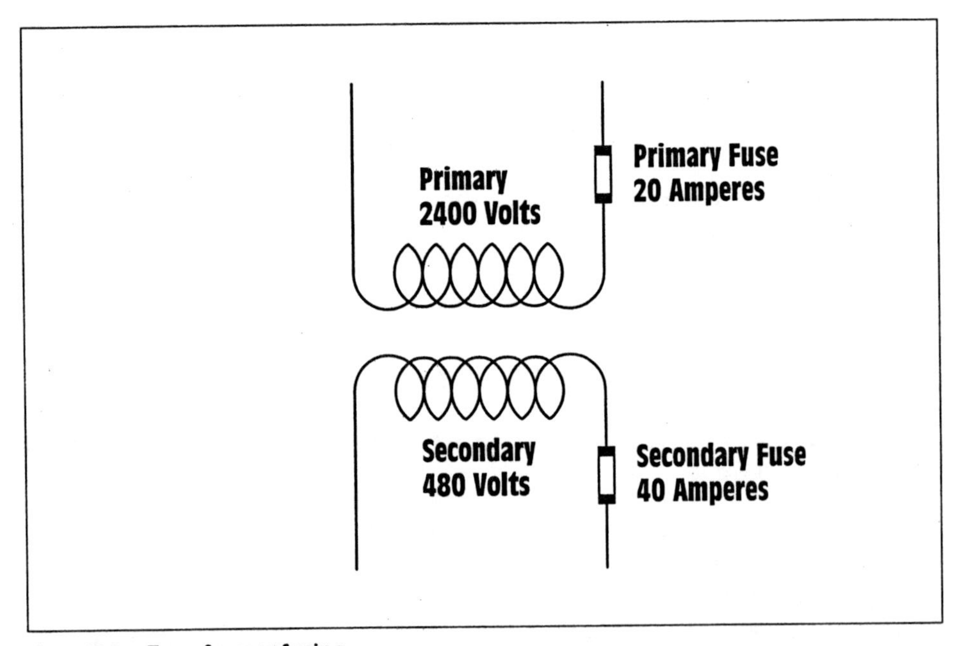

Figure 10-3 Transformer fusing.

Transformers Rated 600 Volts or Less

Fuse protection for transformers rated 600 volts or less is determined by NEC® Sections 450-3(b)(1)&(2). NEC® Section 450-3(b)(1) is concerned with transformers that have primary overcurrent protection only (no secondary protection), and basically states that the primary windings of transformers rated 600 volts or less shall be protected by an overcurrent device rated at not more than 125% of the rated primary current. Exception No. 1 to NEC® Section 450-3(b)(1), however, states that if the primary current is 9 amperes or more, and 125% of this current does not correspond to one of the standard fuse sizes listed in NEC® Section 240-6, *(Figure 10-4)*, that the next higher standard rating can be used.

If the rated primary current is less than 9 amperes, the overcurrent device can be set at *not more than* 167% of that value. If the primary current is less than 2 amperes, the short-circuit protective device can be set at *not more than* 300% of that value.

Notice that if the primary current is 9 amperes or more, it is permissible to increase the fuse size to the next higher standard rating. If the primary current is less than 9 ampere, the next lower fuse size must be used.

NEC® Section 450-3(b)(2) deals with transformers that have short-circuit protection in both the primary and secondary windings. This code basically

240-6. Standard Ampere Ratings.

(a) Fuses and Fixed Trip Circuit Breakers. The standard ampere ratings for fuses and inverse time circuit breakers shall be considered 15, 20, 25, 30, 35, 40, 45, 50, 60, 70, 80, 90, 100, 110, 125, 150, 175, 200, 225, 250, 300, 350, 400, 450, 500, 600, 700, 800, 1000, 1200, 1600, 2000, 2500, 3000, 4000, 5000, and 6000 amperes.

Exception: Additional standard ratings for fuses shall be considered 1, 3, 6, 10, and 601.

(b) Adjustable Trip Circuit Breakers. The rating of an adjustable trip circuit breaker having external means for adjusting the long-time pickup (ampere rating or overload) setting shall be the maximum setting possible.

Exception: Circuit breakers that have removable and sealable covers over the adjusting means, or are located behind bolted equipment enclosure doors, or are located behind locked doors accessible only to qualified personnel, shall be permitted to have ampere ratings equal to the adjusted (set) long-time pickup settings.

(FPN): It is not the intent to prohibit the use of nonstandard ampere ratings for fuses and inverse time circuit breakers.

Figure 10-4 NEC® Section 240-6.

states that if the secondary winding is protected with an overcurrent device that is not rated more than 125% of the rated secondary current, the primary winding does not have to be provided with separate overcurrent protection if the feeder it is connected to is protected with an overcurrent device that is not rated more than 250% of the primary current.

Example 2

Assume a transformer is rated 480/120 volts, and the secondary winding is protected with a fuse that is not greater than 125% of its rated current. Now assume that the rated primary current of this transformer is 8 amperes. If the feeder supplying the primary of the transformer is protected with an overcurrent device rated at 20 amperes or less (8 x 2.50 = 20), the primary of the transformer does not require separate overcurrent protection, *(Figure 10-5)*. NEC® Section 450-3(b)(2) further states that if the transformer is rated at 600 volts or less and has been provided with a thermal-overload device in the

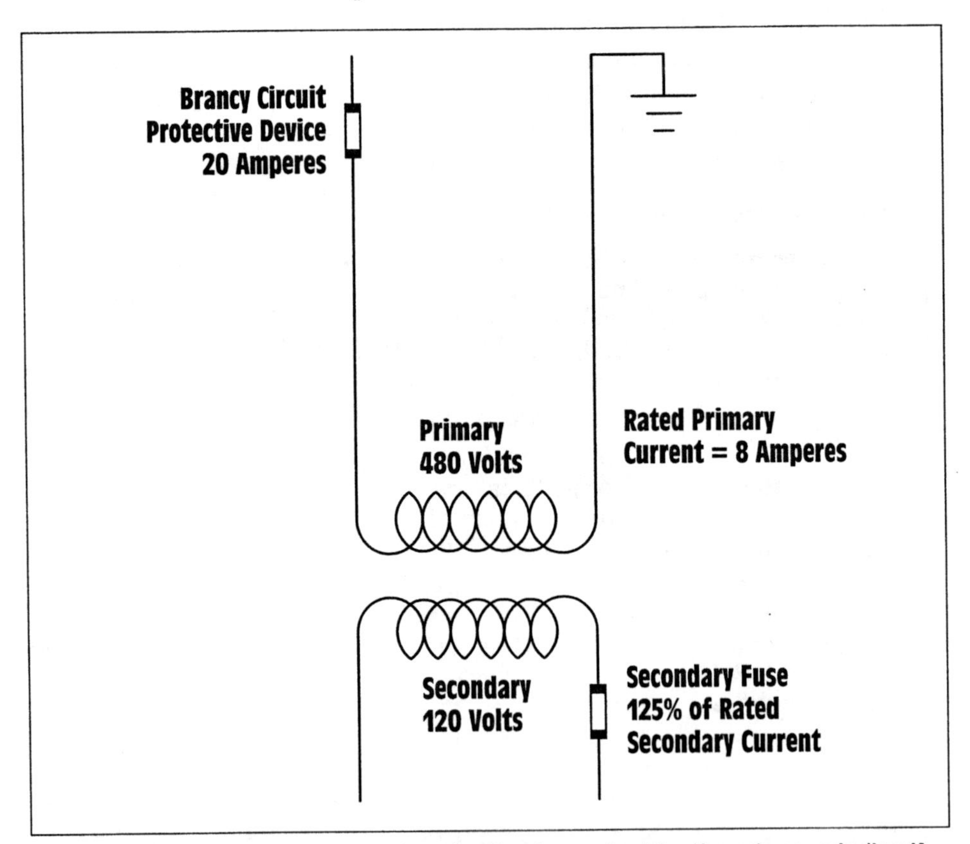

Figure 10-5 No separate overcurrent protection is required for the primary winding if the feeder circuit is protected by an overcurrent device rated not more than 250% of the rated primary current.

primary winding by the manufacturer, no further primary protection is required if the feeder overcurrent-protective device is not greater than 6 times the primary current for a transformer with a rated impedance of not more than 6%, and not more than 4 times the primary rated current for a transformer with a rated impedance greater than 6% but not more than 10%.

Example 3

Assume that a transformer has a primary winding rated at 240 volts and is provided with a thermal-overload device by the manufacturer. Also assume that the primary has a rated current of 3 amperes and an impedance of 4%. To determine if separate overcurrent protection is needed for the primary, multiply the rated primary current by 6 (3 x 6 = 18 amperes). If the branch circuit protective device supplying power to the transformer primary has an overcurrent-protection device rated at 18 amperes or less, no additional protection is required. If the branch circuit overcurrent-protection device is greater than 18 amperes, a separate overcurrent-protection device for the primary will be required.

If separate overcurrent protection is required, it will be computed at 167% of the primary current rating, because the primary current is less than 9 amperes but greater than 2 amperes. In this example, the primary overcurrent-protective device should be rated at

$$3 \times 1.67 = 5.01 \text{ amperes}$$

A 5-ampere fuse should be used to provide primary overcurrent protection.

10-4 Overcurrent Protection for Autotransformers

Section 450-4 (a) of the *National Electrical Code®* states that an autotransformer rated at 600 volts or less must be protected by an overcurrent device in each of the input ungrounded conductors. This section also states that the overcurrent device is to be rated at not more than 125% of the rated full-load input current of the autotransformer. The Code further states that an overcurrent device shall not be installed in series with the shunt winding, *(Figure 10-6)*. The shunt winding is the section of the autotransformer winding that is common to both the primary and secondary.

The Exception to NEC® Section 450-4(a) states that if the input current of an autotransformer is 9 amperes or more, and if 125% of this current does not correspond to one of the standard fuse or circuit breaker sizes listed in NEC® Section 240-6, the next higher size fuse or circuit breaker may be used.

If the input current is rated less than 9 amperes, however, the rating of the fuse or circuit breaker can not be greater than 167% of the full-load input

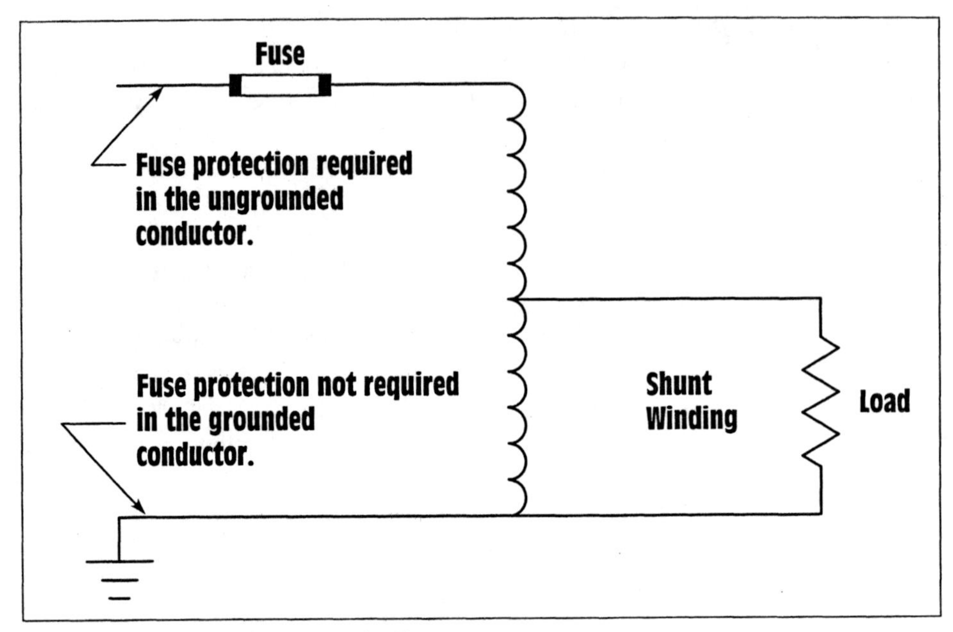

Figure 10-6 Autotransformer protection.

current. Autotransformers with rated voltages over 600 volts are subject to the provisions set forth in Section 450-3(a)(1) of the *National Electrical Code®*.

10-5 Determining Conductor Size for a Transformer

There is more than one method that can be employed to determine the conductor size needed to supply power to a transformer installation. Electrical engineers often determine conductor size by computing the resistivity of the wire when the current, amount of voltage drop, and length of conductor are known. Electricians generally determine the conductor size by referring to the *National Electrical Code®*. This is the reason the conductor size recommended by the manufacturer will often be different than the size determined by using the NEC®. This text uses the *National Electrical Code®* to determine conductor size.

NEC® Section 220-3(a) states that the rating of a branch circuit shall not be less than noncontinuous load plus 125% of the continuous load. Most transformer installations are intended for continuous operation (more than 3 hours). About the only exception is transformers intended for the operation of welding machines. Article 630 of the *National Electrical Code®* should be consulted for special provisions.

Several steps should be followed when selecting a conductor. First determine the amount of current the conductor must carry. This can be done by calculating the rated current of the winding and then increasing this value by 25%.

Next select an insulation type that corresponds to the conditions where the installation is to be made. Factors such as the ambient temperature and whether the conductors will be in a wet or dry location are important. Table 310-13 of the *National Electrical Code*® lists different types of insulation and their characteristics.

Another factor that must be taken into consideration when determining conductor size is the temperature rating of the devices and terminals as specified in NEC® Section 110-14(c). This section states that the conductor be so selected and coordinated as to not exceed the lowest temperature rating of any connected termination, any connected conductor, or any connected device. This means that regardless of the temperature rating of the conductor, the ampacity (current-carrying capacity) must be selected from a column that does not exceed the temperature rating of the termination. The conductors listed in the first column of NEC® Table 310-16, *(Figure 10-7)*, have a temperature rating of 60°C, the conductors in the second column have a rating of 75°C, and the conductors in the third column have a rating of 90°C. Temperature ratings of devices such as circuit breakers, fuses, and terminals are found in UL's (*Underwriters Laboratories*) product directories. Occasionally, the temperature rating may be found on the piece of equipment, but this is the exception and not the rule. As a general rule the temperature rating of most devices will not exceed 75°C.

When the termination temperature rating is not listed or known, NEC® Section 110-14(c) states that for circuits rated at 100 amperes or less, or for #14 AWG through #1 AWG conductors, the ampacity of the wire, regardless of its temperature rating, will be selected from the 60°C column. This does not mean that only those insulation types listed in the 60°C column can be used, but that **ampacity** must be selected from that column. For example, assume that a copper conductor with type XHHW insulation is to be connected to a 50-ampere circuit breaker that does not have a listed temperature rating. According to NEC® Table 310-16, a #8 AWG copper conductor with XHHW insulation is rated to carry 55 amperes of current. Type XHHW insulation is located in the 90°C column, but the temperature rating of the circuit breaker is not known. Therefore, the wire size must be selected from the ampacity ratings in the 60°C column. #8 AWG copper conductor has a current rating of only 40 amperes in the 60°C column. Therefore, a #6 AWG conductor will be used that has a current rating of 55 amperes in the 60°C column.

Table 310-16. Allowable Ampacities of Insulated Conductors Rated 0-2000 Volts, 60° to 90°C (140° to 194°F) Not More Than Three Current-Carrying Conductors in Raceway or Cable or Earth (Directly Buried), Based on Ambient Temperature of 30°C (86°F)

Size	Temperature Rating of Conductor. See Table 310-13.						Size
	60°C (140°F)	75°C (167°F)	90°C (194°F)	60°C (140°F)	75°C (167°F)	90°C (194°F)	
AWG kcmil	TYPES TW†, UF†	TYPES FEPW†, RH†, RHW†, THHW†, THW†, THWN†, XHHW† USE†, ZW†	TYPES TBS, SA SIS, FEP†, FEPB†, MI RHH†, RHW-2, THHN†, THHW†, THW-2†, THWN-2†, USE-2, XHH, XHHW†, XHHW-2, ZW-2	TYPES TW†, UF†	TYPES RH†, RHW†, THHW†, THW†, THWN†, XHHW†, USE†	TYPES TA, TBS, SA, SIS, THHN†, THHW†, THW-2 THWN-2, RHH†, RHW-2, USE-2 XHH, XHHW, XHHW-2, ZW-2	AWG kcmil
	COPPER			ALUMINUM COPPER-CLAD ALUMINUM			
18	....	....	14	....	....	....	....
16	....	....	18	....	....	....	....
14	20†	20†	25†	....	....	....	....
12	25†	25†	30†	20†	20†	25†	12
10	30	35†	40†	25	30†	35†	10
8	40	50	55	30	40	45	8
6	55	65	75	40	50	60	6
4	70	85	95	55	65	75	4
3	85	100	110	65	75	85	3
2	95	115	130	75	90	100	2
1	110	130	150	85	100	115	1
1/0	125	150	170	100	120	135	1/0
2/0	145	175	195	115	135	150	2/0
3/0	165	200	225	130	155	175	3/0
4/0	195	230	260	150	180	205	4/0
250	215	255	290	170	205	230	250
300	240	285	320	190	230	255	300
350	260	310	350	210	250	280	350
400	280	335	380	225	270	305	400
500	320	380	430	260	310	350	500
600	355	420	475	285	340	385	600
700	385	460	520	310	375	420	700
750	400	475	535	320	385	435	750
800	410	490	555	330	395	450	800
900	435	520	585	355	425	480	900
1000	455	545	615	375	445	500	1000
1250	495	590	665	405	485	545	1250
1500	520	625	705	435	520	585	1500
1750	545	650	735	455	545	615	1750
2000	560	665	750	470	560	630	2000

Figure 10-7 NEC® Table 310-16.

If the termination temperature of the device is known, a conductor of that rating or higher may be used. Assume that the 50-ampere circuit breaker in the above example has a known temperature rating of 75°C. The ampacities listed in the 75°C column will be used to select the wire size. A #8 AWG XHHW conductor could now be used in this circuit because NEC® Table 310-16 lists a current rating of 50 amperes for a #8 AWG conductor in the 75°C column.

For circuits rated over 100 amperes, or for conductor sizes larger than #1 AWG, NEC® Section 110-14(c) states that the ampacity ratings listed in the 75°C column may be used to select wire sizes unless a conductor with a 60°C temperature rating has been selected for use. For example, types TW and UF insulation are listed in the 60°C column. If one of these two insulation types has been specified, the wire size must be chosen from the 60°C column regardless of the ampere rating of the circuit.

Example 4

A 25 kVA single-phase transformer has a primary voltage of 480 volts. Copper conductors with type THW insulation are to be used. Determine the conductor size for this installation.

Solution

First determine the full-load rated current of the winding and increase this value by 125%.

$$I = \frac{VA}{E}$$

$$I = \frac{25{,}000}{480}$$

$$I = 52.08 \text{ amperes}$$

$$I_{(Total)} = 52.08 \times 1.25$$

$$I_{(Total)} = 65.1 \text{ amperes}$$

Although type THW insulation is located in the 75°C column, the current is less than 100 amperes. Therefore, in accord with NEC® Section 110-14(c), the conductor must be selected from the 60°C column of NEC® Table 310-16. A #4 AWG conductor will be used.

Summary

1. Fuses or circuit breakers can be used as branch-circuit protective devices.
2. Fuses are generally used for protection in high-voltage circuits.
3. Fuses or circuit breakers must be capable of interrupting the maximum fault current of the circuit.
4. The ability of a fuse or circuit breaker to open under fault current is expressed as the interrupting rating.
5. The interrupting rating of fuses must be plainly marked if it is other than 10,000 amperes.
6. The interrupting rating of circuit breakers must be plainly marked if it is other than 5,000 amperes.
7. The impedance of a transformer is determined by its physical construction.
8. The percent impedance of a transformer can be determined by connecting an ammeter across the low-voltage terminals. Power is applied to the high-voltage terminals until rated current flows in the secondary. The percent impedance is the ratio of the applied voltage compared to the rated voltage.
9. The short-circuit current of a transformer can be determined by dividing the secondary current by the decimal impedance.
10. For transformers rated at 600 volts or less, and having a primary current of 9 amperes or more, the protective device is determined by multiplying the maximum current rating by 125%.
11. For transformers rated at 600 volts or less, and having a primary current less than 9 amperes, the overcurrent-protective device shall be not more than 167% of the rated current of the secondary.
12. When determining the conductor size for a transformer, the 60°C column is generally used for circuits rated less than 100 amperes and the 75°C column is generally used for circuits rated more than 100 amperes.

Review Questions

1. A 20-kVA transformer has secondary voltage of 240 volts, and a listed impedance of 4.2%. What is the short-circuit current?
2. What size overcurrent device should be used to protect the primary of a 480/120 volt 15-kVA transformer.
3. A 150-kVA transformer has a primary voltage of 13,800 volts and a listed impedance of 2.5%. What size fuse should be used to provide primary overcurrent protection?
4. The secondary of the transformer in question 3 has a rated voltage of 2400 volts. What size fuse should be used to provide branch-circuit protection?
5. A 10-kVA transformer has a primary voltage of 208 volts and a secondary voltage of 120 volts. What size circuit breaker should be used to protect the primary winding?
6. A 10-kVA transformer has a primary voltage rating of 240 volts and a secondary voltage rating of 480 volts. The transformer is connected to a circuit protected by a 100 ampere circuit breaker. The secondary is protected by a 25-ampere fuse. Does the primary require separate overcurrent protection?
7. What size conductor should be used to connect the primary of the transformer described in question 6? Assume the conductor is copper with type THW insulation.

11

Transformer Cooling

Objectives

After studying this unit, you should be able to

- Discuss the sources of heat in transformers
- Identify various methods of transformer cooling
- Identify different types of transformer cooling methods
- Describe external devices that aid in transformer cooling

Transformers are very efficient machines with 90% to 99% efficiency. The 1% to 10% losses are mostly from I^2R loss, eddy current loss, and hysteresis loss. These losses are primarily manifested as heat gain in the transformer. If heat is not removed from the transformer it will build up to extremely high temperature and eventually destroy the insulation in the transformer.

11-1 Air-Cooled Transformers

A control transformer is a transformer that supplies voltage to energize motor starters and similar electrical equipment *(Figure 11-1)*. Many small control transformers are shell-type transformers. This design places some of the core material outside the windings, exposing the core material to free air and allows heat to be radiated away from the transformer. The sizes of these transformers limit the heat produced to a few watts, so heat is not a major

Figure 11-1 Control transformer with fuse protection added to the secondary winding. (Courtesy: Hevi-Duty Electric Co.)

problem. For example, a typical 500-VA single-phase control transformer with 95% efficiency would produce only about 25 VA (watts) of heat at maximum rated load. This amount of heat is radiated from the transformer without the need for special cooling provisions.

Larger Control Transformers Supplying a Motor Control Center

Larger control transformers *(Figure 11-2)* feed controls for many starters in a motor control center and might also furnish lighting and convenience outlets for the control-center area. These transformers are usually placed inside the MCC (Motor Control Center) cubicle and have no external case. The core and windings are exposed to free air that is allowed to circulate by convection around the transformer. These cubicles have ventilation openings in the area near the transformer allowing the heated air to escape by convection and ambient air to be admitted. These transformers must be inspected periodically for accumulations of dust and dirt that would impede the flow of cooling air. A vacuum cleaner is used to remove dust and dirt from the **deenergized** transformer. Make sure the ventilation openings in the cubicle are not blocked with dust or debris.

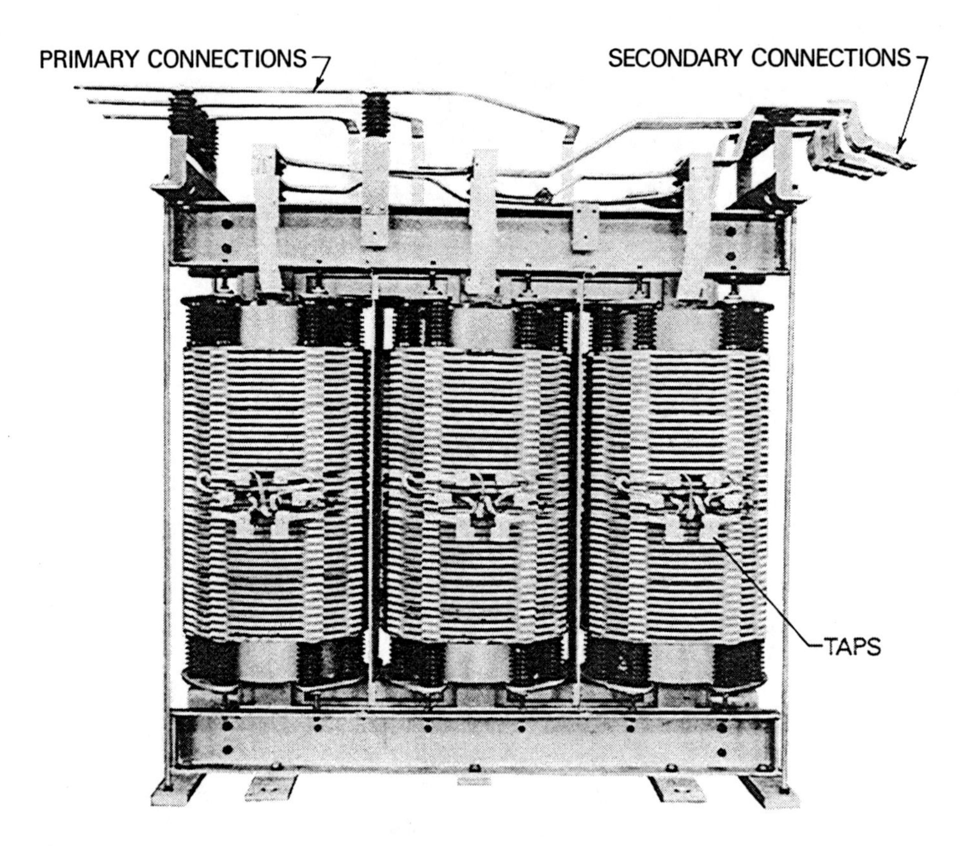

Figure 11-2 Dry-type transformer used in a motor control center.

Small Dry-Type Power Transformers for Commercial and Industrial Use

Small industrial and commercial dry-type transformers are used to provide power for lighting and convenience outlets and other small loads. They are typically installed in utility rooms and are enclosed in a case with ventilation openings. Transformers are subject to accumulations of dust and debris, and must be inspected and cleaned periodically. One hazard with this type of transformer is that because they are usually installed in utility rooms they often have material stored on or around them, blocking the ventilation. Sizes from 25 kVA to 100 kVA are typical.

Large Industrial Dry-Type Transformers for Use in Unit Substations

Large industrial dry-type transformers are used in unit substations *(Figure 11-3)* to reduce distribution voltages, typically 13,800 volts, to voltages used

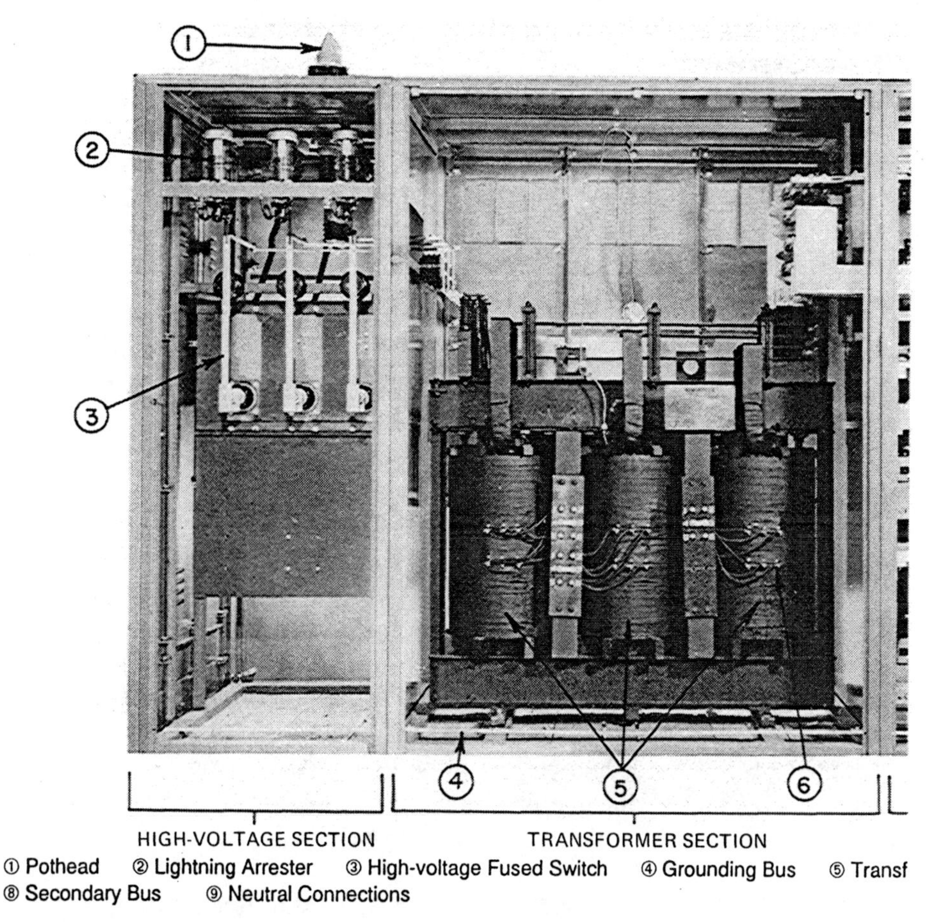

Figure 11-3 A unit substation.

in the industrial unit. These transformers are installed in large cubicles with many square feet of ventilation openings at the top and bottom of the cubicle. Dust and dirt are a problem with these installations, depending on the environment in which the substation is installed. Because of their large size and loads they produce more heat, necessitating more ventilating air circulated through the cubicle, and bringing with it more dirt. Although air filters are often installed to help prevent the entrance of dust and dirt, frequent inspection and cleaning are usually indicated for these transformers. Sizes from 100 kVA through 1 MVA are typical.

Convection and Fan Cooling for Dry-Type Transformers

While most dry-type air-cooled transformers are cooled by natural convection, it is sometimes necessary to design a transformer with forced-air fans for additional cooling in order to reach maximum load rating. In cases where fans are used for cooling of dry-type transformers, it is common practice to install filters in the air inlets to the fans.

11-2 Liquid-Cooled Transformers

Some transformers require increased cooling provided by a material that will absorb heat from the windings and core material. This cooling medium then carries the heat to the outer case or to special cooling devices. Cooling media can be circulated by natural convection or by pumps. In some cases special gasses are used as the cooling medium. More often, a special transformer oil is used that has dual functions of providing insulation for the energized transformer parts, and cooling the parts. The oil must have high dielectric strength and be capable of absorbing heat well. One common oil used for this purpose is known as **Askarel**. A special type of Askarel oil called polychlorinated biphenyl (PCB), is a highly toxic substance now banned in the United States. Before working with older transformers, they must be tested to determine whether the cooling and insulating medium is PCB. If PCB is found, it must be removed and disposed of in a manner approved by the Environmental Protection Agency (EPA). The transformer must then be cleaned and retested for the presence of trace amounts of PCB.

Pole-Mounted Distribution Transformers for Residential and Commercial Use

Transformers mounted on poles and used to supply power to residential and small commercial locations, are filled with special oil. This oil provides insulation between the metal parts of the transformer and the energized windings. It also carries heat away from the windings and core to the housing (tank) of the unit, where it is radiated to the surrounding air. As the oil is heated by the core and windings it becomes less dense and rises in the tank. The oil nearest the outside gives up some of its heat to the housing and outside air and is cooled *(Figure 11-4)*. The cooler oil is more dense and falls to the bottom of the tank. In this way the oil circulates within the tank by natural convection, moving heat away from the heat-producing parts.

Fins or Tubes on Pole-Mounted Transformers

Some larger pole-mounted transformers have fins welded to the outside of the tank to allow more surface area for the transfer of heat to the surround-

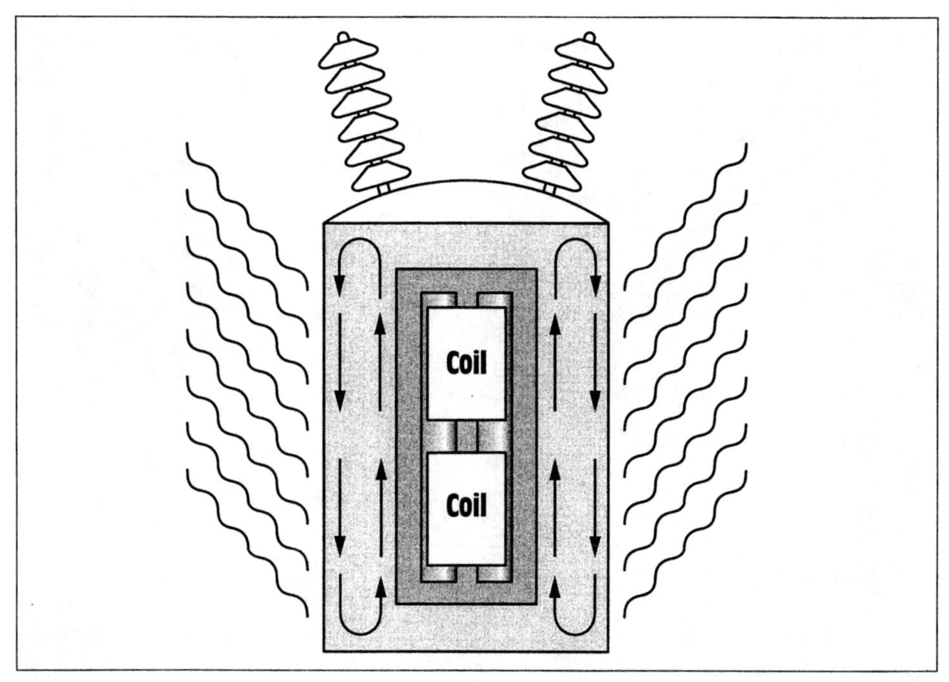

Figure 11-4 Oil circulates inside transformer. Heat is radiated to the air by the case.

ing air *(Figure 11-5)*. Other transformers have hollow tubes connected to the tank at the top and bottom. Oil circulates through these tubes by natural convection and is cooled by air around the tubes *(Figure 11-6)*. Other designs employ hollow plates that increase the surface area and permit a greater degree of cooling *(Figure 11-7)*.

Pad-Mounted Oil-Cooled Transformers

Oil-cooled pad-mounted power transformers range in size from a few hundred kVAs to several hundred MVAs *(Figure 11-8)*. The design and construction of the transformer and its housing permit the insulating and cooling oil to flow around the heat-producing core and windings, removing heat. The heated oil rises by convection and loses some of its heat to the housing of the transformer. The cooled oil goes to the bottom and displaces the hot oil. This circulation moves the heat generated in the transformer to the outside air.

Cooling Fins or Radiators on Pad-Mounted Transformers

As in other transformers some pad-mounted transformers have cooling fins welded to the outside of the housing to aid in the dissipation of heat.

Figure 11-5 Fins increase the surface area and promote cooling.

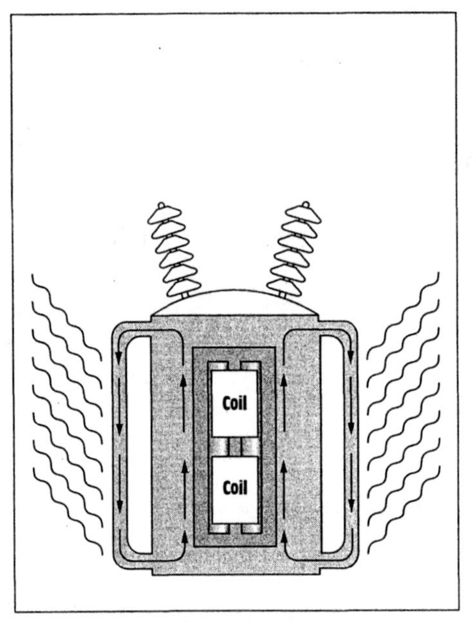

Figure 11-6A & B Oil circulates through hollow tubes to increase cooling.

Figure 11-7 Flat plates increase surface area and permit a faster rate of cooling.

Figure 11-8 Pad mount transformer.

Figure 11-9A & B Pad mount transformers with radiators.

Others have radiators mounted to the outside of the housing through which the heated oil flows by convection *(Figure 11-9)*. Radiators are constructed to give the maximum possible surface area to transfer heat to the surrounding air. Some larger transformers also include thermostatically controlled fans that circulate air through the radiators to further assist in heat removal. The maximum power ratings of large transformers equipped with radiators, fans, or other heat-removing equipment is based on the proper functioning of the heat-removing devices.

Water Cooling of Transformer Oil

Some large transformers, especially those enclosed in vaults where air circulation is not practical, may have their cooling oil circulated by pumps through water-cooled heat exchangers *(Figure 11-10)*. Heat removed from the transformer by oil is transferred to the water to be removed by a cooling tower or other means. It is very important in this type of cooling that internal pressure of the transformer and the oil be maintained at a higher pressure than the water pressure in the heat exchanger to prevent water entering the transformer in case of a leak. If water is allowed to enter the transformer it could destroy the dielectric strength of the insulating liquid, and cause a disastrous short circuit between energized parts.

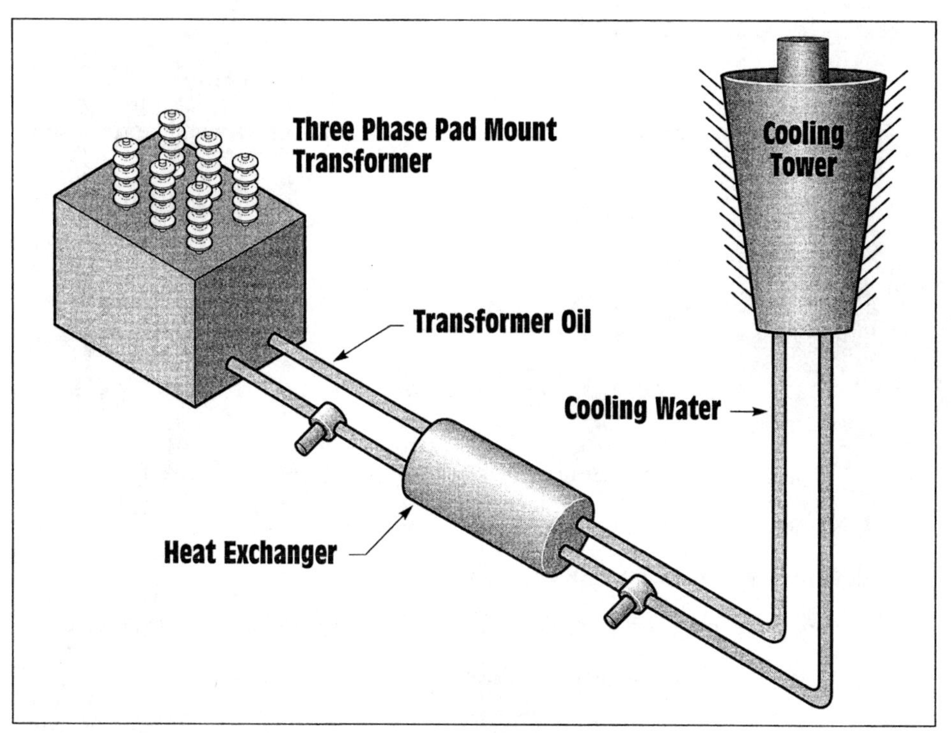

Figure 11-10 Water is used to cool the transformer oil.

Summary

1. Transformers develop heat from I^2R loss, eddy current loss, and hysteresis loss.
2. If heat is not removed from the transformer it will destroy the insulation in the transformer.
3. Small transformers depend on natural convention of the surrounding air to remove heat.
4. Larger dry-heat transformers sometimes use fans to circulate air through the transformer to assist cooling.
5. Oil-filled transformer use special insulating oils to insulate and cool the transformer.
6. Some oil-filled transformers have external radiators through which the oil circulates by convection or pumps.
7. Some oil-filled transformers use external water-cooled heat exchangers to remove heat from the cooling medium.

Review Questions

1. What are the three main sources of heat in a transformer?
2. What is one advantage of shell-type construction in small control transformers?
3. How are larger control transformers supplying a motor control center usually cooled?
4. What is a way to increase cooling on a large dry-type transformer used in a unit substation?
5. What is one special problem with dry-type transformers used in unit substations?
6. What are two types of cooling medium used in transformers?
7. Name two functions of transformer oil.
8. What type of transformer is banned in the USA?
9. Name two ways of increasing the cooling of oil-filled transformers.
10. What hazard is involved in water-cooled transformers?

12

Transformer Maintenance

Objectives

After studying this unit, you should be able to

- List safety procedures used when maintaining transformers
- Discuss the necessity for regular preventive maintenance for transformers
- Describe the procedures to be used for inspection, maintenance, cleaning, and testing of various types of transformers
- Perform maintenance procedures on transformers

Transformers are usually thought of as stationary objects with no moving parts. Because of this misconception, transformers are often neglected and left out of routine preventive-maintenance schedules. This could prove to be a very expensive omission. Transformers must be inspected and maintained on a regular schedule to get maximum performance and life from them. This applies to all transformers, no matter how large or small they are. Environmental conditions such as changing temperatures caused by varying loads and changing ambient temperature affect the operation and life of the transformer. Dust, moisture, and corrosive chemicals in the air surrounding the transformer will greatly affect its operation and life. The type of maintenance procedures and intervals between procedures are governed by the type, size, location, and application of the transformer.

12-1 Safety Procedures

As with any electrical equipment, the primary consideration when working on or near transformers must be the safety of personnel. Before working on any transformer establish whether it is energized and whether the work can be done safely with power on the transformer. Most maintenance procedures will require that power be disconnected and locked or tagged out. On larger transformers with high-voltage connections it is usually advisable to prepare a written switching procedure detailing each step of the process of deenergizing the equipment. By following a written procedure and initialing each step as it is taken, errors in switching can be avoided.

In many larger installations, grounds are placed on each side of the transformer after it is deenergized to protect workers. If these grounds are not removed before the transformer is energized the windings could be severely damaged. A written switching procedure will include the placement and removal of these grounding connections. This will help avoid energizing a transformer with the grounds still in place.

After the power to the transformer has been disconnected, and before doing any work, it is advisable to test all exposed connections for voltage. Use proper test instruments with a voltage rating at least as high as the voltage rating of the connection. This is especially important when there is more than one source of power, as in a double-ended substation *(Figure 12-1)*. Double-ended substations permit power to be supplied from another source in the event of equipment failure. Although the circuit kVA capacity is reduced, power can be maintained until the defective equipment is repaired or replaced. This can, however, cause a backfeed to the secondary side of a transformer that has the primary disconnected. Extreme care must be taken when working with double-ended systems to ensure that power is not being applied to either the primary or secondary windings.

12-2 Entering a Transformer Tank

In some of the maintenance procedures, it is necessary to enter a transformer tank. When this is part of the maintenance procedure, the atmosphere in the tank must be tested for the presence of combustible and/or toxic gases, and also for the presence of sufficient oxygen. Oxygen is normally present in the atmosphere at about 21.2%. If transformer tank concentration is less than 20% it could be a health threat to the worker. If there

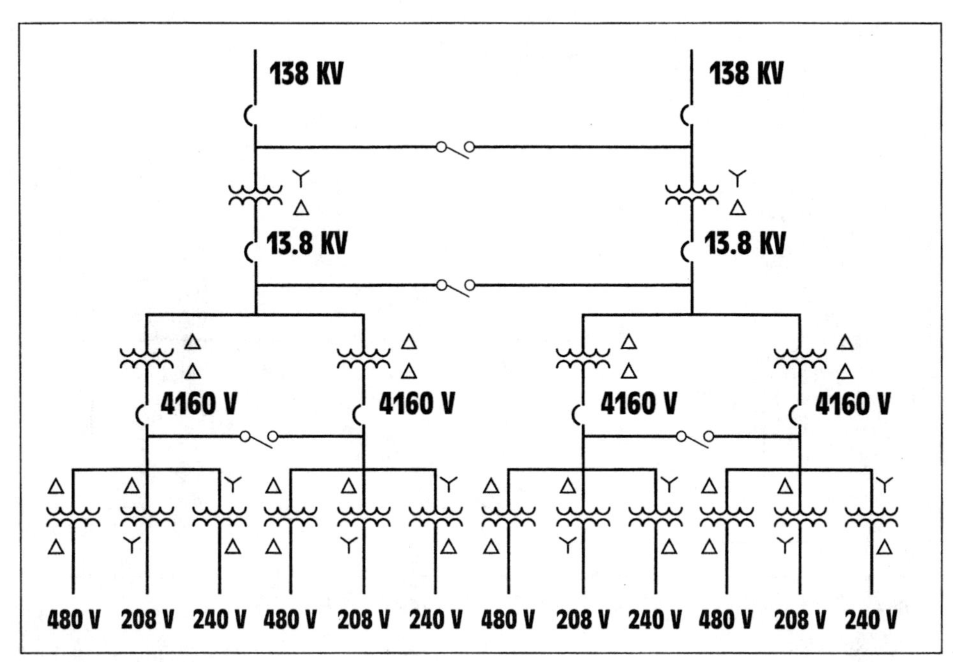

Figure 12-1 Double-ended substations.

are dangerous gases present, or if there is insufficient oxygen in the tank, fresh air ventilation is necessary until safe conditions are met. When anyone is inside the tank, there must be a person outside the entrance to observe the worker in the tank and be alert for any problems.

12-3 Maintenance of Small Control Transformers

Inspection

The first step in any preventive-maintenance procedure is inspection of equipment. Transformer inspection will reveal the presence of rust, corrosion, dirt or dust buildup that should be noted at this time.

Cleaning

The outside of the transformer should be cleaned with an approved solvent or cleaner and rust and corrosion should be removed and the housing painted if necessary.

Tightening

All connections and mounting bolts should be tightened. Any corroded connections should be replaced.

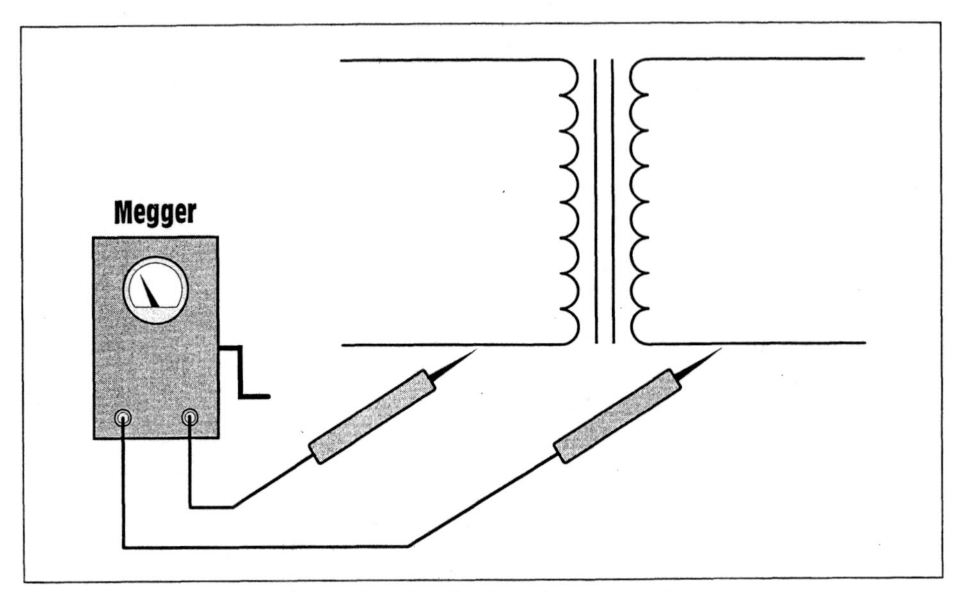

Figure 12-2 A megger is used to test for shorts between windings.

Testing

Small transformers should be tested annually for short circuits and grounds. A megger test between the primary and secondary windings will test the insulation between windings *(Figure 12-2)*, and from each winding to the housing or core will show any insulation weakness in this area *(Figure 12-3)*. Use a megger with voltage ratings close to transformer-winding rating, for example a 500-volt megger would be used to test the insulation on a transformer with a 480-volt rated winding.

After testing the windings with a megger, each winding should be tested for continuity with an ohmmeter *(Figure 12-4)* by connecting the ohmmeter leads across the terminals or leads connected to the ends of each winding. This test will determine if any of the windings are open, but will probably not determine if the windings are shorted. In some instances, the insulation of the wire breaks down and permits the turns to short together. When this occurs, it has the effect of reducing the number of turns for that winding. If these shorted windings do not make contact with the case or core of the transformer, a megger test will not reveal the problem. This type of problem is generally found by connecting the transformer to power and measuring current and voltage values. Excessive current draw or a large deviation from the rated voltage of a winding are good indicators of a shorted winding. When making this test be aware that it is not uncommon for the secondary-

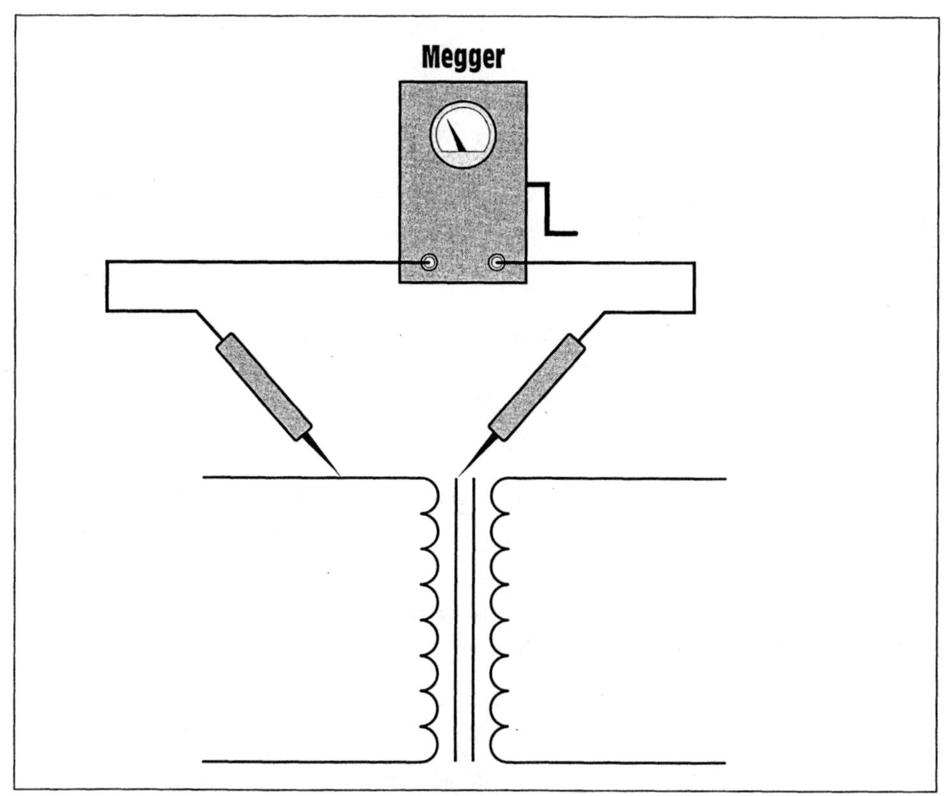

Figure 12-3 Testing for grounds with a megger.

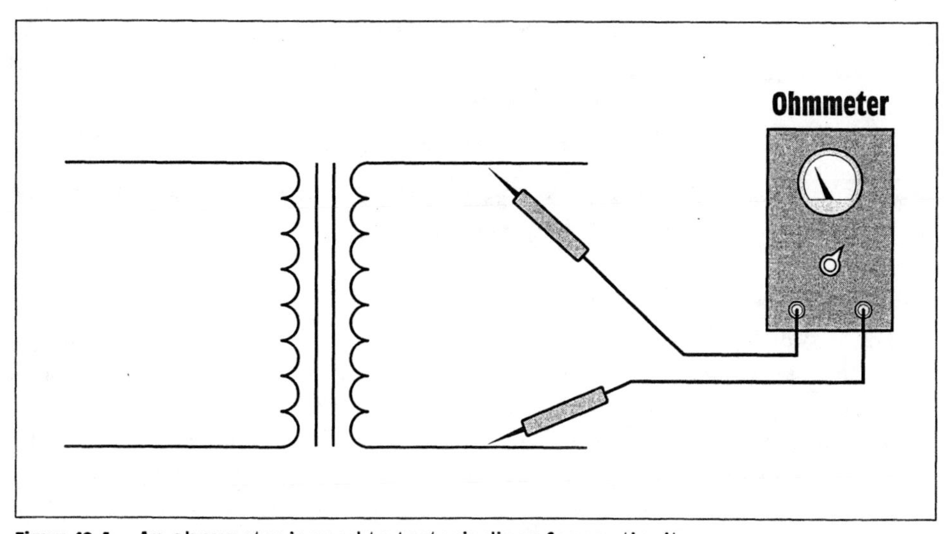

Figure 12-4 An ohmmeter is used to test windings for continuity.

voltage to be higher than the rated value under a no-load condition. Voltage ratings are listed for full load, not no load. It would not be uncommon for a transformer winding rated at 24 volts to measure 28 or 30 volts with no load connected.

12-4 Large Control Transformers Supplying a Motor Control Center

Inspection

Check the housing and windings for dust and dirt accumulation. Check the windings and connecting wires for insulation deterioration, dryness, or evidence of charring. Examine the connections for corrosion or looseness.

Cleaning

Because these transformers are usually located near the floor in the cabinet of the motor control center they are subject to dust and dirt accumulations. Use a vacuum cleaner to remove loose dust from the **deenergized** transformer. Make sure all air passages are open and free of debris Wipe any dirt off with a clean dry cloth. Clean dirt and debris from ventilation openings in the cabinet.

Tightening

Tighten all mounting bolts and electrical connections. Any corroded or burned connections must be replaced.

Testing

As with smaller control transformers, these units should be tested annually with a megger to determine the condition of the insulation. Results should be logged for comparison with previous tests in order to detect any trends that might indicate insulation deterioration.

12-5 Small Dry-Type Transformers for Commercial and Industrial Use

Inspection

These transformers are enclosed in a housing with ventilation openings. They are typically installed in locations that are shared with custodial or other supplies and often have extraneous material stacked around and on them. Inspection requires removing surrounding material and opening the inspection covers. Examine housing and core material for rust or corrosion. Check windings and wiring for dry, cracked insulation. Inspect connections for corrosion or looseness.

Cleaning

Remove all debris from around the transformer housing. Make sure the ventilation openings are not blocked. Use a vacuum cleaner to remove loose dust and dirt from the windings, housing, and area around the transformer. Use a clean, dry cloth to wipe away any remaining dirt. If the housing is rusty or corroded, or if the paint is in poor condition, the housing should be repainted.

Testing

Annual testing should include a megger insulation test between the primary and secondary windings and between the windings and the core material or ground. Record the readings for future reference and comparison.

12-6 Large Industrial Dry-Type Transformers for Use in Unit Substations

Special Safety Precautions

These transformers are frequently installed in double-ended substations and therefore have more than one source of power. Opening the source of power to the primary windings may not deenergize the transformer if the secondary is connected to another transformer secondary through a tie breaker or switch (see Figure 12-1). It is possible for voltage to be stepped up from secondary to primary voltage in this configuration. Be sure all connections to primary and secondary windings are locked or tagged open and checked for voltage before working on any transformer. Be aware that there might be dangerous voltages present in the cubicle for control circuits or space heaters that might not be deenergized when the transformer is deenergized.

Inspection

If possible, remove enough panels from the cubicle enclosing the transformer to allow both sides to be seen and inspected *(Figure 11-3)*. Check for dirt and debris on windings and in air passages. Inspect the support structure for rust and corrosion. Check all coils for distortion of windings and evidence of overheating. Make sure all spacers are in place. Check all bolts in structural members and electrical connections for looseness. Inspect electrical connections and buswork for corrosion or evidence of overheating. If equipment is available, an infrared thermographic inspection of the transformer, connections, and busses, or wiring under normal load conditions, will indicate any trouble spots.

Cleaning

Cleaning schedules for substation transformers will vary according to the environment in which the transformer is installed. Most installations will require annual inspection and cleaning. In dusty, dirty, or corrosive environments the maintenance schedule will be determined by experience. Monthly inspections when the transformer is first installed will indicate how often the unit will have to be taken out of service for maintenance. Cleaning of transformer coils and core and the structure and surrounding areas is done with a vacuum cleaner. Use clean, dry cloths to wipe off any dirt left by the vacuum cleaner. Make sure all air passages are open and clean the ventilation openings in the cubicle panels in the vicinity of the transformer.

Tightening

Tighten all bolts in the support structure. Make sure all electrical connections are tight. If any connections are corroded or show evidence of overheating, they must be cleaned and reconnected. All bus connections must be tightened to the manufacturer's recommended torque.

Testing

Annual testing with a megger will indicate the general condition of the insulation. On transformers with higher voltages special testing equipment and procedures might be required. Follow the manufacturer's recommendations for these transformers. Check the operation of any auxiliary cooling fans and associated temperature switches. Clean or replace ventilation air filters if used. Check calibration of any winding temperature gauges, switches, or alarms. Check operation of space-heating equipment and controls.

12-7 Media-Cooled Transformers

Pole-Mounted Distribution Transformers

Pole-mounted distribution transformers are usually oil filled and require little maintenance. They should be inspected annually for leaks, rust, and corrosion of the housing. If equipped with cooling fins or radiators make sure they are clean and that air flow is not blocked by debris. Inspect the condition of the paint, and repaint if necessary. Inspect bushings for dirt and damage, clean or replace if necessary. Tighten all connections, making sure all mounting bolts are tight and that any arms or brackets are in good condition. Test the condition of insulation with a megger and/or high-voltage test equipment.

12-8 Pad-Mounted Oil-Cooled Transformers

External Inspection

The first step in maintaining transformers is a thorough external inspection. Look for evidence of leaks in the housing or cooling radiators. Inspect the housing for rust, corrosion, or damage and note the general condition of the paint. Inspect bushings for cracks or chips. Look for loose, corroded, or discolored connections. Inspect the housing ground connection to make sure it is tight and corrosion free. Check external gages that indicate temperature and level of the cooling oil, and internal pressure *(Figure 12-5).*

Figure 12-5 External gages indicate level and temperature of the transformer oil and internal pressure of the transformer housing.

Cooling Equipment

If the transformer is equipped with auxiliary cooling equipment it should be checked for proper operation. Check radiator connections to the tank for leaks and make necessary repairs. Cooling fans should be operated manually to be sure they work. Temperature and pressure switches and gauges should be removed and calibrated yearly to assure proper operation. Transformers with a gas blanket (usually dry nitrogen) over the oil, should have the gas pressure checked at least once a week.

Transformer Protective Relaying

Transformers that have a gas blanket on top of the insulating oil have pressure switches that actuate an alarm system if the gas pressure on the blanket drops below a certain point. These switches should be tested frequently along with any temperature or pressure alarm devices on the transformer windings or tank. Protective relaying usually includes overcurrent relays, sudden pressure relays, reverse current relays, and winding and oil over-temperature relays of various types. These devices should be tested and calibrated by qualified technicians on a regularly scheduled basis, but at least once a year.

12-9 Internal Inspection and Maintenance

On larger transformers it will be necessary to open manholes or inspection covers to determine the condition of the windings, connections, and other parts inside the housing. Before removing any covers it is advisable to have new gaskets available for replacement when reclosing. Relieve any internal pressure in the transformer before loosening flange bolts. It is very important that no tools or equipment be left inside the housing. Inventory all tools, parts, and equipment brought to the work area before opening the transformer and before closing it. Anything left in the transformer could cause a short circuit or interfere with normal circulation of the cooling medium and destroy the transformer. Make sure all safety precautions are followed and atmosphere is tested before entering the transformer.

Look for loose, corroded, or discolored connections, distorted or damaged windings, and broken or missing spacers between windings. Check the general condition of the insulation for deterioration. Clean and tighten connections where necessary. Be sure to follow manufacturer's recommendations for torque when tightening connections. Check and tighten mounting bolts. Look for deposits of sludge on windings, core material or other structures. Sludge deposits indicate contamination of the oil and will reduce the dielectric strength of the insulation. Sludge can also act as thermal insulation and decrease the transfer of heat from the internal parts to the cooling oil. While inspecting the internal parts of the transformer it is good practice to observe any evidence of rust on the inside of the housing or covers. This might indicate condensation on these parts which could be caused by a leaky gasket that allows ambient air to be admitted into the housing.

12-10 Insulation Testing

As with any other transformer the dielectric strength of the insulation must be tested at least once a year. Megger testing can be done on the lower-voltage transformers. Hand crank and battery-operated meggers are shown in *Figure 12-6.* Special high-voltage equipment is necessary to test insulation on higher-voltage units. A high-voltage tester, generally referred to as a "HiPot" is shown in *Figure 12-7.* This unit develops a high voltage and measures any current leakage caused by weak or defective insulation.

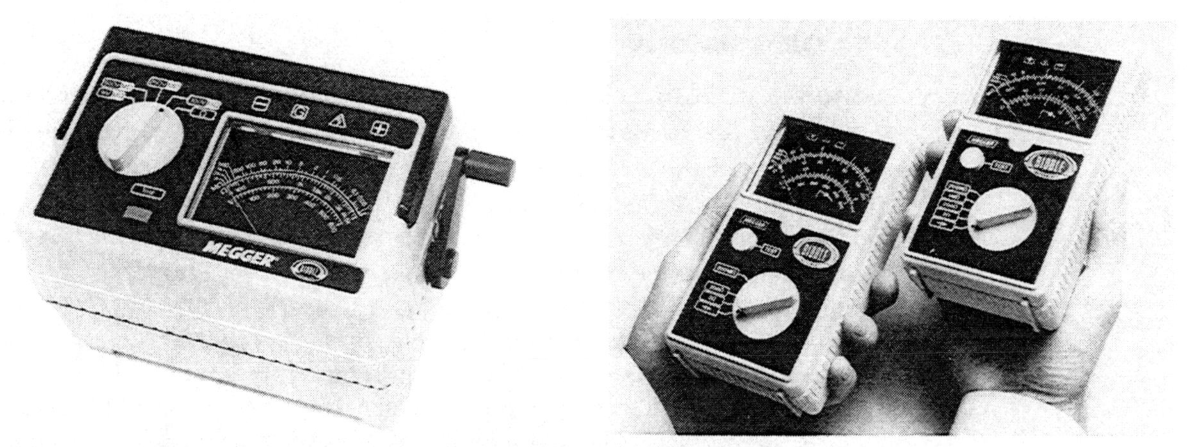

Figure 12-6 Meggers used to test transformer windings. (Courtesy of Biddle Instruments)

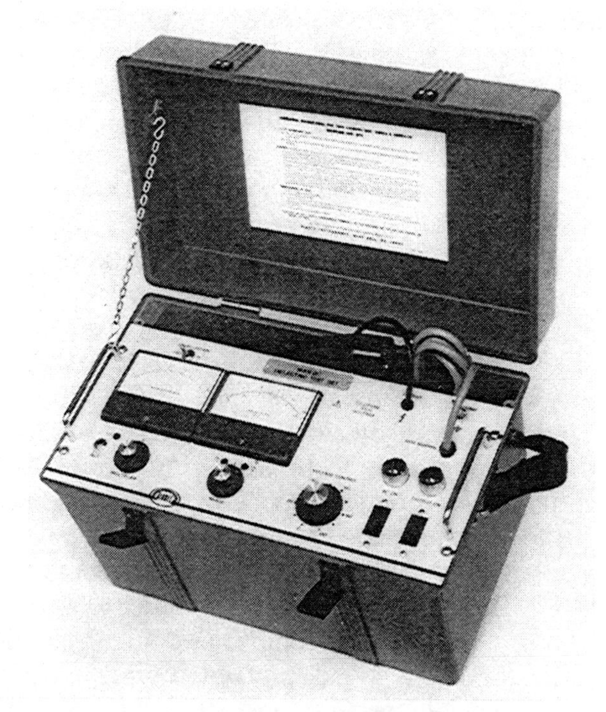

Figure 12-7 High-voltage tester. (Courtesy of Biddle Instruments)

For voltages above 13,800 volts, it is usually advisable to contract out high-voltage insulation tests to a company that specializes in this type of testing and has trained technicians and the proper equipment available. As with any insulation testing a record should be kept of test results to establish trends in insulation dielectric strength.

12-11 Oil Testing

Transformer oil testing should be conducted at least once a year and more frequently in cases of frequent overloads or if there is a history of marginal oil-test results. Oil samples should be in clean, dry containers labeled with the identity of the transformer. After the sample is drawn, it should be allowed to stand for a while to permit any free water to settle to the bottom of the sample. Glass containers make it easier to see any free water in the sample.

Testing for dielectric strength is done in a special device that has a cup for the sample and electrodes placed 0.1 inch apart. Thoroughly clean and dry the sample cup and then rinse it with a portion of the sample. Fill the cup and allow it to settle for at least three minutes to eliminate air bubbles. Turn the device on and gradually increase the voltage until it arcs across the sample. Record the voltage and repeat the test five times on each of three samples from each transformer. Calculate the average of the fifteen tests done in this manner to get the representative dielectric strength of the oil. An average dielectric strength of 26 kV to 29 kV is considered usable, 29 kV to 30 kV is good, under 26 kV is poor and the oil should be replaced or filtered to increase dielectric strength. Special equipment is required to filter transformer oil, and the transformer must be deenergized. This process is usually contracted to companies specializing in transformer maintenance.

Other tests conducted on transformer oil include water content, gas content, and color. A water content of less than 25 parts-per-million is usually acceptable for units operating at voltages up to 228 kV. Excess water can come from condensation or leaks in the housing or cooling system, and will reduce the dielectric strength of the insulation and oil. Filtering is necessary to remove excess water from the oil.

Arcing or overheating can cause combustible gases such as acetylene, hydrogen, methane, and ethane to be formed in the oil. The presence of these gases can be detected only by specialized test equipment and should be done by qualified technicians. Samples should be sent to laboratories specializing in this type of testing. Most transformer consulting firms prefer to have their technicians collect the samples in order to assure uniform sampling procedures. In most cases the companies doing this testing will submit a report listing the conditions found, probable causes, suggested remedies, and suggested frequency of retesting.

Summary

1. Regular preventive maintenance is vital to the performance and life of any transformer.
2. Environmental conditions such as dust, moisture, temperature, and corrosive atmospheres affect the life of a transformer.
3. Personnel safety is the primary consideration when working on or around a transformer.
4. Written switching sheets are advisable when working on substation transformers.
5. Before entering a transformer tank the atmosphere must be tested.
6. The basic steps in transformer maintenance are inspection, cleaning, repair, and testing.
7. Accurate records of test results will indicate any trends in transformer condition.
8. Protective relaying and instrumentation should be included in any transformer maintenance schedule.
9. Dielectric testing of transformer insulation and testing of transformer oil should be done by qualified technicians or companies specializing in this type of testing.

Review Questions

1. What is the primary consideration for personnel performing transformer maintenance?
2. What are the basic steps of transformer maintenance.
3. Why is it important to keep accurate records of inspections and tests?
4. What environmental conditions affect the life of a transformer?
5. What factors govern the type of maintenance procedures and the maintenance intervals for transformers?
6. What tests must be made before entering a transformer tank?
7. What devices and procedures should be used to clean a dry-type transformer?
8. What is a special hazard encountered in large industrial transformers used in unit substations?
9. List some of the protective relays found with large transformers.
10. Before opening manholes on large transformers, what material should be available?
11. Why is it important to inventory tools and equipment before and after working inside a transformer?
12. What effect will sludge deposits have on the internal parts of a transformer?
13. What voltages indicate good oil in the standard dielectric test of oil?
14. What amount of water in transformer oil is considered acceptable for units operating at less than 288 kV?

Laboratory Experiments

These experiments are intended to provide the electrician with hands-on experience dealing with transformers. The transformers used are standard control transformers with two high-voltage windings rated at 240 volts each generally used to provide primary voltages of 480/240, and one low-voltage winding rated at 120 volts. The transformers have a rating of 0.5 kVA. Loads are standard 100-watt lamps that may be connected in parallel or series. It is assumed that the power supply is 208/120-volt three-phase four wire. It is also possibly used with a 240/120-volt three-phase high-leg system, provided adjustments are made in the calculations.

As in industry, these transformers will be operated with full voltage applied to the windings. The utmost caution must be exercised when dealing with these transformers. **These transformers can provide enough voltage and current to seriously injure or kill.** The power should be disconnected before attempting to make or change any connections.

Standard meters such as AC voltmeters, ohmmeters, and AC ammeters will be used to make circuit measurements. If clamp-on-type AC ammeters are used, it is recommended that a 10:1 scale divider, as described in Unit 6, can be used for accuracy.

Experiment 1

Transformer Basics

Objectives

After completing this experiment you will be able to

- Discuss the construction of an isolation transformer
- Determine the winding configuration with an ohmmeter
- Connect a transformer and make voltage measurements
- Compute the turns ratio of the windings

Materials needed

480-240/120 volt, 0.5 kVA control transformer
Ohmmeter
AC voltmeter

The transformer used in this experiment contains two high-voltage windings and one low-voltage winding. The high-voltage windings are labeled H_1 - H_2 and H_3 - H_4. The low-voltage winding is labeled X_1 - X_2.

1. Set the ohmmeter to the Rx1 range and measure the resistance between the following terminals:

H_1 - H_2 ____________ Ω
H_1 - H_3 ____________ Ω
H_1 - H_4 ____________ Ω
H_1 - X_1 ____________ Ω

H_1 - X_2 ____________ Ω
H_2 - H_3 ____________ Ω
H_2 - H_4 ____________ Ω
H_2 - X_1 ____________ Ω
H_2 - X_2 ____________ Ω
H_3 - H_4 ____________ Ω
H_3 - X_1 ____________ Ω
H_3 - X_2 ____________ Ω
H_4 - X_1 ____________ Ω
H_4 - X_2 ____________ Ω
X_1 - X_2 ____________ Ω

2. Using the information provided by the above measurements, which sets or terminals form complete circuits within the transformer?

   ____
   ____
   ____

   These circuits represent the connections to the three separate windings within the transformer.

3. Which of the windings exhibits the lowest resistance and why?

   ____
   ____
   ____
   ____

4. The H_1 - H_2 terminals are connected to one of the high-voltage windings and the H_3 - H_4 terminals are connected to the second high-voltage winding. Each of these windings are rated at 240 volts. When this transformer is connected for 240-volt operation, the two high-voltage windings are connected in parallel to form one winding by connecting H_1 to H_3 and H_2 to H_4, (*Figure Exp. 1-1*). This will provide a 2:1 turns ratio with the low-voltage winding.

   When this transformer is operated with 480 volts connected to the primary, the high-voltage windings are connected in series by connecting H_2 to H_3 and connecting power to H_1 and H_4, (*Figure Exp. 1-2*). This effectively doubles the primary turns, providing a 4:1 turns ratio with the low-voltage winding.

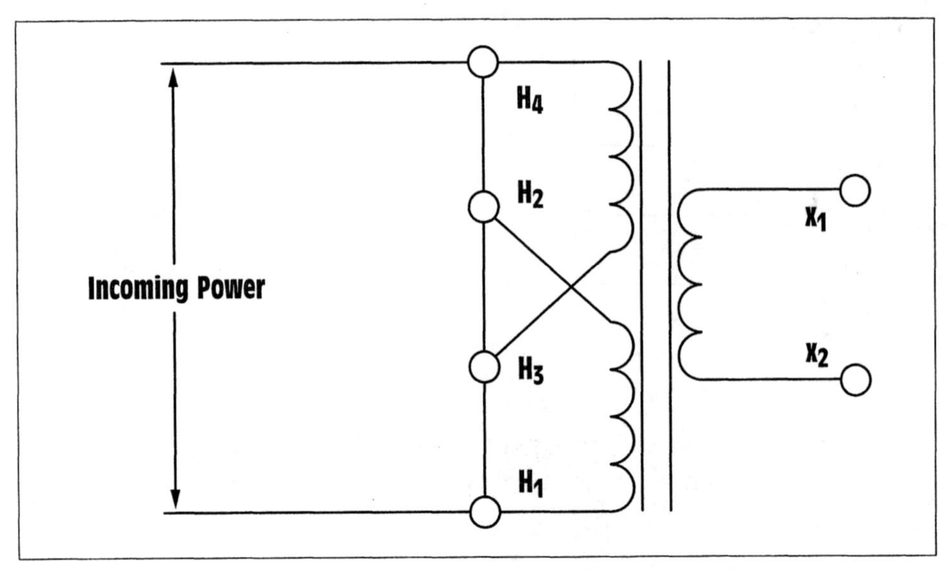

Figure Exp. 1-1 High-voltage windings connected in parallel.

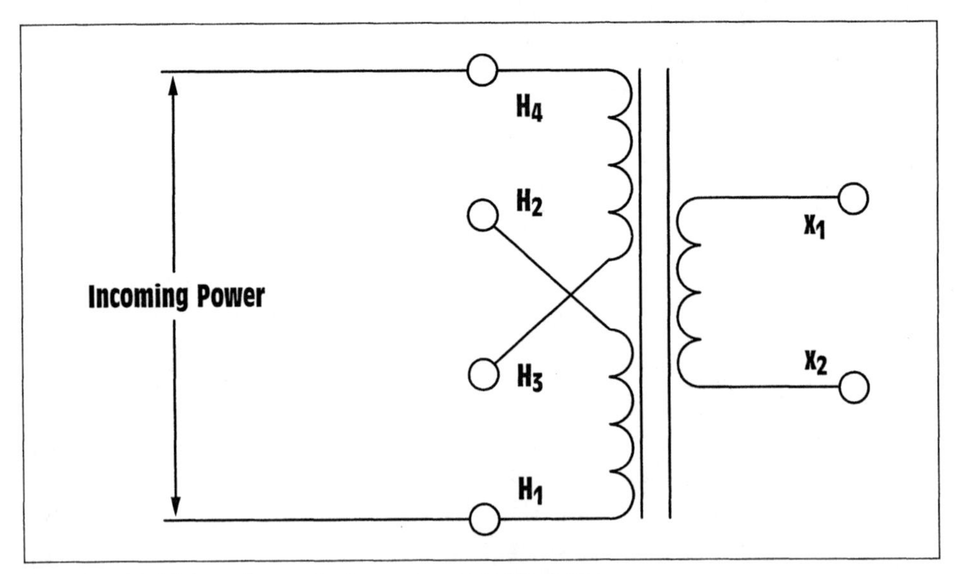

Figure Exp. 1-2 High-voltage windings connected in series.

5. Connect the two high-voltage windings for parallel operation as shown in *Figure Exp. 1-1.* Assuming a voltage of 208 volts is applied to the high-voltage windings, compute the voltage that should be present on the low-voltage winding between terminals X_1 and X_2.

   ____________ volts.

6. Apply a voltage of 208 volts to the transformer and measure the voltage across terminals X_1 and X_2.
 ____________ volts.

7. Notice that the measured voltage is slightly higher than the computed voltage. The rated voltage of a transformer is based on full load. It is normal for the secondary voltage to be slightly higher than rated voltage with no load connected to the transformer. Transformers are generally wound with a few extra turns of wire in the winding that is intended to be used as the load side. This helps overcome the voltage drop when load is added. The slight change in turns ratio generally does not affect the operation of the transformer to a great extent.

8. Disconnect the power from the transformer.

9. Reconnect the high-voltage windings for a series connection as shown in *Figure Exp. 1-2.*

10. Assume a voltage of 208 volts is applied to the high-voltage windings. Compute the voltage across the low-voltage winding.
 ____________ volts.

11. Apply a voltage of 208 volts across the high-voltage windings. Measure the voltage across terminals X_1 - X_2.
 ____________ volts.

12. Disconnect the power from the transformer.

13. Assume that the low-voltage winding is to be used as the primary and the high-voltage winding as the secondary. If the high-voltage windings are connected in series, the turns ratio will become 1:4, which means the secondary voltage will be 4 times greater than the primary voltage. Assume a voltage of 120 volts is connected to terminals X_1 - X_2. If the high-voltage windings are connected in series, compute the voltage across terminals H_1 - H_4.
 ____________ volts

14. Connect a 120-volt source to terminals X_1 - X_2. Make certain the AC voltmeter is set for a higher range than the computed value of voltage. **CAUTION: The secondary voltage in this step will be 480 volts or higher. Use extreme caution when making this measurement.**

15. Turn on the power supply and measure the voltage across terminals H_1 - H_4.

    ____________ volts

    Notice that the voltage is slightly lower than computed. Recall that previously it was stated that a few extra turns of wire are generally added to the winding that is intended to be used as the load side. This transformer is generally used with the low-voltage winding supplying power to the load. Since a few extra turns have been added, the actual turns ratio is probably 3.8:1 or 3.9:1 instead of 4:1.

16. Turn off the power supply.

17. Reconnect the high-voltage windings to form a parallel connection as shown in *Figure Exp. 1-1.*

18. Assume a voltage of 120 volts is connected to the low-voltage winding. Compute the voltage across the high-voltage winding.

    ____________ volts.

19. Connect a 120-volt AC source across terminals X_1 - X_2. Turn on the power supply and measure the voltage across terminals H_1 - H_4.

    ____________ volts.

20. Turn off the power supply and disconnect the transformer. Return the components to their proper place.

Experiment 2

Single-Phase Transformer Calculations

Objectives

After completing this experiment you will be able to

- Discuss transformer excitation current
- Compute values of primary current using the secondary current and the turns ratio
- Compute the turns ratio of a transformer using measured values
- Connect a step-down or step-up isolation transformer

Materials needed

480-240/120 volt, 0.5 kVA control transformer
AC voltmeter
AC ammeter

In this experiment the excitation current of an isolation transformer will be measured. The transformer will then be connected as both a step-down and a step-up transformer. The turns ratio will be determined from measured values and the primary current will be computed and then measured.

1. Connect the high-voltage windings of the transformer in parallel for 240-volt operation.

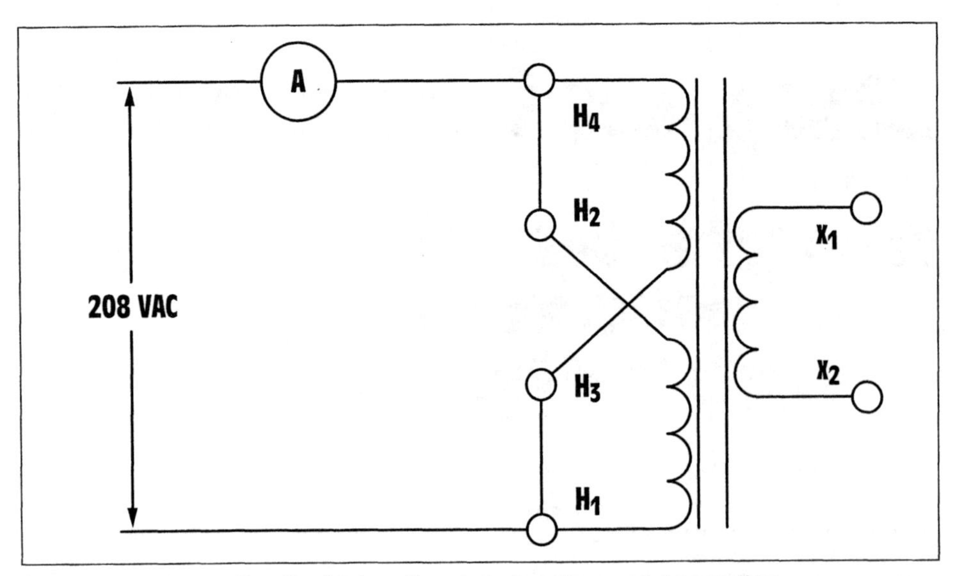

Figure Exp. 2-1 Connecting the high-voltage winding for parallel operation.

2. Connect the high-voltage winding to a 208-volt AC source with an AC ammeter connected in series with one of the lines, *(Figure Exp. 2-1)*. If an inline ammeter is not available and a clamp-on type meter is to be used, form a 10:1 scale divider by wrapping 10 turns of wire around the jaw of the ammeter. This will produce a more accurate reading. When the current value is read, divide the value by 10, by moving the decimal point one place to the left.

3. Turn on the power source and measure the current. This is the **excitation** current of the transformer. The excitation current is the amount of current necessary to magnetize the iron in the transformer, and will remain constant regardless of the load on the transformer.
   ____________ amp

4. Measure the voltage across the low-voltage winding at terminals X_1 - X_2.
   ____________ Volt

5. Compute the turns ratio by dividing the primary voltage by the secondary voltage. Since the primary has the highest voltage the larger number will be placed on the left side of the ratio, such as 3:1 or 4:1.
   ____________ ratio

6. Turn off the power supply.

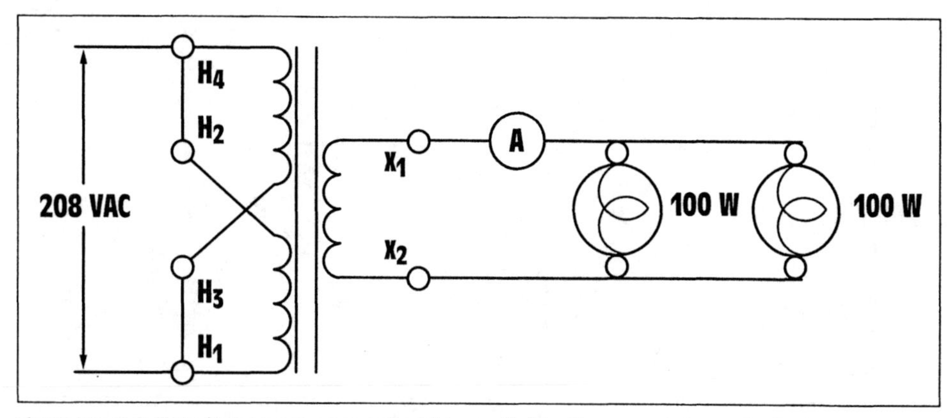

Figure Exp. 2-2 Two lamps are connected in parallel to the secondary winding.

7. Connect 2 100-watt incandescent lamps in parallel with the low-voltage winding of the transformer. Connect an AC ammeter in series with one of the lines, *(Figure Exp. 2-2)*.

8. Turn on the power and measure the current flow in the secondary circuit of the transformer.

    ___________ amp(s)

9. Turn off the power supply.

10. Compute the amount of primary current using the turns ratio. Since the primary voltage is higher, the amount of primary current will be less. Divide the secondary current by the turns ratio, then add the excitation current to this value.

$$I_{(Primary)} = \frac{I_{(Secondary)}}{\text{Turns Ratio}} + \text{Excitation Current}$$

    ___________ $I_{(Primary)}$

11. Reconnect the AC ammeter in one of the primary lines, *(Figure Exp. 2-3)*.

12. Turn on the power supply and measure the primary current. Compare this value with the computed value.

    ___________ $I_{(Primary)}$

13. Reconnect the AC ammeter in the secondary circuit and add two 100-watt incandescent lamps in parallel with the transformer secondary. *(Figure Exp. 2-4)*.

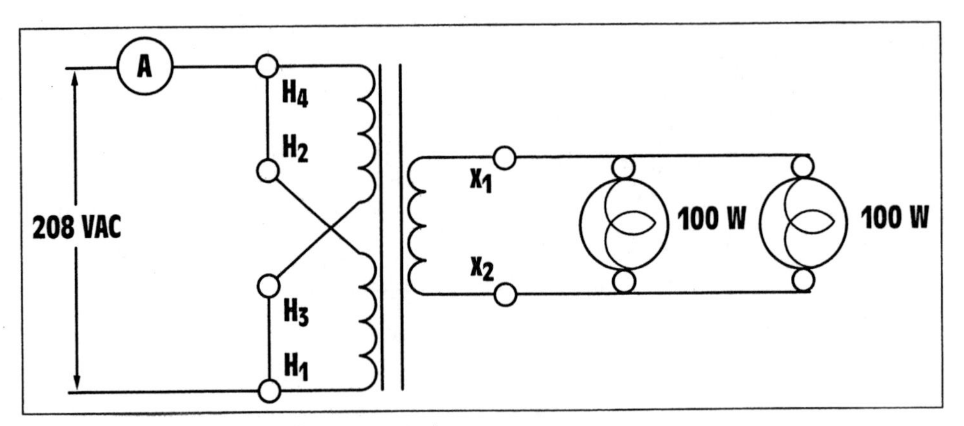

Figure Exp. 2-3 Measuring primary current.

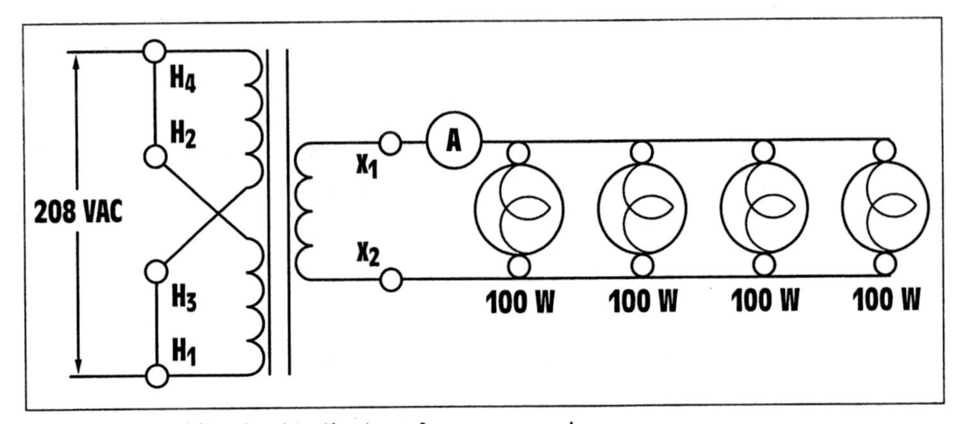

Figure Exp. 2-4 Adding load to the transformer secondary.

14. Turn on the power and measure the secondary current.
    ____________ amp(s)

15. Turn off the power.

16. Compute the amount of current flow that should be in the primary circuit using the turns ratio. Be sure to add the excitation current.
    ____________ $I_{(Primary)}$

17. Reconnect the AC ammeter in series with one of the lines of the primary winding of the transformer.

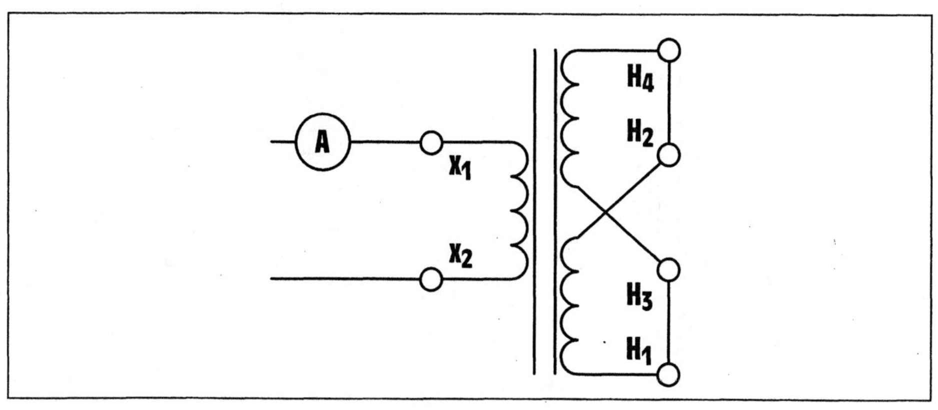

Figure Exp. 2-5 Using the low-voltage winding as the primary.

18. Turn on the power and measure the current flow. Compare the measured value with the computed value.

    ____________ $I_{(Primary)}$

19. Turn off the power supply.

20. Reconnect the transformer by connecting the low-voltage terminals, X_1 - X_2, to a 120-volt AC source. Connect an AC ammeter in series with one of the power lines, *(Figure Exp. 2-5).*

21. Turn on the power and measure the excitation current of the transformer.

    ____________ amp(s)

22. Measure the secondary voltage with an AC voltmeter.

    ____________ volts

23. Determine the turns ratio by dividing the secondary voltage by the primary voltage. Since the primary voltage is lower, the larger number will be placed on the right-hand side of the ratio; 1:3 or 1:4.

    ____________ ratio

24. Turn off the power supply.

25. Connect 2 100-watt incandescent lamps in series. Connect these two lamps in parallel with the high-voltage winding. Connect an AC ammeter in series with one of the secondary leads, *(Figure Exp. 2-6).*

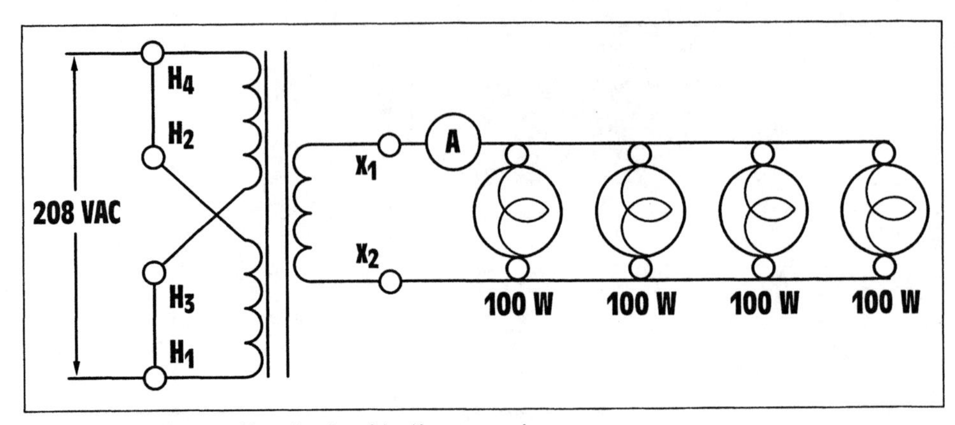

Figure Exp. 2-6 Connecting the load to the secondary.

26. Turn on the power supply and measure the secondary current.
_____________ amp(s)

27. Compute the primary current using the turns ratio. Since the primary voltage is less than the secondary voltage, the primary current will be more than the secondary current. To determine the primary current, multiply the secondary current by the turns ratio and add the excitation current.

$$I_{(Primary)} = I_{(Secondary)} \times \textit{Turns Ratio} + \textit{Excitation Current}$$

_____________ $I_{(Primary)}$

28. Turn off the power supply.

29. Reconnect the AC ammeter in series with the primary side of the transformer.

30. Turn on the power supply and measure the primary current. Compare this value with the computed value.
_____________ $I_{(Primary)}$

31. Turn off the power supply.

32. Reconnect the AC ammeter in series with the secondary winding. Add two more 100-watt lamps that have been connected in series to the sec-

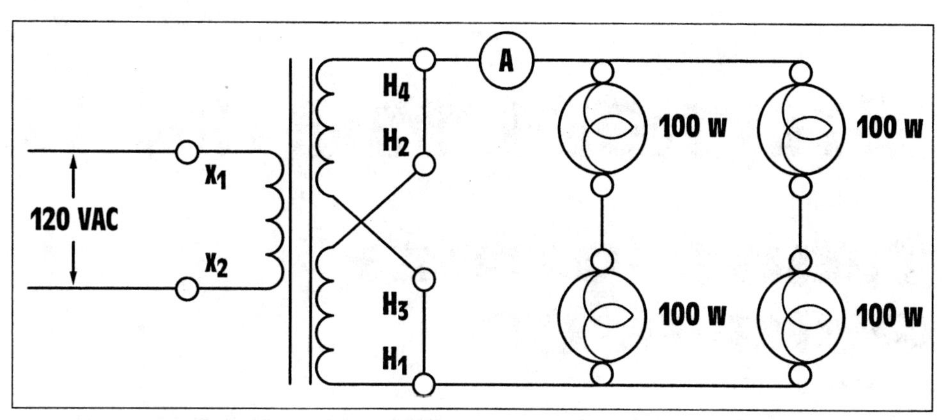

Figure Exp. 2-7 Adding load to the secondary.

ondary circuit. These two lamps should be connected in parallel with the first two lamps, *(Figure Exp. 2-7)*.

33. Turn on the power supply and measure the secondary current.

_____________ amp(s)

34. Turn off the power supply.

35. Compute the amount of current that should flow in the primary circuit.

_____________ $I_{(Primary)}$

36. Reconnect the AC ammeter in series with one of the primary lines.

37. Turn on the power supply and measure the primary current. Compare this value with the computed value.

_____________ $I_{(Primary)}$

38. Turn off the power supply. Disconnect the circuit and return the components to their proper place.

Experiment 3

Transformer Polarities

Objectives

After completing this experiment you will be able to

- Discuss buck and boost connections for a transformer
- Connect a transformer for additive polarity
- Connect a transformer for subtractive polarity
- Determine the turns ratio and calculate current values using measured values

Materials needed

480-240/120 volt, 0.5 kVA control transformer
AC voltmeter
AC ammeter
Four 100-watt lamps

In this experiment a control transformer will be connected for both additive (boost) and subtractive (buck) polarity. Buck and boost connections are made by physically connecting the primary and secondary windings together. If they are connected such that the primary and secondary voltages add, the transformer is additive or boost. If the windings are connected such that the primary and secondary voltages subtract, they are connected subtractive or buck.

In this exercise only one of the high-voltage windings will be used. The other will not be connected.

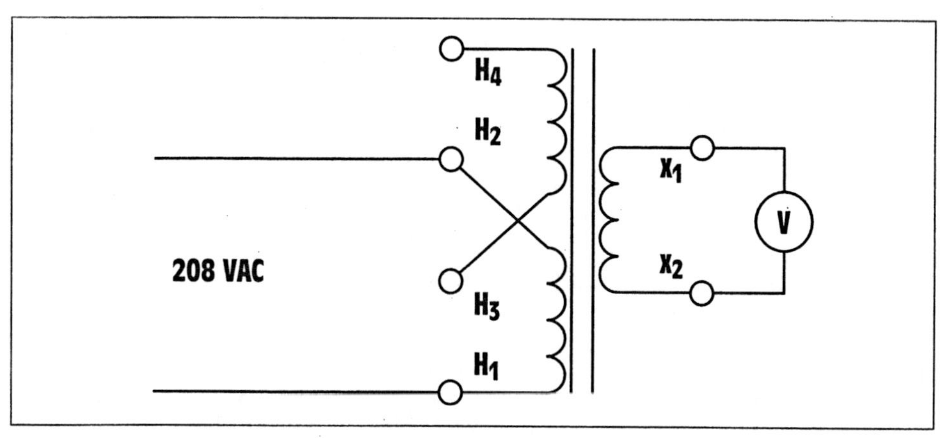

Figure Exp. 3-1 Measuring the secondary voltage.

1. Connect the circuit shown in *Figure Exp. 3-1.*

2. Turn on the power and measure the primary and secondary voltages.
   $E_{(Primary)}$ ______________ volts
   $E_{(Secondary)}$ ______________ volts

3. Turn off the power supply.

4. Determine the turns ratio of this transformer connection by dividing the higher voltage by the lower voltage. Recall that if the primary winding has the higher voltage, the larger number will be placed on the left and 1 will be placed on the right. If the secondary has the higher voltage, a 1 will be placed on the left and the larger number placed on the right.

$$\text{Turns Ratio} = \frac{\text{Higher Voltage}}{\text{Lower Voltage}}$$

   Ratio ____________

5. Connect the circuit shown in *Figure Exp. 3-2* by connecting X_1 to H_1. Connect a voltmeter across terminals X_2 and H_2.

6. Turn on the power supply and measure the voltage across X_2 and H_2.
   ____________ volts

7. Turn off the power supply.

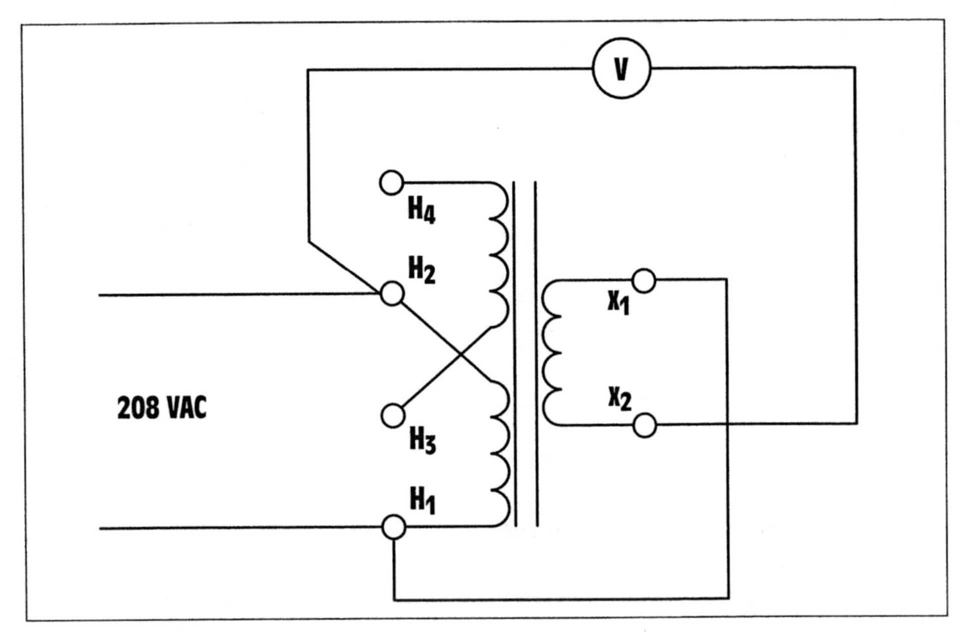

Figure Exp. 3-2 Connecting X_2 to H_2.

8. Determine the turns ratio of this transformer connection.
   Ratio ____________

9. If the measured voltage is the difference between the applied voltage and the secondary voltage, the transformer is connected subtractive polarity or buck. If the measured voltage is the sum of the applied voltage and the secondary voltage, the transformer is connected additive or boost. Is the transformer connected buck or boost?

   ____

10. Connect an AC ammeter in series with one of the power-supply lines.

11. Turn on the power supply and measure the excitation current of the transformer.
    $I_{(Exc)}$ ____________ amp(s)

12. Turn off the power supply.

13. Reconnect the transformer as shown in *Figure Exp. 3-3* by connecting X_2 to H_1. Connect an AC voltmeter across terminals X_1 and H_2.

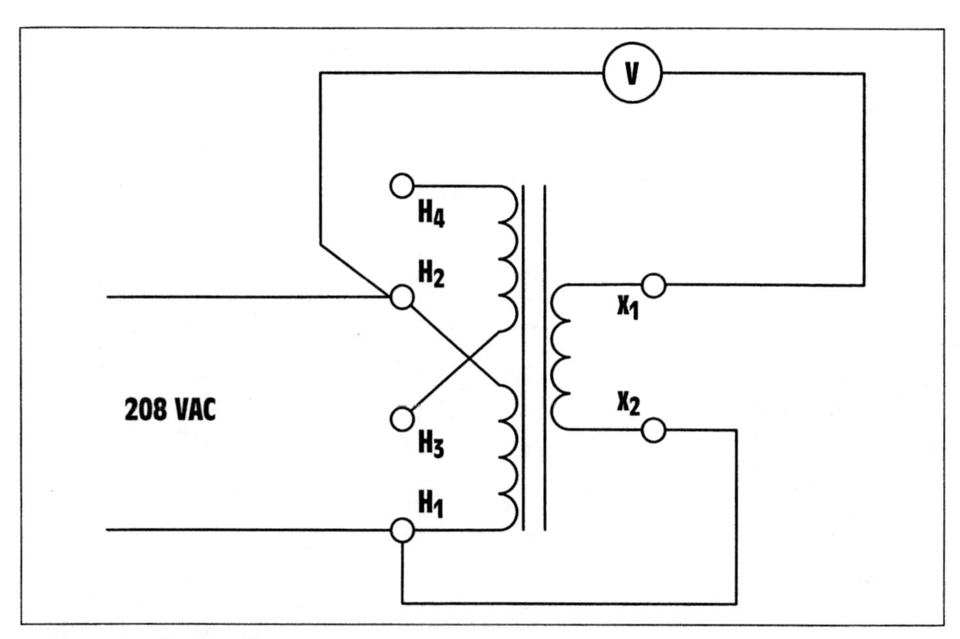

Figure Exp. 3-3 Connecting X_1 to H_2.

14. Turn on the power supply and measure the voltage across terminals X_1 and H_2.

    ____________ volts

15. Turn off the power supply.

16. Is the transformer connected buck or boost?

    ____

17. Determine the turns ratio of this transformer connection.

    Ratio ____________

18. Connect an AC ammeter in series with one of the primary leads.

19. Turn on the power supply and measure the excitation current of this connection.

    $I_{(Exc)}$ ____________amp(s)

20. Compare the value of excitation current for the buck and boost connections. Is there any difference between these two values?

    ____

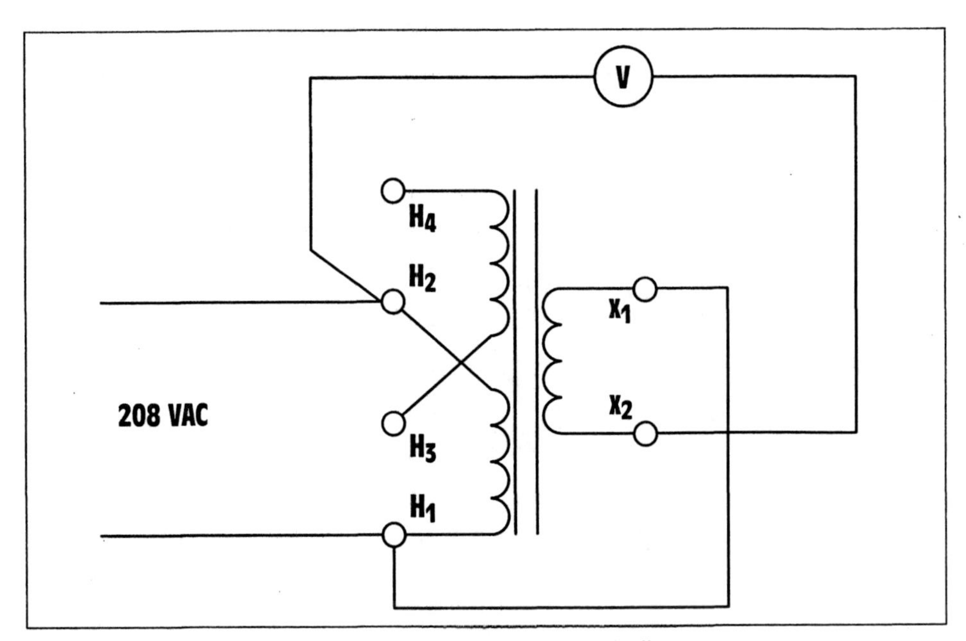

Figure Exp. 3-4 Placing polarity dots on the transformer windings.

21. Turn off the power supply.

22. *Figure Exp. 3-4* shows the proper location for the placement of polarity dots. Recall that polarity dots are used to indicate which windings of a transformer have the same polarity at the same time. To better understand how the dots are placed, redraw the two transformer windings in a series connection as shown in *Figure Exp. 3-5.* Place a dot beside one of the high-voltage terminals. In this example, a dot has been placed beside the H_1 terminal. Next draw an arrow pointing to the dot. To place the second dot, draw an arrow in the same direction as the first arrow. This arrow should point to the dot that is to be placed beside the secondary terminal. Since terminal X_2 is connected to H_1, the arrow must point to terminal X_1.

23. Reconnect the transformer for subtractive polarity. If two ammeters are available, place one ammeter in series with one of the primary leads and the second ammeter in series with the secondary lead that is not connected to the H_1 terminal. Connect a 100-watt lamp in the secondary circuit, and connect a voltmeter in parallel with the lamp *Figure Exp. 3-6.*

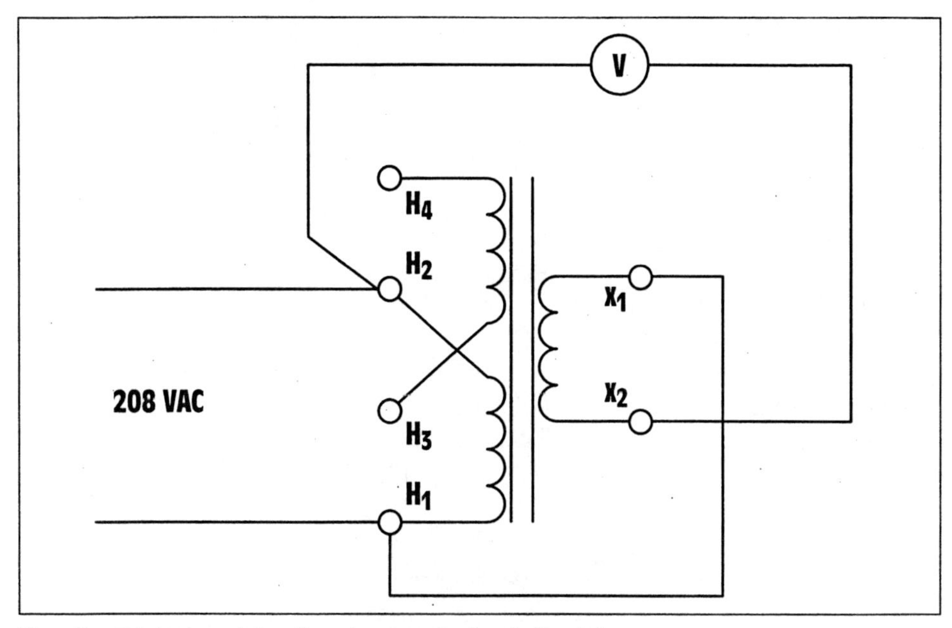

Figure Exp. 3-5 Determining the placement of polarity dots.

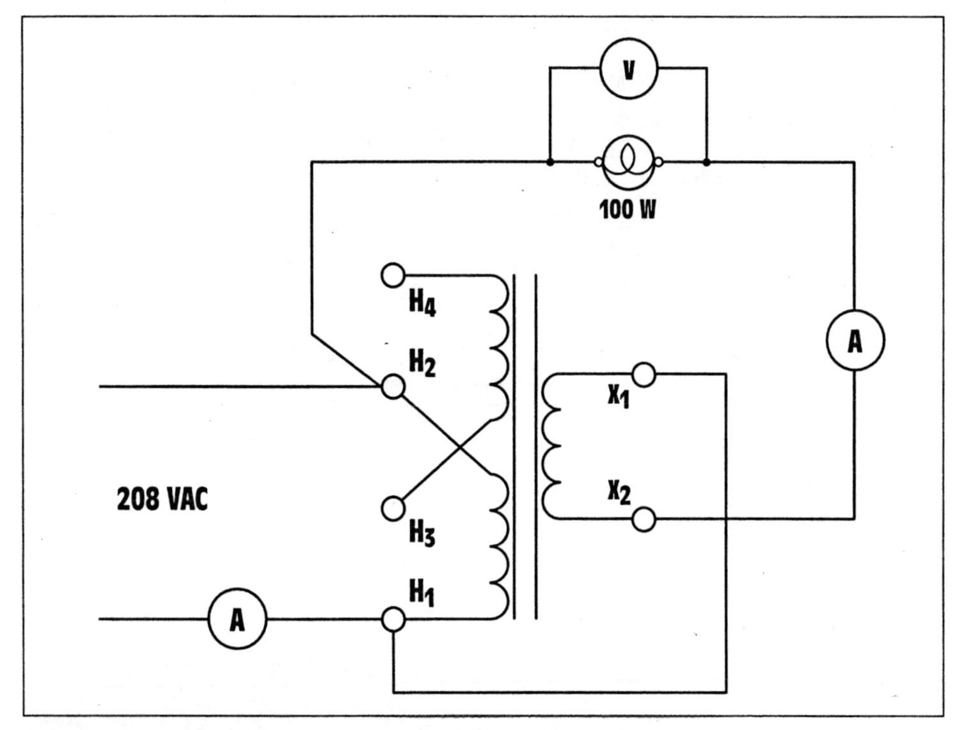

Figure Exp. 3-6 Connecting load to a subtractive polarity transformer.

24. Turn on the power supply and measure the secondary current.

    $I_{(Secondary)}$ ______________ amp(s)

25. Measure the secondary voltage. Since the lamp is the only load connected to the secondary, the voltage drop across the lamp will be the secondary voltage.

    $E_{(Secondary)}$ ______________ volts

26. Calculate the amount of primary current using the measured value of secondary current and the turns ratio. Be sure to use the turns ratio for this connection as determined in step 8. Since the primary voltage is greater than the secondary voltage, the primary current should be less. Therefore, divide the secondary current by the turns ratio and then add the excitation current measured in step 11.

$$I_{(Primary)} = \frac{I_{(Secondary)}}{\text{Turns Ratio}} + I_{(Exc)}$$

    $I_{(Primary)}$ ______________ amp(s)

27. If necessary, turn off the power supply and connect an AC ammeter in series with one of the primary leads.

28. Turn on the power supply and measure the primary current. Compare this value with the calculated value.

    $I_{(Primary)}$ ______________ amp(s)

29. Turn off the power supply.

30. Connect another 100-watt lamp in parallel with the first as shown in *Figure Exp. 3-7.* Reconnect the AC ammeter in series with the secondary winding if necessary.

31. Turn on the power supply and measure the amount of secondary current.

    $I_{(Secondary)}$ ______________ amp(s)

32. Calculate the primary current.

    $I_{(Primary)}$ ______________ amp(s)

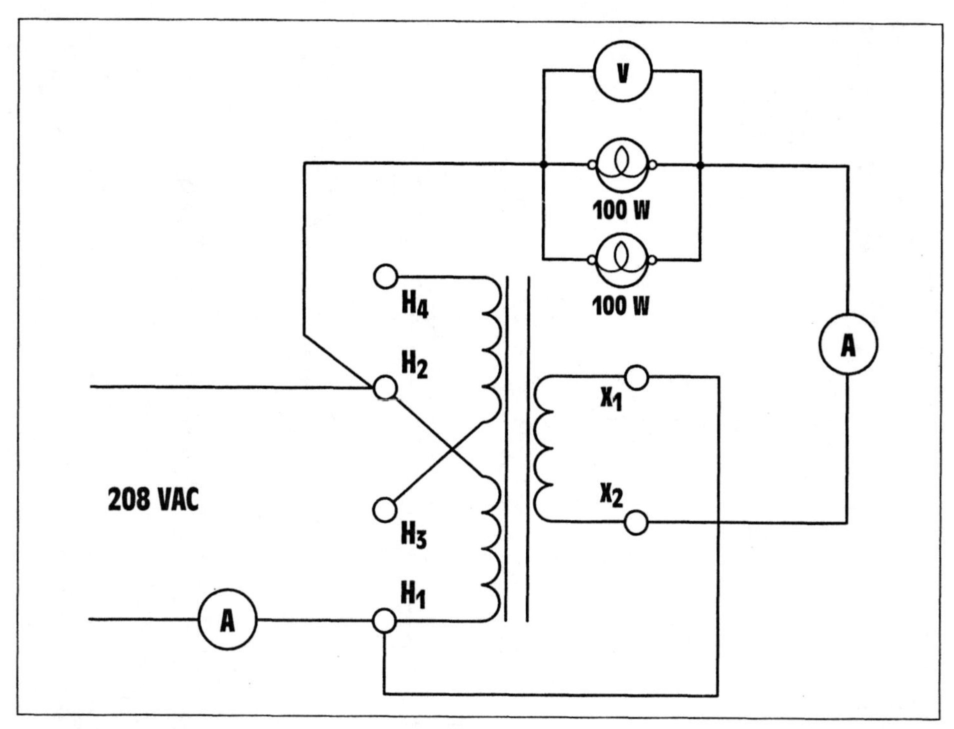

Figure Exp. 3-7 Adding load to the transformer.

33. If necessary, turn off the power supply and connect the AC ammeter in series with one of the primary leads.

34. Turn on the power supply and measure the primary current. Compare this value with the computed value.
$I_{(Primary)}$ ____________ amp(s)

35. Turn off the power supply.

36. Reconnect the transformer for the boost connection by connecting terminal X_2 to H_1. If two ammeters are available, connect one AC ammeter in series with one of the power-supply leads and the second AC ammeter in series with the secondary. Connect four 100-watt lamps in series with terminals X_1 and H_2 as shown in *Figure Exp. 3-8.* Connect an AC voltmeter across terminals X_2 and H_1.

37. Turn on the power and measure the secondary current.
$I_{(Secondary)}$ ____________ amp(s)

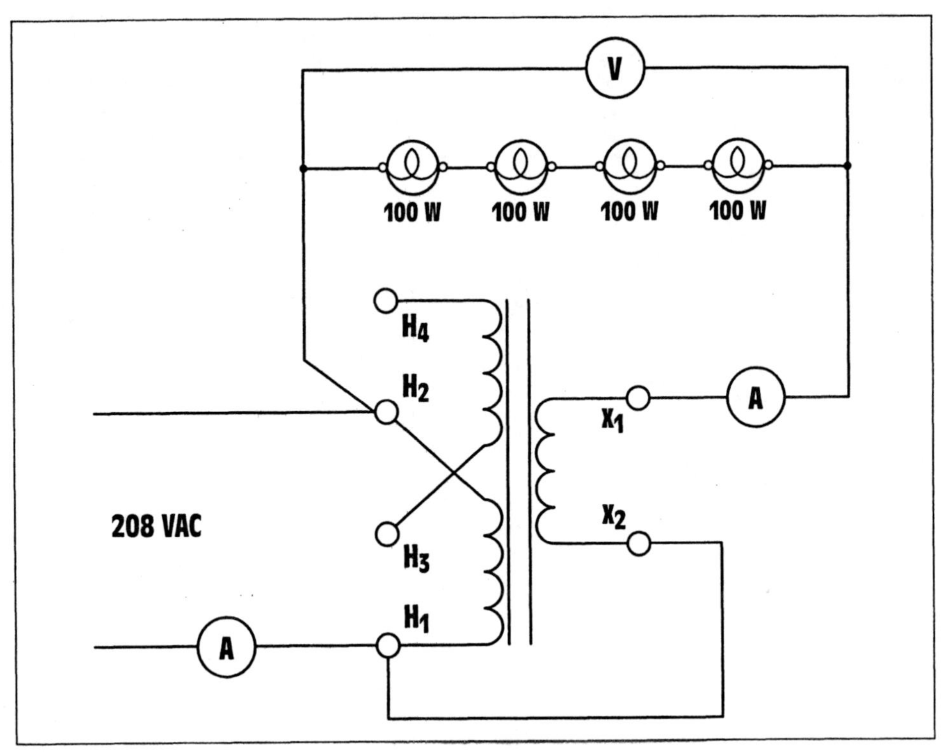

Figure Exp. 3-8 Connecting load to the boost connection.

38. Turn off the power supply.

39. Compute the primary current using the turns ratio. Be sure to use the turns ratio for this connection as determined in step 17. Since the primary voltage in this connection is less than the secondary voltage, the primary current will be greater. To calculate the primary current, multiply the secondary current by the turns ratio and then add the excitation current.

$$I_{(Primary)} = (I_{(Secondary)} \times \text{Turns Ratio}) + I_{(Exc)}$$

$I_{(Primary)}$ ______________ amp(s)

40. If necessary, connect the AC ammeter in series with one of the power supply leads.

41. Turn on the power supply.

42. Measure the primary current. Compare this value with the calculated value.

    $I_{(Primary)}$ ____________ amp(s)

43. Turn off the power supply.

44. Disconnect the circuit and return the components to their proper place.

Experiment 4

Autotransformers

Objectives

After completing this experiment you will be able to

- Discuss the operation of an autotransformer
- Connect a control transformer as an autotransformer
- Calculate the turns ratio from measured voltage values
- Calculate primary current using the secondary current and the turns ratio
- Connect an autotransformer as a step-down transformer
- Connect an autotransformer as a step-up transformer

Materials needed

480-240/120 volt, 0.5 kVA control transformer
AC voltmeter
AC ammeter
Four 100-watt lamps

In this experiment, the control transformer will be connected for operation as an autotransformer. The-low voltage winding will not be used in this experiment. The two high-voltage windings will be connected in series to form one continuous winding. The transformer will be connected as both a step-down and a step-up transformer.

1. Series connect the two high voltage windings by connecting terminals H_2 and H_3 together. The H_1 and H_4 terminals will be connected to a source

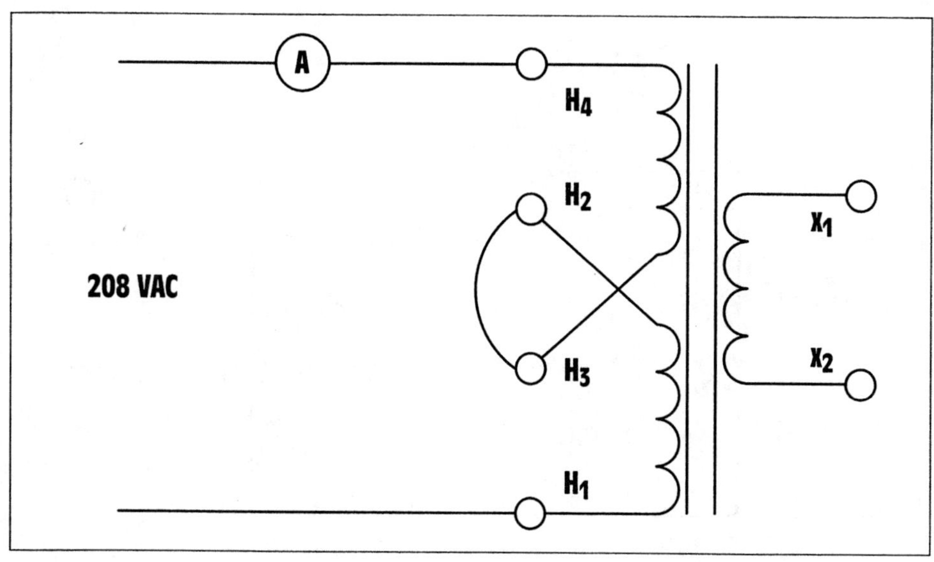

Figure Exp. 4-1 Connecting the high-voltage windings as an autotransformer.

of 208 VAC. Connect an ammeter in series with one of the power-supply lines, *(Figure Exp. 4-1).*

2. Turn on the power supply and measure the excitation current. (The current will be small and it may be difficult to determine this current value.)
   $I_{(Ecx)}$ ____________ amp(s)

3. Measure the primary voltage across terminals H_1 and H_4.
   $E_{(Primary)}$ ____________ volts

4. Measure the secondary voltage across terminals H_1 and H_2. (Note: it is also possible to use terminals H_3 and H_4 as the secondary winding.)
   $E_{(Secondary)}$ ____________ volts

5. Determine the turns ratio of this transformer connection.

$$\text{Turns Ratio} = \frac{\text{Higher Voltage}}{\text{Lower Voltage}}$$

   Ratio ____________

6. Turn off the power supply.

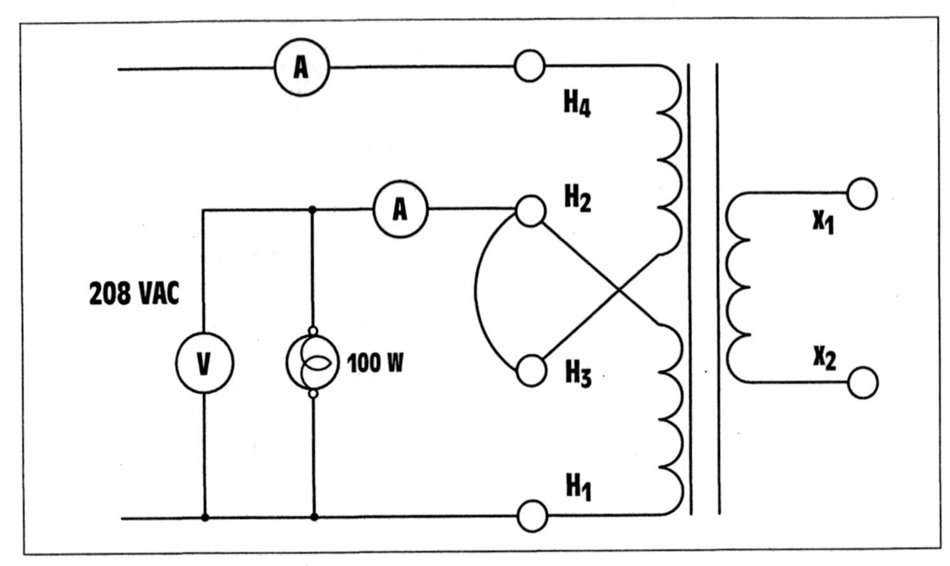

Figure Exp. 4-2 Connecting a load to the autotransformer.

7. Connect an AC ammeter in series with the H_2 terminal and a 100-watt lamp as shown in *Figure Exp. 4-2.* The secondary winding of the transformer will be between terminals H_2 and H_1.

8. Turn on the power supply and measure the amount of current flow in the secondary winding.
   $I_{(Secondary)}$ ____________ amp(s)

9. Measure the voltage drop across the secondary winding with an AC voltmeter.
   $E_{(Secondary)}$ ____________ volts

10. Turn off the power supply.

11. Calculate the primary current using the turns ratio. Since the primary voltage is greater than the secondary voltage, the primary current will be less than the secondary current. To determine the primary current, divide the secondary current by the turns ratio and add the excitation current.

$$I_{(Primary)} = \frac{I_{(Secondary)}}{\text{Turns Ratio}} + I_{(Exc)}$$

$I_{(Primary)}$ ____________ amp(s)

12. If necessary, reconnect the AC ammeter in series with one of the power-supply leads.

13. Turn on the power and measure the primary current. Compare this value with the computed value.
    $I_{(Primary)}$ ______________ amp(s)

14. Turn off the power supply.

15. Connect another 100 watt lamp in parallel with the existing lamp, *(Figure Exp. 4-3)*.

16. If necessary, reconnect the AC ammeter in series with the secondary winding of the transformer.

17. Turn on the power supply and measure the secondary current.
    $I_{(Secondary)}$ ______________ amp(s)

18. Calculate the primary current using the turns ratio.
    $I_{(Primary)}$ ______________ amp(s)

19. Turn off the power supply.

20. If necessary, reconnect the AC ammeter in series with one of the power-supply leads.

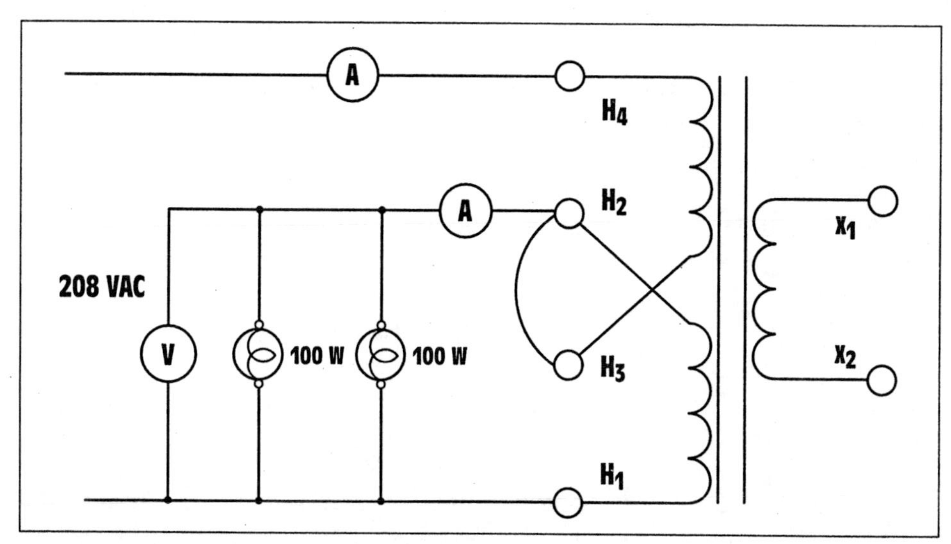

Figure Exp. 4-3 Adding load to the autotransformer.

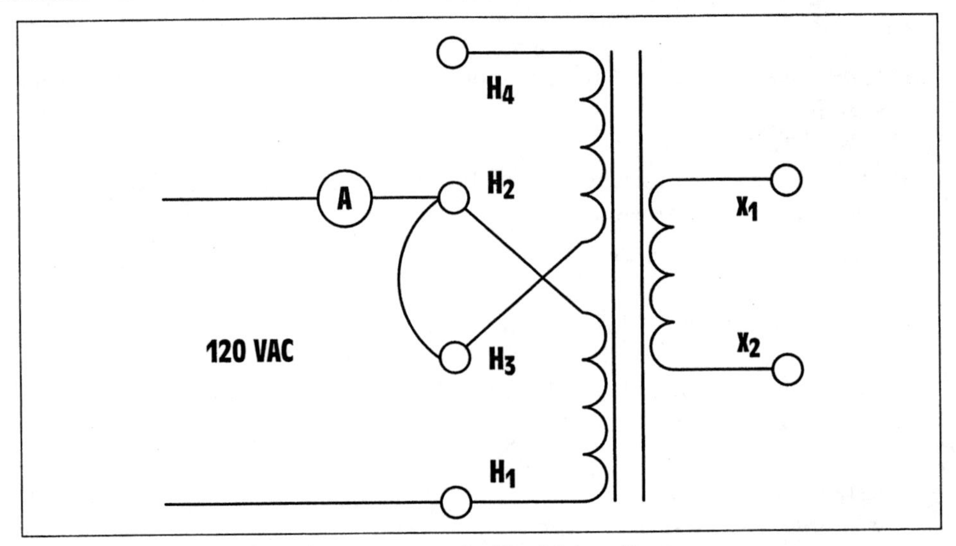

Figure Exp. 4-4 The autotransformer connected for high voltage.

21. Turn on the power supply and measure the primary current. Compare this value with the computed value.
    $I_{(Primary)}$ ____________ amp(s)

22. Turn off the power supply.

23. Reconnect the circuit as shown in *Figure Exp. 4-4.* Terminals H_1 and H_2 will be connected to a source of 120 VAC. Connect an AC ammeter in series with terminal H_2. The entire winding between terminals H_1 and H_4 will be used as the secondary.

24. Turn on the power and measure the excitation current of this transformer connection.
    $I_{(Exc)}$ ____________ amp(s)

25. Measure the voltage across terminals H_4 and H_1.
    ____________ volts

26. Compute the turns ratio of this transformer connection.
    Ratio ____________

27. Turn off the power supply.

28. Connect an AC ammeter and four 100-watt lamps in series with terminals H_4 and H_1, *(Figure Exp. 4-5).*

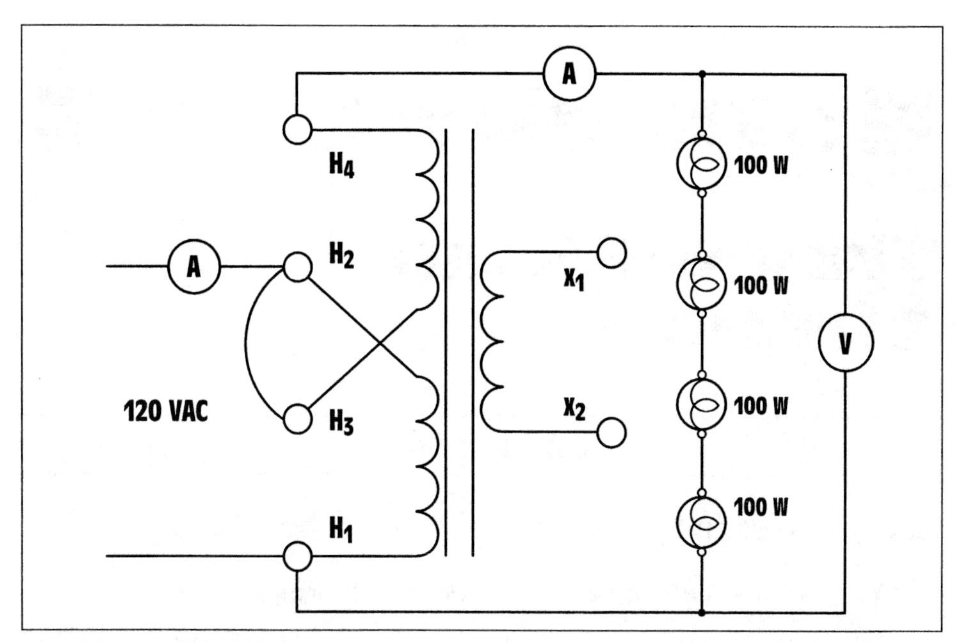

Figure Exp. 4-5 Adding load to the secondary winding.

29. Turn on the power and measure the secondary current.
 $I_{(Secondary)}$ ______________ amp(s)

30. Turn off the power supply.

31. If necessary, connect the AC ammeter in series with one of the primary leads.

32. Compute the value of primary current using the turns ratio and the measured value of secondary current.
 $I_{(Primary)}$ ______________ amp(s)

33. Turn on the power and measure the primary current. Compare this value with the computed value.
 $I_{(Primary)}$ ______________ amp(s)

34. Measure the voltage across terminals H_1 and H_4.
 $E_{(Secondary)}$ ______________ volts

35. Turn off the power supply.

36. Disconnect the circuit and return the components to their proper place.

Experiment 5

Three-Phase Circuits

Objectives

After completing this experiment you will be able to

- Connect a wye connected three-phase load
- Calculate and measure voltage and current values for a wye-connected load
- Connect a delta-connected load
- Calculate and measure voltage and current values for a delta-connected load

Materials needed

AC voltmeters
AC ammeters
Six 100-watt lamps

In this experiment, six 100-watt lamps will be connected to form different three-phase loads. Two lamps will be connected in series to form three separate loads. These loads will be connected to form wye or delta connections.

1. Connect two 100-watt lamps in series to form three separated load banks. Connect the load banks in wye by connecting one end of each bank together to form a center point, *(Figure Exp. 5-1)*. It is assumed that this load is to be connected to a 208-VAC, three-phase line. Connect an AC ammeter in series with the line supplying power to the load.

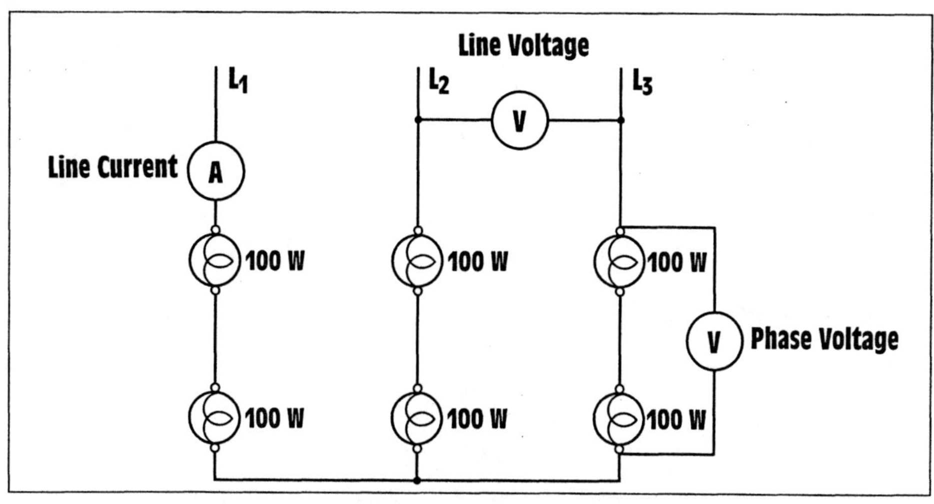

Figure Exp. 5-1 Measuring the line current and voltages of a wye connection.

2. Turn on the power and measure the line voltage supplied to the load.
   $E_{(Line)}$ ______________ volts

3. Calculate the value of phase voltage for a wye-connected load.

$$E_{(Phase)} = \frac{E_{(Line)}}{\sqrt{3}}$$

   $E_{(Phase)}$ ______________ volts

4. Measure the phase voltage and compare this value to the computed value.
   $E_{(Phase)}$ ______________ volts

5. Measure the line current.
   $I_{(Line)}$ ______________ amp(s)

6. Turn off the power supply.

7. In a wye-connected system, the line current and phase current are the same. Reconnect the circuit as shown in *Figure Exp. 5-2.*

8. Turn on the power and measure the phase current.
   $I_{(Phase)}$ ______________ amp(s)

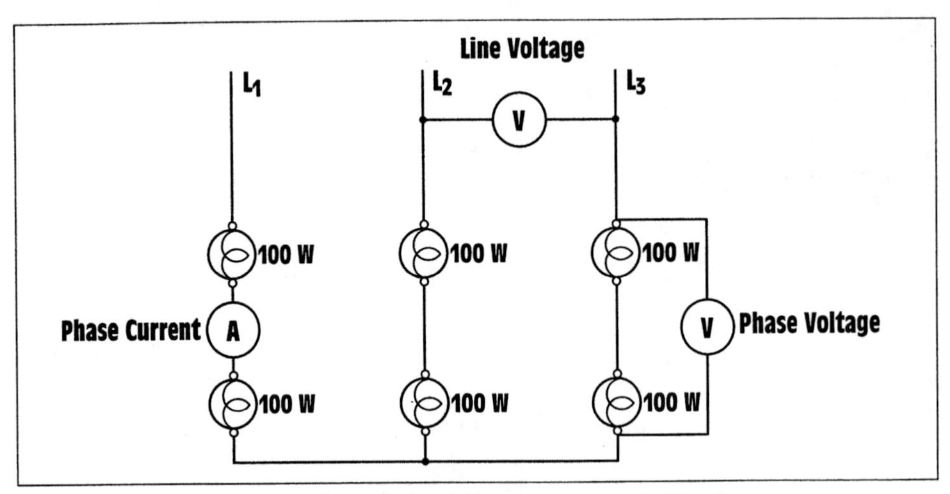

Figure Exp. 5-2 Measuring the phase current in a wye-connected load.

9. Turn off the power supply.

10. Reconnect the three banks of lamps to form a delta-connected load *(Figure Exp. 5-3)*.

11. Turn on the power and measure the line voltage supplied to the load.
    $E_{(Line)}$ ____________ volts

12. Measure the phase value of voltage.
    $E_{(Phase)}$ ____________ volts

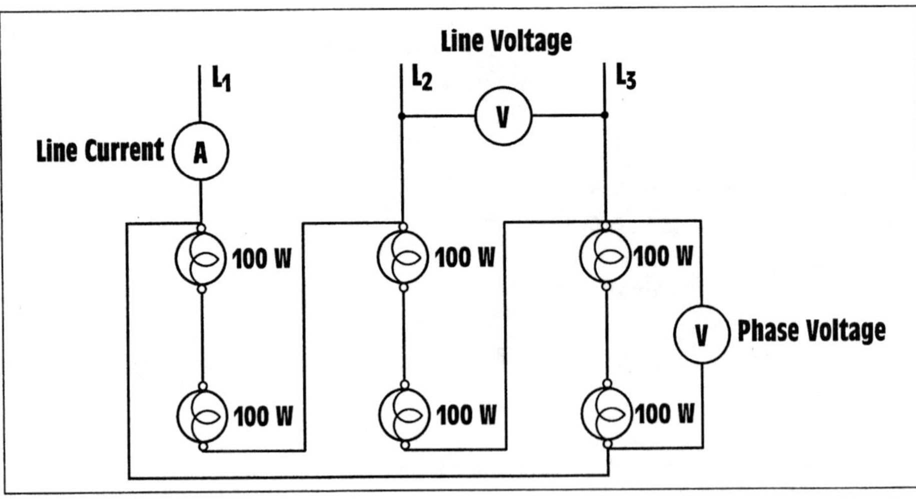

Figure Exp. 5-3 Measuring the voltage and line current values of a delta-connected load.

13. Are the line and phase voltage values the same or different?

____

14. Measure the line current.
$I_{(Line)}$ ______________ amp(s)

15. Turn off the power supply.

16. In a delta-connected system, the phase current will be less than the line current by a factor of 1.732. Calculate the phase-current value for this connection.

$$I_{(Phase)} = \frac{I_{(Line)}}{1.732}$$

$I_{(Phase)}$ ______________ amp(s)

17. Reconnect the circuit as shown in *Figure Exp. 5-4.*

18. Turn on the power supply and measure the phase current. Compare this value with the computed value.
$I_{(Phase)}$ ______________ amp(s)

19. Turn off the power supply.

20. Disconnect the circuit and return the components to their proper place.

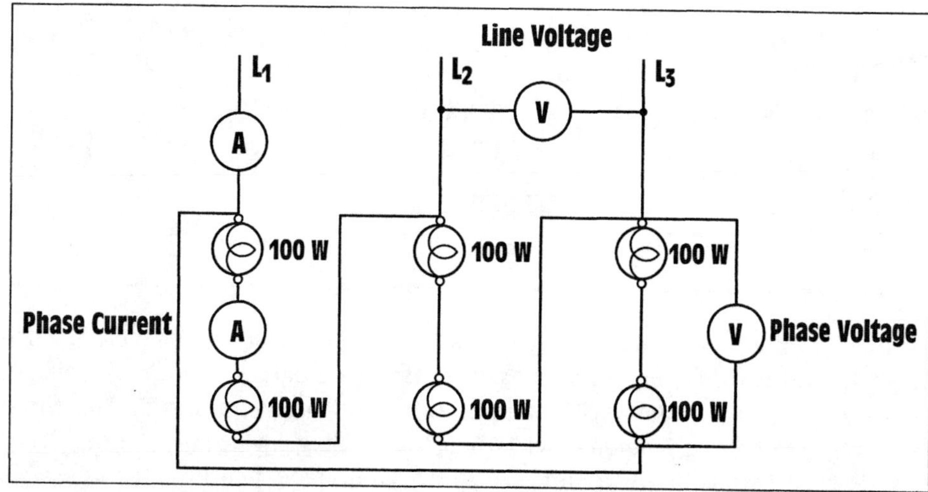

Figure Exp. 5-4 Measuring the phase current of a delta-connected load.

Experiment 6

Three-Phase Transformers

Objectives

After completing this experiment you will be able to

- Connect three single-phase transformers to form a three-phase bank
- Connect transformer windings in a delta configuration
- Connect transformer windings in a wye configuration
- Compute values of voltage, current, and turns ratio for different three-phase connections
- Compute the values for an open-delta-connected transformer bank

Materials needed

Three 480-240/120 volt, 0.5 kVA, control transformers
AC voltmeter
AC ammeter
Six 100-watt lamps

In this experiment, three single-phase control transformers will be connected to form different three-phase transformer banks. Values of voltage, current, and turns ratios will be computed and then measured. The three transformers will be operated with their high-voltage windings connected in parallel for low-voltage operation. The high-voltage windings are used as the primary for each connection.

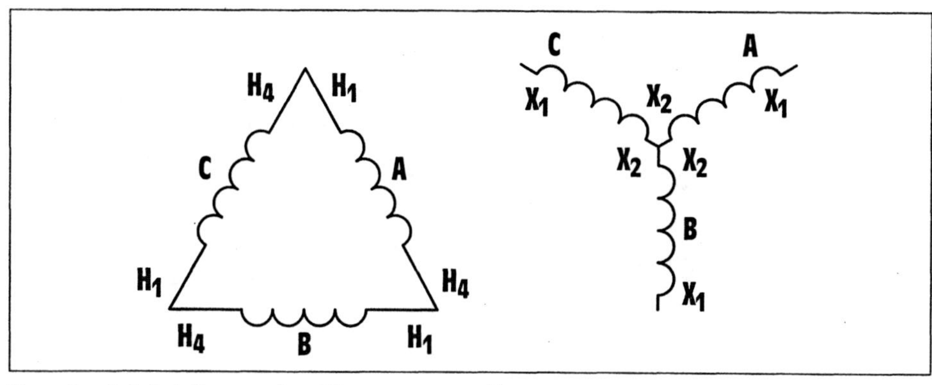

Figure Exp. 6-1 A delta-wye transformer connection.

Delta-Wye Connection

A delta-wye connected three-phase transformer bank has its primary windings connected in a delta configuration and its secondary windings connected in a wye configuration, *(Figure Exp. 6-1)*. Notice that the three primary windings have been labeled A, B, and C. The H_1 terminal of transformer A is connected to the H_4 terminal of transformer C. The H_4 terminal of transformer A is connected to the H_1 terminal of transformer B, and the H_4 terminal of transformer B is connected to the H_1 terminal of transformer C. The secondary windings form a wye by connecting all the X_2 terminals together.

1. Connect the circuit shown in *Figure Exp. 6-2.* Notice that the three transformers have been labeled A, B, and C. The H_1 terminal of transformer A is connected to the H_4 terminal of transformer C. The H_4 terminal of transformer A is connected to the H_1 terminal of transformer B, and the H_4 terminal of transformer B is connected to the H_1 terminal of transformer C. This is the same connection shown in the schematic drawing of *Figure Exp. 6-1.* Also notice that the X_2 terminal of each transformer is connected together to form a wye-connected secondary. The 100-watt lamp loads form a wye connection also.

2. Turn on the power supply and measure the phase voltage of the secondary. Since the secondary windings of the three transformers form the phases of the wye, the phase voltage can be measured across the X_1 - X_2 terminals of any transformer.

   $E_{(Phase\ Secondary)}$ ________________ volts.

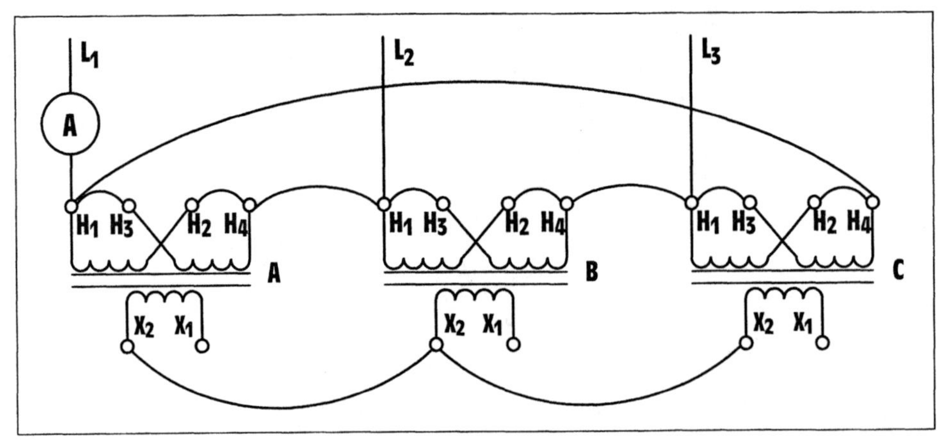

Figure Exp. 6-2 Transformer bank with a delta-connected primary and wye-connected secondary.

3. Calculate the line-to-line (line) voltage of the secondary.

$$E_{(Line)} = E_{(Phase)} \times \sqrt{3}$$

$E_{(Line\ Secondary)}$ ______________ volts

4. Measure the line voltage of the secondary by connecting an AC voltmeter across any two of the X_1 terminals.
   $E_{(Line\ Secondary)}$ ______________ volts

5. Measure the excitation current flowing in the primary winding. The excitation current will remain constant as long as the primary windings remain connected in a delta configuration. Since this measurement indicates the **line** value of current for this delta connection, it will later be added to the **line** value of computed current.
   $I_{(Exc)}$ ______________ amp(s)

6. Turn off the power supply.

7. Connect three 100-watt lamps to the secondary of the transformer bank. These lamps will be connected in wye to form a three-phase load for the transformer. If available, connect a second AC ammeter in series with one of the secondary leads, *(Figure Exp. 6-3)*.

8. Turn on the power supply and measure the secondary current.
   $I_{(Line\ Secondary)}$ ______________ amp(s)

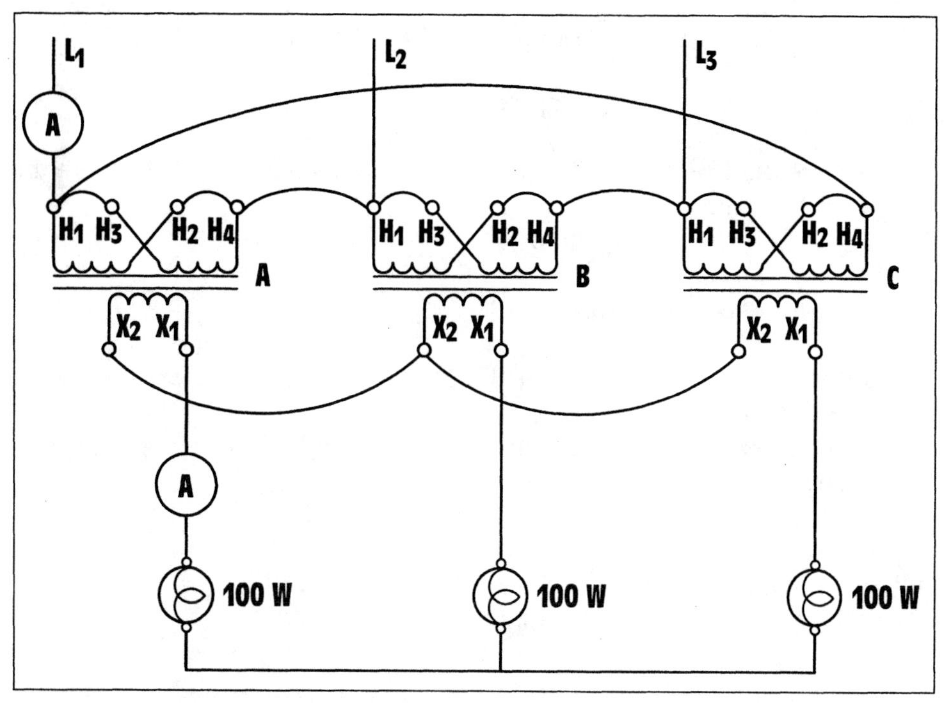

Figure Exp. 6-3 Adding load to the connection.

9. The measured value is the **line** current value. Since the secondary is connected in a wye configuration, the phase-current value will be the same as the line-current value. When calculating the primary current using secondary current and the turns ratio, **the phase-current value must be used**. Calculate the phase current of the primary using the turns ratio. Since the phase-voltage value of the primary is greater than the phase-voltage value of the secondary, the phase current of the primary will be less than the phase current of the secondary. To determine the primary phase-current value, divide the phase current of the secondary by the turns ratio.

$$I_{(Primary)} = \frac{I_{(Secondary)}}{\text{Turns Ratio}}$$

$I_{(Phase\ Primary)}$ ______________ amp(s)

10. Calculate the line current of the primary. Since the primary is connected as a delta, the line current will be greater than the phase current by a factor of 1.732. Be sure to add the line value of the excitation current in this calculation.

$$I_{(Line\ Primary)} = (I_{(Phase\ Primary)} \times 1.732) + I_{(EXC)}$$

$I_{(Line\ Primary)}$ ______________ amp(s)

11. Measure the line current of the primary and compare this value with the calculated value.
    $I_{(Line\ Primary)}$ ______________ amp(s)

12. Turn off the power supply.

13. Add a 100-watt lamp in parallel with each of the three existing loads, *(Figure Exp. 6-4)*.
14. Turn on the power supply and measure the line voltage of the secondary.
    $E_{(Line\ Secondary)}$ ______________ volts

15. Measure the line current of the secondary.
    $I_{(Line\ Secondary)}$ ______________ amp(s)

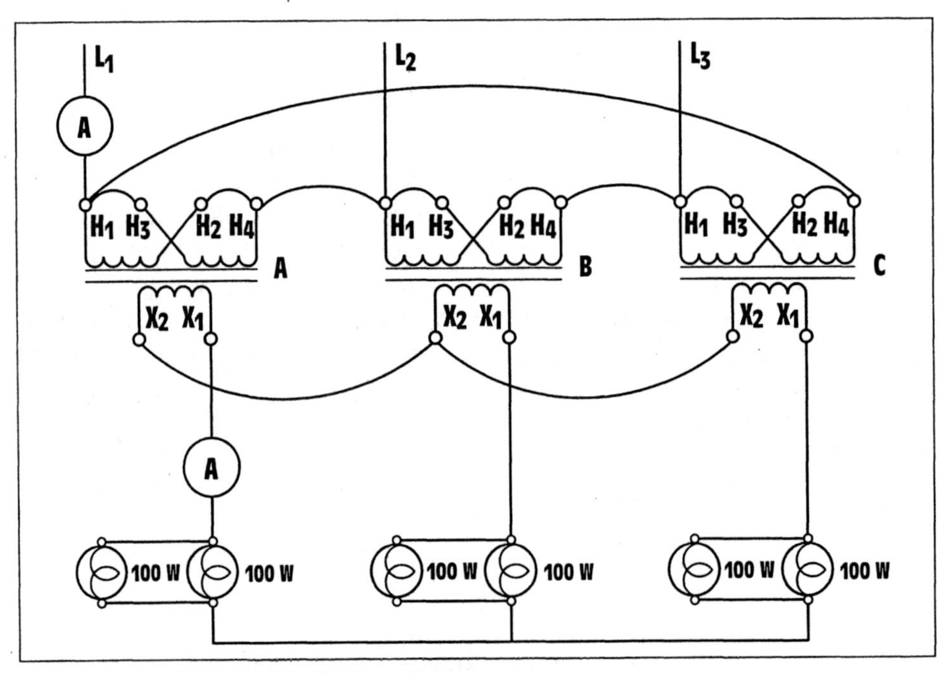

Figure Exp. 6-4 Adding load to the transformers.

16. Calculate the phase-current value of the primary using the phase-current value of the secondary and the turns ratio.
$I_{(Phase\ Primary)}$ ________________ amp(s)

17. Calculate the line current value of the primary.
$I_{(Line\ Primary)}$ ________________ amp(s)

18. Measure the line current of the primary and compare this value with the computed value.
$I_{(Line\ Primary)}$ ________________ amp(s)

19. Turn off the power supply.

Delta-Delta Connection

The three transformers will now be reconnected to form a delta-delta connection. The schematic diagram for a delta-delta connection is shown in *Figure Exp. 6-5.*

20. Reconnect the transformers as shown in *Figure Exp. 6-6.* In this connection, the primary windings remain connected in a delta configuration, but the secondary windings have been reconnected from a wye to a delta.

21. Turn on the power supply and measure the phase voltage of the secondary. The phase voltage can be measure across any set of X_1 - X_2 terminals.
$E_{(Phase\ Secondary)}$ ________________ volts

22. Since the secondary is connected as a delta, the line-voltage value

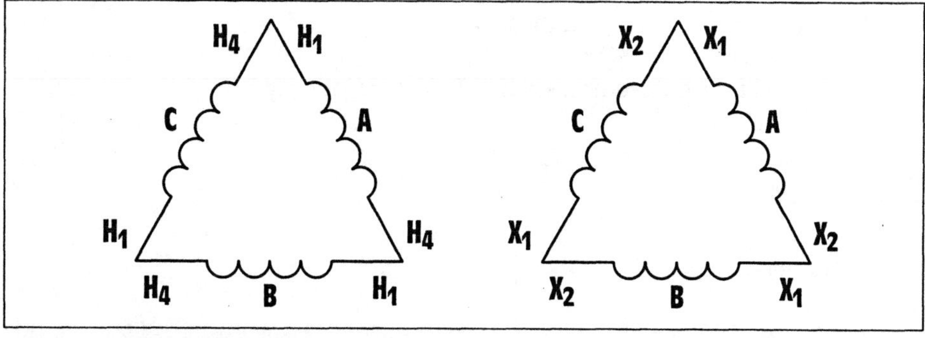

Figure Exp. 6-5 Delta-delta transformer connection.

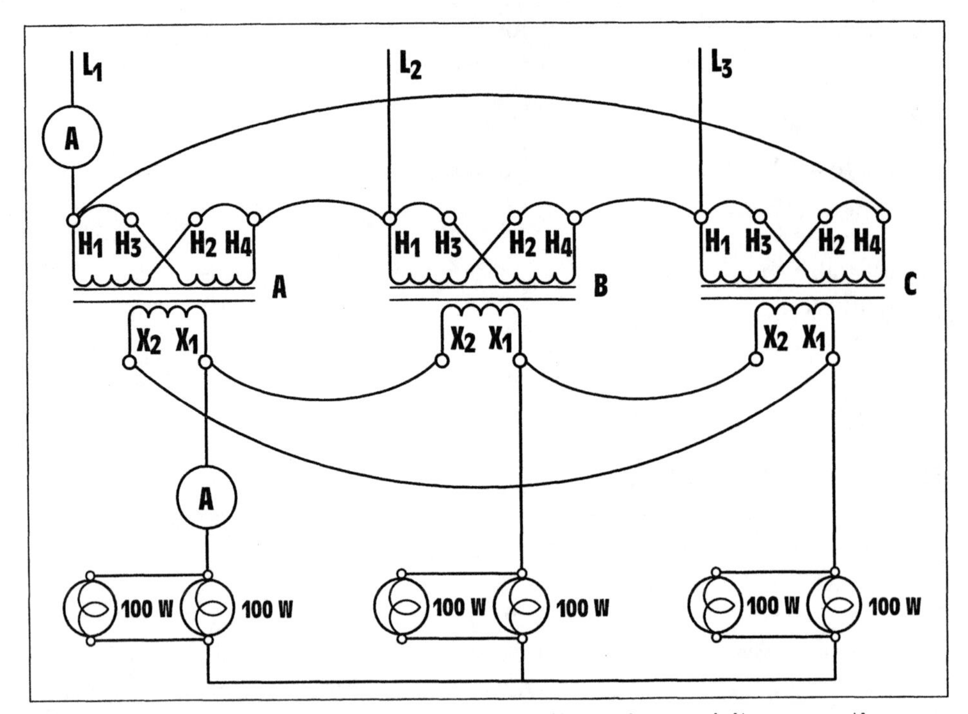

Figure Exp. 6-6 Transformers with a delta-connection primary, delta-connection secondary, and wye-connected load.

should be the same as the phase-voltage value. Measure the line-to-line voltage of the secondary. The line voltage can be measured between any two X_1 terminals.

$E_{(\text{Line Secondary})}$ ______________ volts

23. Measure the line-current value of the secondary.

$I_{(\text{Line Secondary})}$ ______________ amp(s)

24. Calculate the phase-current value of the secondary.

$$I_{(\text{Phase})} = \frac{I_{(\text{Line})}}{1.732}$$

$I_{(\text{Phase Secondary})}$ ______________ amp(s)

25. Using the phase current of the secondary and the turns ratio, calculate the phase current of the primary.

$I_{(\text{Phase Primary})}$ ______________ amp(s)

26. Compute the line current of the primary.
    $I_{(Line\ Primary)}$ ____________ amp(s)

27. Measure the line current of the primary and compare this value to the computed value.
    $I_{(Line\ Primary)}$ ____________ amp(s)

28. Turn off the power supply.

29. Reconnect the lamps to form a delta-connected load instead of a wye-connected load. Each phase should have two 100-watt lamps connected in parallel as shown in *Figure Exp. 6-7.*

30. Turn on the power supply and measure the line voltage of the secondary.
    $E_{(Line\ Secondary)}$ ____________ volts

31. Measure the line current of the secondary.
    $I_{(Line\ Secondary)}$ ____________ amp(s)

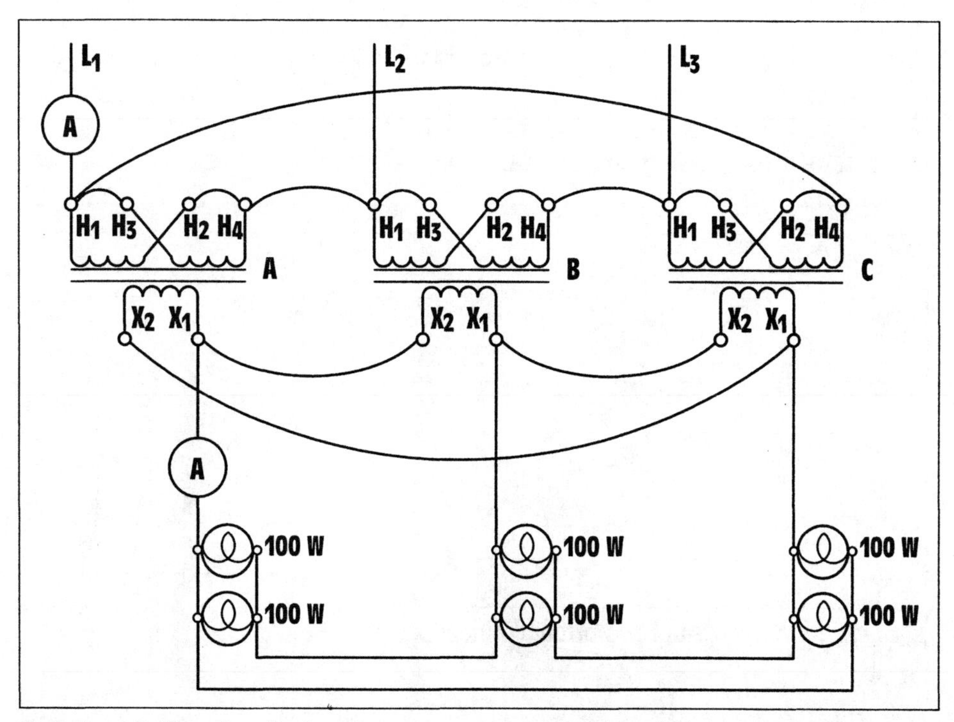

Figure Exp. 6-7 Changing the load from a wye connection to a delta connection.

32. Calculate the value of secondary-phase current.
    $I_{(Phase\ Seconary)}$ ______________ amp(s)

33. Calculate the phase-current value of the primary using the secondary phase current and the turns ratio.
    $I_{(Phase\ Primary)}$ ______________ amp(s)

34. Calculate the line-current value of the primary.
    $I_{(Line\ Primary)}$ ______________ amps(s)

35. Measure the primary line current and compare this value with the computed value.
    $I_{(Line\ Primary)}$ ______________ amps(s)

36. Turn off the power supply.

Wye - Delta Connection

In the next part of the experiment, the three transformers will be reconnected to form a wye-delta transformer bank. The schematic drawing of the connection is shown in *Figure Exp. 6-8*. Notice that all of the H_4 terminals have been joined together to form the wye connection. Power will be applied to the H_1 terminals. The secondary winding will remain in a delta connection.

37. Reconnect the transformers as shown in *Figure Exp. 6-9*. For the first part of this experiment, be sure that no load is connected to the secondary.

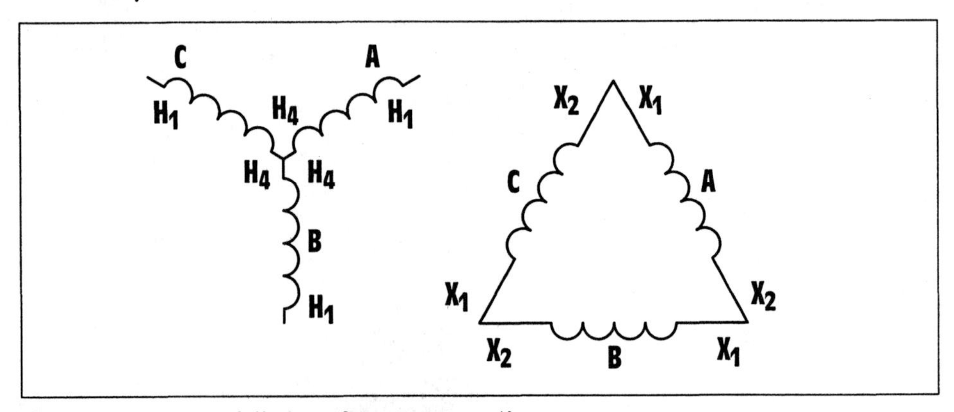

Figure Exp. 6-8 A wye-delta transformer connection.

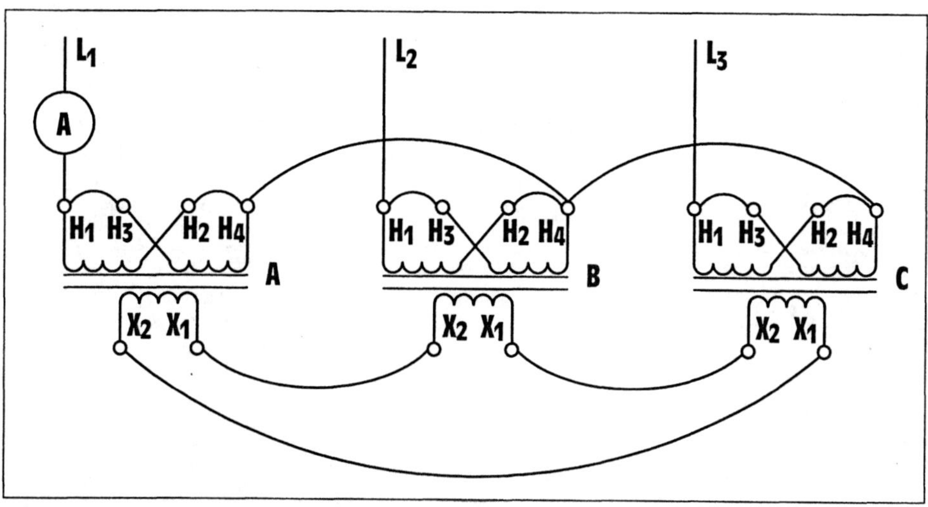

Figure Exp. 6-9 Transformer bank with a wye-connected primary and delta-connected secondary.

38. Turn on the power supply and measure the phase voltage of the primary.

    $E_{(Phase\ Primary)}$ ______________ volts

39. Measure the phase voltage of the secondary.

    $E_{(Phase\ Secondary)}$ ______________ volts

40. Compute the turns ratio of this transformer connection.

$$\text{Turns Ratio} = \frac{\text{Higher Voltage}}{\text{Lower Voltage}}$$

    Ratio ______________

41. Measure the excitation current of this connection. As long as the primary remains connected in a wye configuration, this excitation current will remain constant.

    $I_{(Exc)}$ ______________ amps(s)

42. Turn off the power supply.

43. Reconnect the delta connected lamp bank to the secondary of the transformer as shown in *Figure Exp. 6-10.*

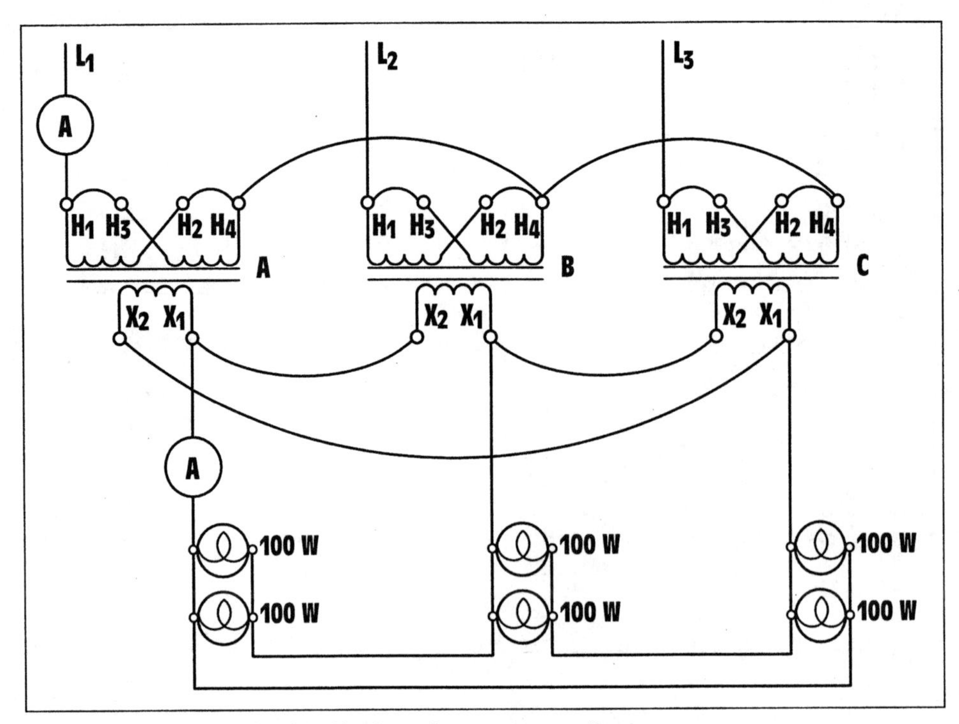

Figure Exp. 6-10 Adding load to the transformer connection.

44. Turn on the power supply and measure the line voltage of the secondary.

    $E_{(Line\ Secondary)}$ ______________ volts

45. Measure the line current of the secondary.

    $I_{(Line\ Secondary)}$ ______________ amps(s)

46. Calculate the phase current of the secondary.

$$I_{(Phase)} = \frac{I_{(Line)}}{1.732}$$

    $I_{(Phase\ Secondary)}$ ______________ amps(s)

47. Using the turns ratio and the secondary phase current, compute the phase current of the primary.

    $I_{(Phase\ Primary)}$ ______________ amps(s)

48. In a wye connection, the line current and the phase current are the same. To determine the line current for a transformer connection, however, the excitation current must be added to the line-current value. Compute the value of primary line current.
$I_{(Line\ Primary)}$ ____________ amps(s)

49. Measure the primary line current and compare this value to the computed value.
$I_{(Line\ Primary)}$ ____________ amps(s)

50. Turn off the power supply.

Wye-Wye Connection

The next section of this experiment deals with transformers connected in a wye-wye connection. The schematic diagram for this connection is shown in *Figure Exp. 6-11.* Notice that both the primary and secondary windings are connected to form a wye connection.

51. Connect the circuit shown in *Figure Exp. 6-12.*

52. Turn on the power supply and measure the phase voltage of the secondary.
$E_{(Phase\ Secondary)}$ ____________ volts

53. Calculate the line voltage value of the secondary.

$$E_{(Line)} = E_{(Phase)} \times 1.732$$

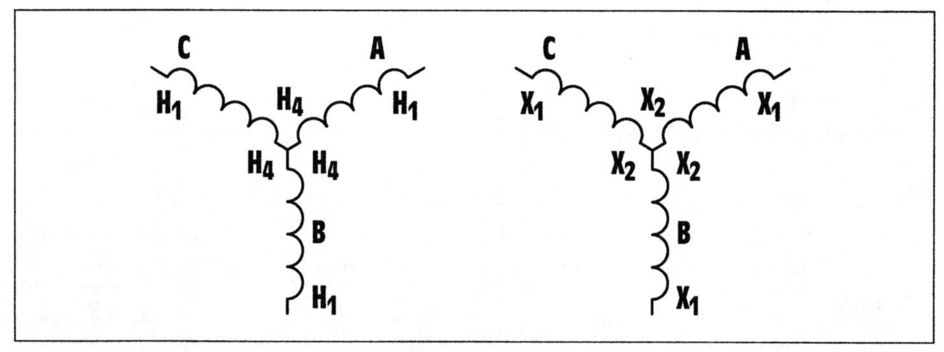

Figure Exp. 6-11 A wye-wye three-phase transformer connection.

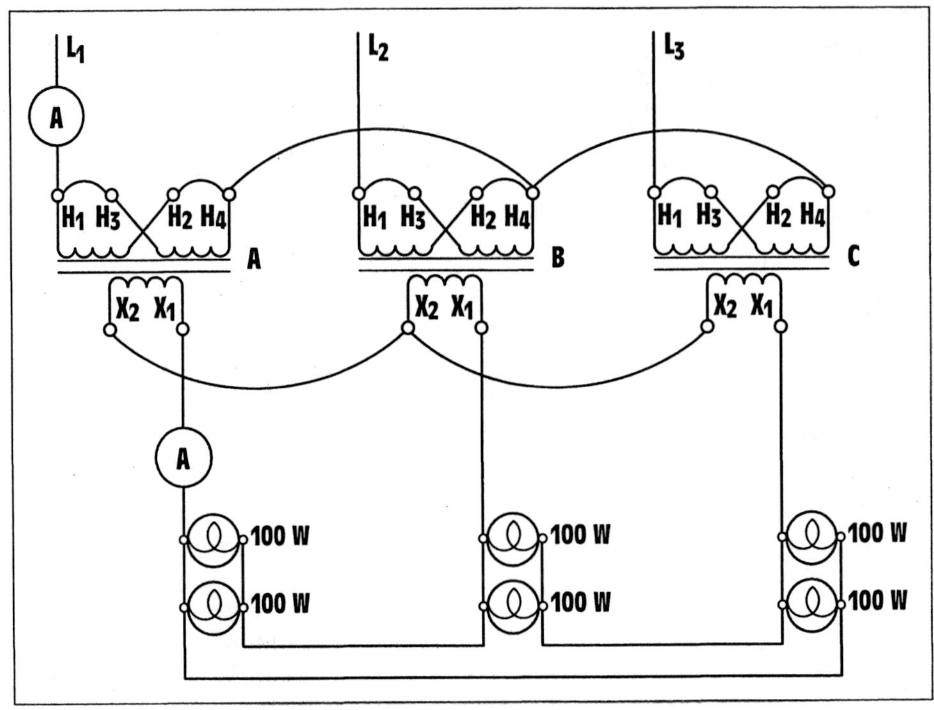

Figure Exp. 6-12 Transformers with a wye-connected primary, wye-connected secondary and delta-connected load.

$E_{(Line\ Secondary)}$ ______________ volts

54. Measure the line voltage of the secondary and compare this value to the computed value.
$E_{(Line\ Secondary)}$ ______________ volts

55. Measure the line current of the secondary.
$I_{(Line\ Secondary)}$ ______________ amps(s)

56. Since the secondary is now connected in a wye configuration, the phase current will be the same as the line current. Compute the value of primary phase current using the secondary phase current and the turns ratio.
$I_{(Phase\ Primary)}$ ______________ amps(s)

57. Since the primary is connected in a wye configuration also, the line current will be the same as the phase current plus the excitation current. Compute the total line-current value for the primary.

$I_{(Line\ Primary)}$ ______________ amps(s)

58. Measure the line current value and compare it to the computed value.
$I_{(Line\ Primary)}$ ______________ amps(s)

59. Turn off the power supply.

Open-Delta Connection

The last connection to be made is the open delta. The open-delta connection requires the use of only two transformers to supply three-phase power to a load. The schematic diagram for an open-delta connection is shown in *Figure Exp. 6-13*. It should be noted that the open-delta connection can provide only about 87% of the combined kVA capacity of the two transformers.

60. Connect the circuit shown in *Figure Exp. 6-14*.

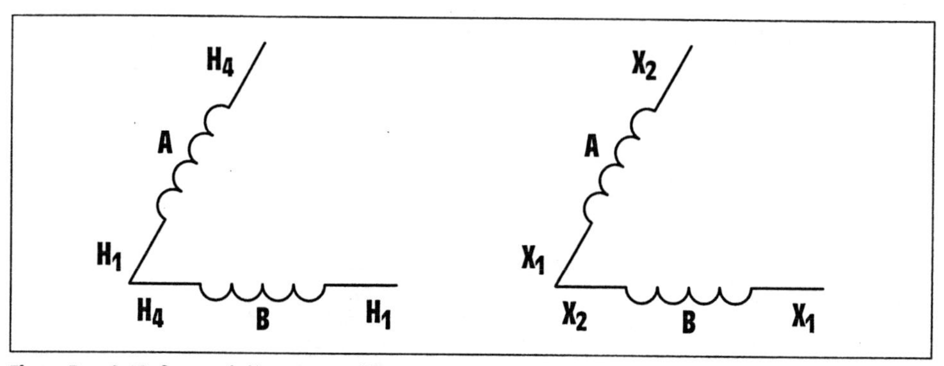

Figure Exp. 6-13 Open-delta connection.

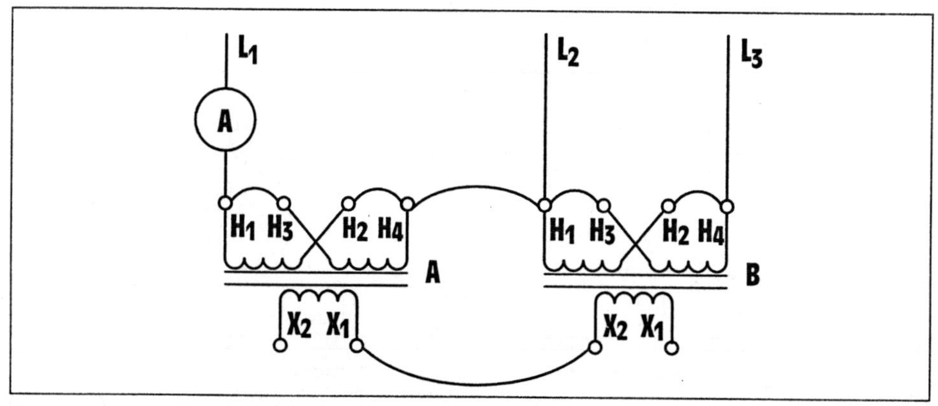

Figure Exp. 6-14 Two transformers connected in an open delta.

61. Turn on the power supply and measure the phase voltage of the primary.
    $E_{(Phase\ Primary)}$ ____________ volts

62. Measure the phase voltage of the secondary.
    $E_{(Phase\ Secondary)}$ ____________ volts

63. Calculate the turns ratio of this transformer connection.
    Ratio ____________

64. Measure the line-to-line voltage between all three of the secondary line terminals. Is there any variation in the voltages?
    ____

65. Turn off the power supply.

66. Connect three 100-watt lamps to form a delta connection. Connect these lamps to the line terminals of the secondary, *(Figure Exp. 6-15)*.

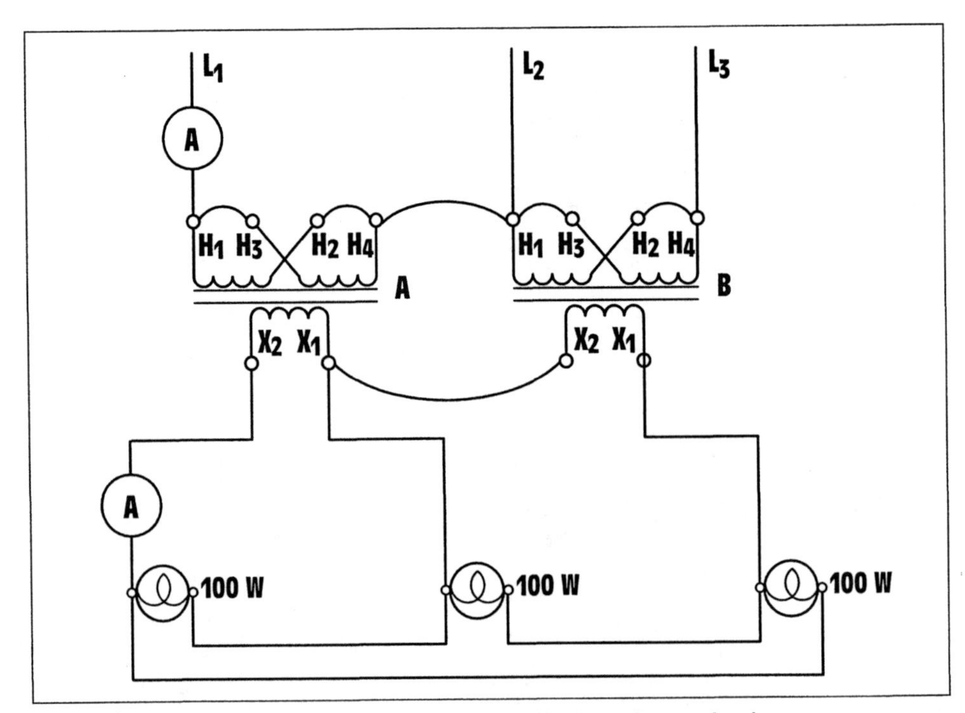

Figure Exp. 6-15 Connecting a three-phase load to the transformer bank.

67. Turn on the power supply and measure the line voltage between each of the three lines. Are the voltage values the same?

____

68. Measure the line current of the secondary.
$I_{(Line\ Secondary)}$ _____________ amps(s)

69. The phase-current value for an open delta is calculated in the same way as a closed delta. Calculate the phase-current value for the secondary.

$$I_{(Phase)} = \frac{I_{(Line)}}{1.732}$$

$I_{(Phase\ Secondary)}$ _____________ amps(s)

70. Using the secondary phase current and the turns ratio, calculate the phase-current value for the primary.
$I_{(Phase\ Primary)}$ _____________ amps(s)

71. Calculate the line current value.

$$I_{(Line)} = (I_{(Phase)} \text{ X } 1.732) + I(EXC)$$

$I_{(Line\ Primary)}$ _____________ amps(s)

72. Measure the line current of the primary and compare this value to the computed value.
$I_{(Line\ Primary)}$ _____________ amps(s)

73. Turn off the power supply.

74. Disconnect the circuit and return the components to their proper place.

Answers to Practice Problems

Unit 4 Single Phase Isolation Transformers

1.

E_P 120	E_s 24
I_P 1.6	I_s 8
N_P 300	N_s 60
Ratio: 5:1	Z = 3 Ω

2.

E_P 240	E_s 320
I_P 0.853	I_s 0.643
N_P 210	N_s 280
Ratio: 1:1.333	Z = 500 Ω

3.

E_P 64	E_s 160
I_P 33.333	I_s 13.333
N_P 32	N_s 80
Ratio: 1:2.5	Z = 12 Ω

4.

E_P 48	E_s 240
I_P 3.333	I_s 0.667
N_P 220	N_s 1,100
Ratio: 1:5	Z = 360 Ω

5.

E_P 35.848	E_s 182
I_P 16.5	I_s 3.25
N_P 87	N_s 450
Ratio: 1:5.077	Z = 56 Ω

6.

E_P 480	E_s 916.346
I_P 1.458	I_s 0.764
N_P 275	N_s 525
Ratio: 1:1.909	Z = 1.2 kΩ

7.

E_P = 208	E_{s1} = 320	E_{s2} = 120	E_{s3} = 24
I_P = 11.93	I_{s1} = 0.0267	I_{s2} = 20	I_{s3} = 3
N_P = 800	N_{s1} = 1231	N_{s2} = 462	N_{s3} = 92
	$Ratio_1$ 1:1.54	$Ratio_2$ 1.73:1	$Ratio_3$ 1:8.67
	R_1 = 12 kΩ	R_2 = 6 Ω	R_3 = 8 Ω

8.

E_P = 277	E_{s1} = 480	E_{s2} = 208	E_{s3} = 120
I_P = 8.93	I_{s1}= 2.4	I_{s2} = 3.47	I_{s3} = 5
N_P = 350	N_{s1} = 606	N_{s2} = 263	N_{s3} = 152
	$Ratio_1$ 1:1.73	$Ratio_2$ 1.33:1	$Ratio_3$ 2.31:1
	R_1 = 200 Ω	R_2 = 60 Ω	R_3 = 24 Ω

Unit 5 Autotransformers

1.

A-B 89.1	A-C 148.5	A-D 252.5	A-E 297.1	B-C 59.4	B-D 163.4
B-E 208	C-D 104	C-E 148.6	D-E 44.6		

2. I_p = 3.1 amperes I_s = 4.3 amperes

3. 280/60 = 4.67:1

4. 168 volts (480/400 = 1.2 volts per turn) (1.2 x 140 turns = 168 volts)

5. 1.35 amperes (325/240 = 1.35)

Unit 7 Three-Phase Circuits

1.

$EP_{(A)}$ 138.57	$EP_{(L)}$ 240
$IP_{(A)}$ 34.64	$IP_{(L)}$ 20
$E_{L(A)}$ 240	$E_{L(L)}$ 240
$I_{L(A)}$ 34.64	$I_{L(L)}$ 34.64
P 14,399.16	$Z_{(Phase)}$ 12 Ω

2.

$EP_{(A)}$ 4,160	$EP_{(L)}$ 2401.85
$IP_{(A)}$ 23.11	$IP_{(L)}$ 40.03
$E_{L(A)}$ 4,160	$E_{L(L)}$ 4,160
$I_{L(A)}$ 40.03	$I_{L(L)}$ 40.03
P 288,420.95	$Z_{(Phase)}$ 60 Ω

3.

$EP_{(A)}$ 323.33	$E_{P(L1)}$ 323.33	$E_{P(L2)}$ 560
$IP_{(A)}$ 185.91	$I_{P(L1)}$ 64.67	$I_{P(L2)}$ 70
$E_{L(A)}$ 560	$E_{P(L1)}$ 560	$E_{L(L2)}$ 560
$I_{L(A)}$ 185.91	$I_{L(L1)}$ 64.67	$I_{L(L2)}$ 121.24
P 180,317.83	$Z_{(Phase)}$ 5 Ω	$Z_{(Phase)}$ 8 Ω

4.

$EP_{(A)}$ 277.14	$E_{P(L1)}$ 277.14	$E_{P(L2)}$ 480	$E_{P(L3)}$ 277.14
$IP_{(A)}$ 33.49	$I_{P(L1)}$ 23.1	$I_{P(L2)}$ 30	$I_{P(L3)}$ 27.71
$E_{L(A)}$ 480	$E_{L(L1)}$ 480	$E_{L(L2)}$ 480	$E_{L(L3)}$ 480
$I_{L(A)}$ 33.49	$I_{L(L1)}$ 23.1	$I_{L(L2)}$ 51.96	$I_{L(L3)}$ 27.71
VA 27,843.39	$R_{(Phase)}$ 12 Ω	$X_{L(Phase)}$ 16 Ω	$X_{C(Phase)}$ 10 Ω
	P 19,204.42	$VARs_L$ 43,197.47	$VARs_C$ 23,037

Unit 8 Three-Phase Transformers

1.

E_P 2401.8	E_P 440	E_P 254.04
I_P 7.67	I_P 41.9	I_P 72.58
E_L 4160	E_L 440	E_L 440
I_L 7.67	I_L 72.58	I_L 72.58
Ratio 5.46:1	$Z = 3.5\ \Omega$	

2.

E_P 4157.04	E_P 240	E_P 138.57
I_P 1.15	I_P 20	I_P 34.64
E_L 7200	E_L 240	E_L 240
I_L 1.15	I_L 34.64	I_L 34.64
Ratio 17.32:1	$Z = 4\ \Omega$	

3.

E_P 13,800	E_P 277	E_P 480
I_P 6.68	I_P 332.54	I_P 192
E_L 13,800	E_L 480	E_L 480
I_L 11.57	I_L 332.54	I_L 332.54
Ratio 49.76:1	$Z = 2.5\ \Omega$	

4.

E_P 23,000	E_P 120	E_P 208
I_P 0.626	I_P 120.08	I_P 69.33
E_L 23,000	E_L 208	E_L 208
I_L 1.08	I_L 120.08	I_L 120.08
Ratio 191.66:1	$Z = 3\ \Omega$	

Glossary

AC (alternating current) current that reverses its direction of flow periodically. Reversals generally occur at regular intervals.
across-the-line a method of motor starting which connects the motor directly to the supply line on starting or running. (Also known as Full-Voltage Starting.)
air gap space between two magnetically related components
alternator a machine used to generate alternating current by rotating conductors through a magnetic field
ambient temperature the temperature surrounding a device.
American Wire Gauge (AWG) a measurement of the diameter of wire The gage scale was formally known as the Brown and Sharp scale. The scale has a fixed constant of 1.123 between gage sizes.
ammeter instrument used to measure the flow of current
amortisseur winding squirrel-cage winding on the rotor of a synchronous motor used for starting purposes only
ampacity maximum current rating of a wire or device
ampere unit of measure for the rate of current flow
ampere-turns basic unit for the measurement of magnetism (The product of turns or wire and current flow.)
amp-hour unit of measure for describing the capacity of a battery
amplifier device used to increase a signal
amplitude highest value reached by a signal, voltage, or current
analog voltmeter voltmeter that uses a meter movement to indicate the voltage value. Analog meters use a pointer and scale.
anode positive terminal of an electrical device
apparent power value found by multiplying the applied voltage and total current of an AC circuit. Apparent power should not be confused with true power or watts.
applied voltage amount of voltage connected to a circuit or device
armature rotating member of a motor or generator. The armature generally contains windings and a commutator.
armature reaction twisting or bending of the main magnetic field of a motor or generator. Armature reaction is proportional to armature current.
ASA American Standards Association
Askarel special type of dielectric oil used to cool electrical equipment such as transformers. Some types of Askarel contains polychlorinated biphenyl (PCB).

atom smallest part of an element that contains all the properties of that element.

attenuator device that decreases the amount of signal voltage or current

automatic self-acting, operation by its own mechanical or electrical mechanism

autotransformer transformer that uses only one winding for both primary and secondary

back voltage induced voltage in the coil of an inductor or generator that opposes the applied voltage

base semiconductor region between the collector and emitter of a transistor. The base controls the current flow through the collector-emitter circuit

battery device used to convert chemical energy into electrical energy. A group of voltaic cells connected together in a series or parallel connection.

bias DC voltage applied to the base of a transistor to preset its operating point

bimetallic strip strip made by bonding two unlike metals together that expand at different temperatures when heated. This causes a bending or warping action.

branch circuit portion of a wiring system that extends beyond the circuit-protective device such as a fuse or circuit breaker.

breakdown torque maximum amount of torque that can be developed by a motor at rated voltage and frequency before an abrupt change in speed occurs

bridge circuit circuit that consists of four sections connected in series to form a closed loop

bridge rectifier device constructed with four diodes that converts both positive and negative cycles of AC voltage into DC voltage. The bridge rectifier is one type of full-wave rectifier.

brush sliding contact, generally made of carbon, used to provide connection to rotating parts of machines.

bus way enclosed system used for power transmission that is voltage and current rated

capacitance electrical size of a capacitor

capacitive reactance (X_L) current-limiting property of a capacitor in an alternating-current circuit

capacitor device made with two conductive plates separated by an insulator or dielectric

capacitor-start motor single-phase induction motor that uses a capacitor connected in series with the start winding to increase starting torque

charging current current flowing from an electrical source to a capacitor

cathode negative terminal of an electrical device

center-tapped transformer transformer that has a wire connected to the electrical midpoint of its winding. Generally the secondary winding is tapped.
choke inductor designed to present an impedance to AC current or to be used as the current filter of a DC-power supply.
circuit electrical path between two points
circuit breaker device designed to open under an abnormal amount of current flow. The device is not damaged and may be used repeatedly. They are rated by voltage, current, and horsepower.
clock timer time-delay device that uses an electric clock to measure delay time
collapse (of a magnetic field) occurs when a magnetic field suddenly changes from its maximum value to a zero value
collector semiconductor region of a transistor that must be connected to the same polarity as the base
commutating field field used in direct-current machines to help overcome the problems of armature reaction. The commutating field connects in series with the armature and is also known as the interpole winding.
commutator strips or bars of metal insulated from each other and arranged around an armature. They provide connection between the armature windings and the brushes. The commutator is used to insure proper direction of current flow through the armature windings.
comparator device or circuit that compares two like quantities such as voltage levels
compensating winding winding embedded in the main field poles of a DC machine. The compensating winding is used to help overcome armature reaction.
compound DC machine generator or motor that uses both series and shunt-field windings. DC machines may be connected long-shunt compound, short-shunt compound, cumulative compound, or differential compound.
continuity complete path for current flow
conduction level point at which an amount of voltage or current will cause a device to conduct
conductor device or material that permits current to flow through it easily
contact conducting part of a relay that acts as a switch to connect or disconnect a circuit or component
conventional current flow theory that considers current to flow from the most positive source to the most negative source
copper losses power loss due to current flowing through wire. Copper loss is proportional to the resistance of the wire and the square of the current.

core magnetic material used to form the center of a coil or transformer. The core may be made of a nonmagnetic conductor (air core), iron, or some other magnetic material.

core losses power loss in the core material due to eddy-current induction and hysteresis loss

cosine In trigonometry, it is the ratio of the adjacent side of the angle and the hypotenuse.

coulomb quantity of electrons equal to 6.25×10^{18}

counter EMF (CEMF) voltage induced in the armature of a DC motor which opposes the applied voltage and limits armature current

counter torque magnetic force developed in the armature of a generator which makes the shaft difficult to turn. Counter torque is proportional to armature current and is a measure of the electrical energy produced by the generator.

current rate of flow of electrons

current rating amount of current flow a device is designed to withstand

current relay relay that is operated by a predetermined amount of current flow. Current relays are often used as one type of starting relay for air conditioning and refrigeration equipment.

cycle one complete AC waveform

D'Arsonval meter meter movement using a permanent magnet and a coil of wire which is the basic meter movement used in many analog-type voltmeters, ammeters, and ohmmeters.

DC (direct current) current that does not reverse its direction of flow

delta connection circuit formed by connecting three electrical devices in series to form a closed loop. It is used most often in three-phase connections.

diac bidirectional diode

diamagnetic material which will not conduct magnetic lines of flux. Diamagnetic materials have a permeability rating less than that of air (1).

dielectric electrical insulator

dielectric breakdown point at which the insulating material separating two electrical charges permits current to flow between the two charges. Dielectric breakdown is often caused by excessive voltage, excessive heat, or both.

digital device device that has only two states of operation, on or off.

digital logic circuit elements connected in such a manner as to solve problems using components that have only two states of operation

digital voltmeter voltmeter that uses direct reading numerical display as opposed to a meter movement

diode two-element device that permits current to flow through it in only one direction

direct current current that flows in only one direction
disconnecting means (disconnect) device or group of devices used to disconnect a circuit or device from its source of supply
domain group of atoms aligning themselves north and south to create a magnetic material
dot notation dots placed beside transformer windings on a schematic to indicate relative polarity between different windings
DVM abbreviation for digital voltmeter
dynamic braking (1) Using a DC motor as a generator to produce counter torque and thereby produce a braking action. (2) Applying direct current to the stator winding of an AC induction motor to cause a magnetic braking action.
eddy current circular induced current contrary to the main currents. Eddy currents are a source of heat and power loss in magnetically operated devices.
electrical interlock when the contacts of one device or circuit prevent the operation of some other device or circuit.
electric controller device or group of devices used to govern in some predetermined manner the operation of a circuit or piece of electrical apparatus
electrodynamometer machine used to measure the torque developed by a motor or engine for the purpose of determining output horsepower
electrolysis decomposition of a chemical compound or metals caused by an electric current
electrolyte chemical compound capable of conducting electric current by being broken down into ions. Electrolytes can be acids or alkalines.
electron one of the three major parts of an atom. The electron carries a negative charge.
electronic control control circuit which uses solid-state devices as control components
electrostatic field of force that surrounds a charged object. The term is often used to describe the force of a charged capacitor.
element (1) One of the basic building blocks of nature. An atom is the smallest part of an element. (2) One part of a group of devices.
EMF abbreviation for electromotive force
emitter semiconductor region of a transistor that must be connected to a polarity different than the base
enclosure mechanical, electrical, or environmental protection for components used in a system
eutectic alloy metal with a low and sharp melting point used in thermal overload relay

excitation current direct current used to produce electromagnetism in the fields of a DC motor or generator, or in the rotor of an alternator or synchronous motor

farad basic unit of capacitance

feeder circuit conductor between the service equipment or the generator switchboard of an isolated plant and the branch circuit overcurrent-protective device

ferromagnetic material that will conduct magnetic lines of force easily, such as iron (ferris). Ferromagnetic materials have a permeability much greater than that of air (1).

field loss relay (FLR) current relay connected in series with the shunt field of a direct-current motor. The relay causes power to be disconnected from the armature in the event that field current should drop below a certain level.

filter device used to remove the ripple produced by a rectifier

flashing the field method used to produce residual magnetism in the pole pieces of a DC machine. It is done by applying full voltage to the field winding for a period of not less than 30 seconds.

flat compounding setting the strength of the series field in a DC generator so that the output voltage will be the same at full load as it is at no load

flux magnetic lines of force

flux density number of magnetic lines contained is a certain area. The area measurement depends on the system of measurement.

frequency number of complete cycles of AC voltage that occur in one second

full-load torque amount of torque necessary to produce the full horsepower of a motor at rated speed

fuse device used to protect a circuit or electrical device from excessive current. Fuses operate by melting a metal link when current becomes excessive.

gain increase in signal power produced by an amplifier

galvanometer meter movement requiring microamperes to cause a full-scale deflection. Many galvanometers utilize a zero center which permits them to measure both positive and negative values.

gate (1) A device that has multiple inputs and a single output. There are five basic types of gates, the and, or, nand, nor, and inverter. (2) One terminal of some electronic devices such as SCRs, Triacs, and field-effect transistors.

gauss unit of measure in the CGS system. One gauss equals one maxwell per-square-centimeter.

generator device used to convert mechanical energy into electrical energy
giga metric prefix meaning one billion ($x10^9$)
gilbert basic unit of magnetism in the CGS system
heat sink metallic device designed to increase the surface area of an electronic component for the purpose of removing heat at a faster rate
henry basic unit of inductance
hermetic completely enclosed. Air tight.
hertz international unit of frequency
holding contacts contacts used for the purpose of maintaining current low to the coil of a relay
holding current amount of current needed to keep an SCR or Triac turned on
horsepower measure of power for electrical and mechanical devices
hydrometer device used to measure the specific gravity of a fluid, such as the electrolyte used in a battery
hypotenuse longest side of a right triangle
hysteresis loop graphic curve that shows the value of magnetizing force for a particular type of material
hysteresis loss power loss in a conductive material due to molecular friction. Hysteresis loss is proportional to frequency.
impedance total opposition to current flow in an electrical circuit.
incandescent ability to produce light as a result of heating an object.
induced current current produced in a conductor by the cutting action of a magnetic field
inductive reactance (X_L) current-limiting property of an inductor in a alternating-current circuit
inductor coil
input voltage amount of voltage connected to a device or circuit
insulator material used to electrically isolate two conductive surfaces
interlock device used to prevent some action from taking place in a piece of equipment or circuit until some other action has occurred
interpole small pole piece placed between the main field poles of a DC machine to reduce armature reaction
ion charged atom
isolation transformer transformer whose secondary winding is electrically isolated from its primary winding
joule basic unit of electrical energy. A joule is to the amount of power used when 1 amp flows through 1 ohm for 1 second. A joule is equal to a watt/second.
jumper short piece of conductor used to make connection between components or a break in a circuit

junction diode made by joining together two pieces of semiconductor material

kick-back diode used to eliminate the voltage spike induced in a coil by the collapse of a magnetic field

kilo metric measure prefix meaning thousand ($x10^3$)

kinetic energy energy of a moving object such as the energy of a flywheel in motion

lamination one thickness of the sheet material used to construct the core material for transformers, inductors, and alternating-current motors

LED (light-emitting diode) diode that produces light when current flows through it

Leyden jar glass jar used to store electrical charges in the very early days of electrical experimentation. The Leyden jar was constructed by lining the inside and outside of the jar with metal foil. The Leyden jar was a basic capacitor.

limit switch mechanically operated switch which detects the position or movement of an object

linear when used in comparing electrical devices or quantities; one unit is equal to another

load center generally the service entrance. A point from which branch circuits originate.

locked-rotor current amount of current produced when voltage is applied to a motor and the rotor is not turning

locked-rotor torque amount of torque produced by a motor at the time of starting

lockout mechanical device used to prevent the operation of some component

long-shunt compound connection of field windings in a DC machine where the shunt field is connected in parallel with both the armature and series field

low-voltage protection magnetic relay circuit so connected that a drop in voltage causes the motor starter to disconnect the motor from the line

magnetimotive force magnetic force produced by current flowing through a conductor or coil

magnetic contactor contactor operated electromechanically

magnetic field space in which a magnetic force exists

maintaining contact also known as a holding or sealing contact. It is used to maintain the coil circuit in a relay-control circuit. The contact is connected in parallel with the start push button.

manual controller controller operated by hand at the location of the controller

maxwell measure of magnetic flux in the CGS system
mica mineral used as an electrical insulator
micro metric prefix meaning one millionth ($x10^{-6}$)
micro-farad measurement of capacitance
microprocessor small computer. The central processing unit is generally made from a single integrated circuit.
mil unit for measuring the diameter of a wire equal to one thousandth of an inch
mill-foot standard for measuring the resistivity of wire. A mill-foot is the resistance of a piece of wire one mill in diameter and one foot in length.
milli metric prefix for one thousandth ($x10^{-3}$)
mode state of condition
motor device used to convert electrical energy into rotating motion
motor controller device used to control the operation of a motor
multi-speed motor motor that can be operated at more than one speed
nano metric prefix meaning one billionth ($x10^{-9}$)
negative one polarity of a voltage, current, or charge
NEMA National Electrical Manufacturers Association
NEMA ratings electrical control device ratings of voltage, current, horsepower, and interrupting capability given by NEMA.
neutron one of the principle parts of an atom. The neutron has no charge and is part of the nucleus.
noninductive load electrical load that does not have induced voltages caused by a coil. Noninductive loads are generally considered to be resistive, but can be capacitive.
nonreversing device that can be operated in only one direction.
normally closed contact of a relay that is closed when the coil is deenergized
normally open contact of a relay that is open when the coil is deenergized
off-delay timer timer that delays changing its contacts back to their normal position when the coil is deenergized
on-delay timer timer that delays changing the position of its contacts when the coil is energized
ohm unit of measure for electrical resistance
ohmmeter device used to measure resistance
operational amplifier (OP AMP) integrated circuit used as an amplifier
opto-isolator device used to connect different sections of a circuit by means of a light beam
oscillator device used to change DC voltage into AC voltage
oscilloscope voltmeter that displays a wave form of voltage in proportion to its amplitude with respect to time

out-of-phase condition in which two components do not reach their positive or negative peaks at the same time

over compounded condition of a DC generator when the series field is too strong. It is characterized by the output voltage being greater at full load than it is at no load.

over excited condition that occurs when the DC current supplying excitation current to the rotor of a synchronous motor is greater than necessary

overload relay relay used to protect a motor from damage due to overloads. The overload relay senses motor current and disconnects the motor from the line if the current is excessive for a certain length of time.

panelboard metallic or nonmetallic panel used to mount electrical controls, equipment, or devices.

parallel circuit circuit that contains more than one path for current flow.

paramagnetic material that has a permeability slightly greater than that of air (1).

peak-inverse/peak-reverse voltage rating of a semiconductor device that indicates the maximum amount of voltage in the reverse direction that can be applied to the device

peak-to-peak voltage amplitude of AC voltage measured from its positive peak to its negative peak

peak voltage amplitude of voltage measured from zero to its highest value

phase shift change in the phase relationship between two quantities of voltage or current

permalloy alloy used in the construction of electromagnets (approximately 78% nickel and 21% iron)

permeability measurement of a materials ability to conduct magnetic lines of flux. The standard is air which has a permeability of 1.

photoconductive material that changes its resistance due to the amount of light

photovoltaic material or device which produces a voltage in the presence of light

pico unit of metric measure for one trillionth ($x10^{-12}$)

piezoelectric production of electricity by applying pressure to a crystal

pilot device control component designed to control small amounts of current. Pilot devices are used to control larger control components.

pneumatic timer device that used the displacement of air in a bellows or diaphragm to produce a time delay

polarity characteristic of a device that exhibits opposite quantities within itself: positive and negative

potentiometer variable resistor with a sliding contact that is used as a voltage divider

power factor comparison of the true power (watts) to the apparent power (volt amps) in an AC circuit
power rating rating of a device that indicates the amount of current flow and voltage drop that can be permitted
pressure switch device that senses the presence or absence of pressure and causes a set of contacts to open or close
primary winding winding of a transformer to which power is applied
proton one of the three major parts of an atom. The proton has a positive charge.
primary cell voltaic cell that cannot be recharged
prime mover device supplying the turning force necessary for turning the shaft of a generator or alternator. (Steam turbine, diesel engine, water wheel, etc.)
printed circuit board on which a predetermined pattern of printed connections has been made
push button pilot control device operated manually by being pushed or pressed
reactance opposition to current flow in an AC circuit offered by pure inductance or pure capacitance
rectifier device or circuit used to change AC voltage into DC voltage
regulator device that maintains a quantity at a predetermined level
relay magnetically-operated switch that may have one or more sets of contacts
reluctance resistance to magnetism
remote control controls the functions of some electrical device from a distant location
residual magnetism amount of magnetism left in an object after the magnetizing force has been removed
resistance opposition to current flow in an AC or DC circuit
resistance-start induction-run motor one type of split-phase motor that uses the resistance of the start winding to produce a phase shift between the current in the start winding and the current in the run winding
resistor device used to introduce some amount of resistance into an electrical circuit
retentivity material's ability to retain magnetism after the magnetizing force has been removed
rheostat variable resistor
RMS value value of AC voltage that will produce as much power when connected across a resistor as a like amount of DC voltage
rotor rotating member of an alternating-current machine
saturation maximum amount of magnetic flux a material can hold

schematic electrical drawing showing components in their electrical sequence without regard for physical location

SCR (silicon-controlled rectifiers) semiconductor device than can be used to change AC voltage into DC voltage. The gate of the SCR must be triggered before the device will conduct current.

sealing contacts contacts connected in parallel with the start button and used to provide a continued path for current flow to the coil of the contactor when the start button is released. See also Holding Contacts and Maintaining Contacts.

secondary cell voltaic cell which can be recharged

secondary winding winding of a transformer to which the load is connected

semiconductor material that contains four valence electrons and is used in the production of solid-state devices

sensing device pilot device that detects some quantity and converts it into an electrical signal

series circuit circuit that contains only one path for current flow

series field winding of large wire and few turns designed to be connected in series with the armature of a DC machine

series machine direct-current motor or generator that contains only a series-field winding connected in series with the armature

service conductors and equipment necessary to deliver energy from the electrical supply system to the premises served

service factor allowable overload for a motor indicated by a multiplier which, when applied to a normal horsepower rating, indicates the permissible loading

shaded pole motor AC induction motor that develops a rotating magnetic field by shading part of the stator windings with a shading loop

shading loop large copper wire or band connected around part of a magnetic pole piece to oppose a change of magnetic flux

short circuit electrical circuit that contains no resistance to limit the flow of current

shunt field coil wound with small wire having many turns designed to be connected in parallel with the armature of a DC machine

shunt machine DC motor or generator that contains only a shunt field connected in parallel with the armature

sine-wave voltage voltage waveform whose value at any point is proportional to the trigonometric sine of the angle of the generator producing it

slip difference in speed between the rotating magnetic field and the speed of the rotor in an induction motor

slip rings circular bands of metal placed on the rotating part of a machine. Carbon brushes riding in contact with the slip rings provide connection to the external circuit.
snap-action quick opening and closing action of a spring-loaded contact
solenoid magnetic device used to convert electrical energy into linear motion
solenoid valve valve operated by an electric solenoid
solid-state device electronic component constructed from semiconductor material
specific gravity ratio of the volume and weight of a substance as compared to an equal volume and weight of water. Water has a specific gravity of 1.
split-phase motor type of single-phase motor that uses resistance or capacitance to cause a shift in the phase of the current in the run winding and the current in the start winding. The three primary types of split phase motors are: resistance start induction run, capacitor start induction run, and permanent split-capacitor motor.
squirrel-cage rotor rotor of an AC induction motor constructed by connecting metal bars together at each end
star connection see wye connection
starter relay used to connect a motor to the power line
stator stationary winding of an AC motor
step-down transformer transformer that produces a lower voltage at its secondary than is applied to its primary
step-up transformer transformer that produces a higher voltage at its secondary than is applied to its primary
surge transient variation in the current or voltage at a point in the circuit. Surges are generally unwanted and temporary.
switch mechanical device used to connect or disconnect a component or circuit
synchronous speed speed of the rotating magnetic field of an AC induction motor
synchroscope instrument used to determine the phase-angle difference between the voltage of two alternators
temperature relay relay that functions at a predetermined temperature. Generally used to protect some other component from excessive temperature.
terminal fitting attached to a device for the purpose of connecting wires to it
tesla unit of magnetic measure in the MKS system. (1 Tesla = 1 Weber per-square-meter.)

thermistor resistor that changes its resistance with a change of temperature

thyristor electronic component that has only two states of operation; on or off

time constant amount of time required for the current flow through an inductor or for the voltage applied to a capacitor to reach 63.2% of its total value

torroid doughnut-shaped electromagnet. Donut-shaped in the fact that it has a hole through which conductors are wound.

torque turning force developed by a motor

transducer device that converts one type of energy into another type of energy. Example: A solar cell converts light into electricity.

transformer electrical device that changes one value of AC voltage into another value of AC voltage

transistor solid-state device made by combining three layers of semiconductor material together. A small amount of current flow through the base emitter can control a larger amount of current flow through the collector emitter.

triac bidirectional thyristor used to control AC voltage

troubleshoot locate and eliminate problems in a circuit

turns ratio (Transformer) The ratio of the number of primary turns of wire as compared to the number of secondary turns

under compounded condition of a direct-current generator when the series field is too weak. The condition is characterized by the fact that the output voltage at full load will be less than the output voltage at no load

unity power factor power factor of 1 (100%). Unity power factor is accomplished when the applied voltage and circuit current are in phase with each other.

valence electrons electrons located in the outer orbit of an atom

variable resistor resistor whose resistance value can be varied between its minimum and maximum values

varistor resistor that changes its resistance value with a change of oltage

vector lines have a specific length and direction

voltage electrical measurement of potential difference, electrical pressure, or electromotive force (EMF)

voltage drop amount of voltage required to cause an amount of current to flow through a certain resistance

voltage rating rating that indicates the amount of voltage that can be safely connected to a device

voltage regulator device or circuit that maintains a constant value of voltage

voltaic cell device that converts chemical energy into electrical energy

voltmeter instrument used to measure a level of voltage

volt-ohm-milliammeter (VOM) test instrument designed to measure voltage, resistance, or milliamperes

watt measure of true power

waveform shape of a wave as obtained by plotting a graph with respect to voltage and time

weber measure of magnetic lines of flux in the MKS system. (1 Weber = 100 million [$x10^8$] lines of flux)

windage loss losses encountered by the armature or rotor of a rotating machine caused by the friction of the surrounding air

wiring diagram electrical diagram used to show components in their approximated physical location with connecting wires

wound-rotor motor three-phase motor containing a rotor with windings and slip rings. This rotor permits control of rotor current by connecting external resistance in series with the rotor winding.

wye connection connection of three components made in such a manner that one end of each component is connected. This connection is generally used to connect devices to a three-phase power system.

zener diode diode that has a constant voltage drop when operated in the reverse direction. Zener diodes are commonly used as voltage regulators in electronic circuits.

INDEX

CPSIA information can be obtained
at www.ICGtesting.com
Printed in the USA
FFOW03n0028190816
26931FF